AF389405

Vibration Control and Structure
Health Monitoring

Vibration Control and Structure Health Monitoring

Editors

Zhao-Dong Xu
Siu-Siu Guo
Jinkoo Kim

MDPI • Basel • Beijing • Wuhan • Barcelona • Belgrade • Manchester • Tokyo • Cluj • Tianjin

Editors
Zhao-Dong Xu
Southeast University
China

Siu-Siu Guo
Xi'an University of
Architecture and Technology
China

Jinkoo Kim
Sungkyunkwan University
Korea

Editorial Office
MDPI
St. Alban-Anlage 66
4052 Basel, Switzerland

This is a reprint of articles from the Special Issue published online in the open access journal *Actuators* (ISSN 2076-0825) (available at: https://www.mdpi.com/journal/actuators/special_issues/vibration_control_SHM).

For citation purposes, cite each article independently as indicated on the article page online and as indicated below:

LastName, A.A.; LastName, B.B.; LastName, C.C. Article Title. *Journal Name* **Year**, *Volume Number*, Page Range.

ISBN 978-3-0365-6336-7 (Hbk)
ISBN 978-3-0365-6337-4 (PDF)

Contents

About the Editors

Zhao-Dong Xu

Zhao-Dong Xu is a Professor at Southeast University; Director of the International Joint Laboratory for Smart Disaster Prevention of major Infrastructure, Yangtze River Scholar Professor of the Ministry of Educationl; a National Outstanding Youth Distinguished Professor; and Winner of the First Scientific Exploration Award. He is mainly engaged in research on structural vibration control and health monitoring, intelligent prevention of seismic and multi-disaster damage to structures, and intelligent materials and structures. He chaired the 863 Program, the Major Research Program of the National Natural Science Foundation, key special projects for the National Key Research and Development Program, and five projects for the National Natural Science Foundation, etc. He is now in charge of the key special projects of the National Key Research and Development Program of the Smart City, the Key Projects of the National Natural Science Foundation, and more. As a leading professor, he won the second prize of National Technology Invention, the first prize of Technology Invention of the Ministry of Education, the Gold Prize of the Geneva International Invention Expo, the first prize of Science and Technology of Jiangsu Province, and the first prize of Technology Invention of Building Materials of China, among other.

Siu-Siu Guo

Siu-Siu Guo is a Professor at Xi'an University of Architecture and Technology. Her research interests include structural stochastic response and dynamic reliability. She has presided over two National Natural Science Foundation projects and participated in a number of domestic and foreign cooperative scientific research projects. She was awarded as one of the 100 Young talent People of Shaanxi Province and as a science and technology new star of Shaanxi Province. One of her published papers was selected as a key scientific article by Advances in Engineering. One of her conference papers was awarded a Marie Curie ITN award and an ISMA Student fellowship sponsored by the EC via the FP7 Marie Curie ITN projects VECOM and MID-FREQUENCY. One of her conference papers also won the Best poster award prize.

Jinkoo Kim

Jinkoo Kim is a Professor at Sungkyunkwan University and President of the Korea Earthquake Engineering Society. His research interests include slotted steel plate shock absorbers, friction shock absorbers, structural reinforcement and structural damage. Kim received a bachelor's degree from the National Seoul University in 1986, a master's degree from the National Seoul University in 1988, a master's degree from the Massachusetts Institute of Technology in 1992, and a doctorate degree from the Massachusetts Institute of Technology in 1995. He has presided over five projects for the National Natural Science Foundation of Korea. He has published more than 150 journal papers in international core journals, including 117 indexed by SCI; has received more than 1300 citations by others; and owns 17 authorized patents.

Editorial

Some Recent Developments in the Vibration Control and Structure Health Monitoring

Siu-Siu Guo [1,*] and Jinkoo Kim [2]

[1] School of Civil Engineering, Xi'an University of Architecture and Technology, Xi'an 710055, China
[2] Department of Civil and Architectural Engineering, Sungkyunkwan University, Jangan-Gu, Suwon 03063, Republic of Korea
* Correspondence: siusiuguo@xauat.edu.cn

Keywords: vibration control; structural health monitoring; vibration mitigation; vibration analysis

Citation: Guo, S.-S.; Kim, J. Some Recent Developments in the Vibration Control and Structure Health Monitoring. *Actuators* **2023**, *12*, 11. https://doi.org/10.3390/act12010011

Received: 9 December 2022
Revised: 22 December 2022
Accepted: 22 December 2022
Published: 26 December 2022

Vibration is a common phenomenon when a structure is exposed to mechanical or environmental actions. It may inflict great cost to lives and to the economy. In order to reduce this, and to monitor resulting damages, vibration control and structural health monitoring have become increasingly important. In particular, the past decade has witnessed emerging thoughts, perspectives, ideas, and methods in vibration control and structural health monitoring, inspired by big data, supercomputing, and AI, as well as from new advances in mechanics, mathematics, materials, and related multi-disciplinary fusions. Significant contributions have been made, such as that of Xu [1], who established a spectral energy theory of frequency response function in the distribution area and proposed a series of methods to accurately describe different structural damages and collapse processes, for instance the perturbation distribution area energy method, the energy catastrophe collapse analysis method, and the regional energy flow density method. For the first time ever, he carried out a 3D shaking table test on a long-span reticulated structure with multi-dimensional earthquake isolation and mitigation devices [2]. He was also the pioneer [2–6] of conducting large-scale shaking table tests on a variety of structures with isolation and mitigation devices, and relevant research results have been applied to several representative construction projects, such as the Haikou Meilan International Airport of China in 2017. However, there are still key challenges. These are due to, on the one hand, the increasing frequency and intensity of natural and man-made disasters in recent years, such as earthquakes, tropical cyclones, floods, and industrial accidents, and on the other hand, the increasing size, complexity, multi-field and intra-and-between-system coupling effects, and the enhanced performance demands of structures and infrastructure systems. These factors have motivated frontier research topics in the field of vibration control and structural health monitoring, including: theories and computational methods of, and materials and devices for, vibration control; the application of vibration mitigation or isolation techniques; the development of structural health monitoring equipment and methods of damage detection and localization for structures; data sensing and processing in structural health monitoring; safety diagnosis and assessment of structures; and interdisciplinary approaches and applications for structural health monitoring, vibration analysis, tests and applications in relative fields.

The Special Issue "vibration control and structural health monitoring", is an opportunity to share knowledge, experience, and information in the field of vibration control and structural health monitoring. In this Special Issue, nine original research papers and one review paper have been included. These papers concerned the vibration control and structural health monitoring in areas of theory and computational methods of vibration control materials and devices for vibration control, tests and applications of vibration mitigation or isolation techniques, vibration analysis, and tests and applications in relative fields.

Zhang and Mou et al. [7] proposed, by investigating a five-story reinforced concrete frame structure, a multi-dimensional elastic–plastic calculation model for frame structures with a magnetorheological damper. The results show that the proposed calculation model can effectively simulate the multi-dimensional seismic response of the structure with and without the magnetorheological damper.

Torsional vibration of eccentric structures would result in and accelerate seismic damage to structures. Yang and Guo [8] established numerical models of spatial eccentric structures with multiple MR dampers and conducted a time–history analysis. Numerical results demonstrate that the control system with multiple MR dampers can significantly attenuate the torsional vibrations of eccentric structures, and thus possess significant application prospects. This study used a mixed ferromagnetic particle coated with carbon nanotubes and graphene, overcoming the technical bottleneck of easy sedimentation and poor re-dispersibility of magnetorheological fluid, which was developed for the first time by Xu [9]. Xu [2] also developed magnetorheological dampers with the world's largest damping force 47.2 t, with integrated magnetorheological control systems, and carried out dynamic tests on building structures and long-span bridges with this intelligent control system.

In order to solve the problem of slow response, poor precision, and weak anti-interference ability in hydraulic servo position controls, Guo and Zha et al. [10] designed a Kalman genetic optimization PID controller. Compared with traditional PID control algorithms, the PID algorithm optimized by generic algorithm improves the system's response speed and control accuracy; the Kalman filter solves the problem of amplitude fluctuations and reduces the influence of external disturbances on the hydraulic servo system. This study is based on the real-time online control algorithm for earthquake catastrophes, solving the problems of time delay and nonlinear time-varying of magnetorheological control, proposed for the first time by Xu [9].

Terrains with inclination, settlements, displacements, and other phenomena lead transmission towers to collapse. Wang and Liu et al. [11] designed a device to monitor the settlement of a transmission tower. The experimental results show that the third to fifth natural frequencies decreased significantly, especially when the tower legs are adjacent to the excitation position.

Machine tool ram affects the ability of the machine to achieve stable cutting conditions. There are various solutions for increasing its stiffness and damping. Novotny and Cervenka et al. [12] proposed an innovative two-axial electromagnetic actuator for controlled vibration dampers with high dynamic force values. This position measurement concept will enable possible use in the field of vibration suppression of vertical rams of large machine tools.

Ying and Ni [13] proposed a method to study vibration response and amplitude frequency characteristics of a controlled nonlinear meso-scale beam under periodic loading. The proposed method includes a general analytical expression of harmonic balance solution to periodic vibration, and an updated cycle iteration algorithm for amplitude frequency relation of periodic response. Numerical results demonstrate a good convergency and accuracy of the proposed method.

Shen and Huang [14] conducted a sensitivity experiment to study the dependence of the shear strain on the seismic properties of bearings (lead rubber bearing and super-high damping rubber bearing). Test results showed that temperature is the most dominant factor, whilst the shear modulus is the least contributing factor.

Road roughness plays an important role in road maintenance and ride quality. Zhang and Hou at al. [15] proposed a road-roughness estimation method based on the frequency response function of a vehicle, which can be estimated directly using the measured response. The results show that road roughness can be estimated using the proposed method with acceptable accuracy and robustness.

Ge and Huang et al. [16] developed a cylindrical viscoelastic damper and investigated its mechanical and damping performance. The experimental results demonstrate that the cylindrical viscoelastic damper presents a full hysteretic curve in a wide frequency range.

This kind of cylindrical viscoelastic damper can be simulated by a multi-molecular-chain model of viscoelastic materials and dampers with eight parameters proposed by Xu [17], which can easily calculate the effect of temperature, frequency, and displacement amplitude. Among the existing mathematical models worldwide, only the finite element model can reflect the influence of these three factors at the same time; however, this model is very complex and involves the determination of nearly 20 parameters.

Aabid and Parveez et al. [18] conduced a review on piezoelectric-material-based structural control and health monitoring techniques. They reported fundamental modeling and applications in engineering fields, explored new approaches and hypotheses and discussed the challenges and opportunities for future work.

Although only ten papers are included in this Special Issue, new advances and developments in vibration control and structural health monitoring have been presented as much as possible. They stimulate frontier research topics in various fields, including civil engineering, mechanical engineering, hydraulic engineering, offshore and marine engineering, and aeronautics and aerospace engineering, amongst others. Given that it is still a challenging topic, new thoughts, perspectives, ideas, and methods, as well as new advances in materials and devices, are still sought to promote research on vibration control and structural health monitoring.

We are striving to enhance the influence of the journal *Actuator* and maintain its leading position in the field of vibration control and structural health monitoring. We are grateful for the contributions from authors who make this Special Issue successful.

Author Contributions: All authors listed have made substantial, direct, and intellectual contribution to the work and approved it for publication. All authors have read and agreed to the published version of the manuscript.

Funding: This research was funded by National Key Research and Development Programs of China under Grant 2019YFE0121900 and Jiangsu Province International Science and Technology Cooperation Project (Grant No. BZ2022037).

Data Availability Statement: The datasets generated and/or analyzed during the current study are available from the corresponding author upon reasonable request.

Acknowledgments: We would like to acknowledge all the authors, reviewers, editors, and publishers who have supported this Research Topic.

Conflicts of Interest: The authors declare that they have no conflict of interest.

References

1. Xu, Z.D.; Wu, Z.S. Energy damage detection strategy based on acceleration responses for long-span bridge structures. *Eng. Struct.* **2007**, *29*, 609–617. [CrossRef]
2. Xu, Z.D. Horizontal shaking table tests on structures using innovative earthquake mitigation devices. *J. Sound Vib.* **2009**, *325*, 34–48. [CrossRef]
3. Xu, Z.D.; Tu, Q.; Guo, Y.F. Experimental study on vertical performance of multidimensional earthquake isolation and mitigation devices for long-span reticulated structures. *J. Vib. Control* **2012**, *18*, 1971–1985. [CrossRef]
4. Xu, Z.D.; Huang, X.H.; Lu, L.H. Experimental study on horizontal performance of multi-dimensional earthquake isolation and mitigation devices for long-span reticulated structures. *J. Vib. Control* **2012**, *18*, 941–952. [CrossRef]
5. Xu, Z.D.; Wang, S.A.; Xu, C. Experimental and numerical study on long-span reticulate structure with multidimensional high-damping earthquake isolation devices. *J. Sound Vib.* **2014**, *333*, 3044–3057. [CrossRef]
6. Xu, Z.D.; Chen, B.B. Experimental and Numerical Study on Magnetorheological Fluids Based on Mixing Coated Magnetic Particles. *J. Mater. Civ. Eng.* **2016**, *28*, 04015198. [CrossRef]
7. Zhang, X.C.; Mou, C.C.; Zhao, J.; Guo, Y.Q.; Song, Y.M.; You, J.Y. A multidimensional elastic-plastic calculation model of the frame structure with magnetorheological damper. *Actuator* **2022**, *11*, 362. [CrossRef]
8. Yang, Y.; Guo, Y.Q. Numerical simulation and torsional vibration mitigation of spatial eccentric structures with multiple magnetorheological dampers. *Actuators* **2022**, *11*, 235. [CrossRef]
9. Xu, Z.D.; Guo, Y.Q. Neuro-fuzzy control strategy for earthquake-excited nonlinear magnetorheological structures. *Soil Dyn. Earthq. Eng.* **2008**, *28*, 717–727. [CrossRef]
10. Guo, Y.Q.; Zha, X.M.; Shen, Y.Y.; Wang, Y.N.; Chen, G. Research on PID position control of a hydraulic servo system based on Kalman genetic optimization. *Actuators* **2021**, *11*, 162. [CrossRef]

11. Wang, L.H.; Liu, C.F.; Zhu, X.W.; Xu, Z.X.; Zhu, W.W.; Zhao, L. Active vibration-based condition monitoring of a transmission line. *Actuators* **2021**, *10*, 309. [CrossRef]
12. Novotny, L.; Cervenka, J.; Sulitka, M.; Sveda, J.; Janota, M.; Kupka, P. Design of two-axial actuator for controlled vibration damper for large rams. *Actuator* **2021**, *10*, 199. [CrossRef]
13. Ying, Z.G.; Ni, Y.Q. Vibrational amplitude frequency characteristics analysis of a controlled nonlinear meso-scale beam. *Actuator* **2021**, *10*, 180. [CrossRef]
14. Shen, C.Y.; Huang, X.Y.; Chen Y., Y.; Ma, Y.H. Factors affecting the dependency of shear strain of LRB and SHDR: Experimental study. *Actuator* **2021**, *10*, 98. [CrossRef]
15. Zhang, Q.X.; Hou, J.L.; Duan, Z.D.; Jankowshi, L.; Hu, X.Y. Road roughness estimation based on the vehicle frequency response function. *Actuator* **2021**, *10*, 89. [CrossRef]
16. Ge, T.; Huang, X.H.; Guo, Y.Q.; He, Z.W.; Hu, Z W. Investigation of mechanical and damping performance of cylindrical viscoelastic dampers in wide frequency range. *Actuator* **2021**, *10*, 71. [CrossRef]
17. Xu, Z.D.; Ge, T.; Liu, J. Experimental and theoretical study of high energy dissipation viscoelastic dampers based on acrylate rubber matrix. *J. Eng. Mech. ASCE* **2020**, *146*, 04020057. [CrossRef]
18. Aabid, A.; Parveez, B.; Raheman, M.; Ibrahim, Y.; Anjum, A.; Hrairi, M.; Parveen, N.; Zayan, J.M. A review of piezoelectric material-based structural control and health monitoring techniques for engineering structures: Challenges and opportunities. *Actuator* **2021**, *10*, 101. [CrossRef]

actuators

MDPI

Article

A Multidimensional Elastic–Plastic Calculation Model of the Frame Structure with Magnetorheological Damper

Xiangcheng Zhang [1], Changchi Mou [1], Jun Zhao [1,*], Yingqing Guo [2], Youmin Song [3] and Jieyong You [3]

[1]   School of Mechanics and Safety Engineering, Zhengzhou University, Zhengzhou 450001, China
[2]   College of Mechanical and Electronic Engineering, Nanjing Forestry University, Nanjing 210037, China
[3]   China Construction Fifth Engineering Bureau Co., Ltd., Changsha 410004, China
*   Correspondence: zhaoj@zzu.edu.cn

Abstract: To analyze the multidimensional elastic–plastic response of the frame structure with magnetorheological (MR) dampers under strong seismic excitations, the test of the MRD was performed, the location matrix of the MRD in the frame structure was derived, and the multidimensional elastic–plastic calculation models of the frame structure with and without an MRD were established based on the three-segment variable stiffness beam. Taking a five-story reinforced concrete (RC) frame structure as an example, the multidimensional elastic–plastic calculation models were developed by MATLAB software and the dynamic time history analyses were performed under strong seismic excitations. The results show that under the seismic wave, after the MRD is installed in the structure, the maximum horizontal displacements of the top-story node of the structure in X and Y directions is reduced by 51.87% and 39.59%, respectively, and the maximum horizontal accelerations are reduced by 36.67% and 47.86%. The maximum displacements and the story drift ratios of each story of the structure are significantly reduced, and the reduction in the maximum accelerations of each story is small relatively. In the frame structure without an MRD, plastic hinges appear at the ends of most columns, and the structure is characterized by a column hinge yield mechanism. The maximum residual displacement angles of the column end in X and Y directions which reach 1.628×10^{-3} rad and 2.101×10^{-3} rad, respectively. After setting the MRD, the number of plastic hinges in X and Y directions at the column end are both reduced by 37.5%, and the residual displacement angle at some column ends are reduced to 0. The results show that the complied calculation model programs of the frame structure can effectively simulate the multi-dimensional seismic response of the structure with and without MRD.

Keywords: magnetorheological damper; reinforced concrete frame; three-section variable stiffness; multidimensional elastic–plastic; time history analysis

Citation: Zhang, X.; Mou, C.; Zhao, J.; Guo, Y.; Song, Y.; You, J. A Multidimensional Elastic–Plastic Calculation Model of the Frame Structure with Magnetorheological Damper. *Actuators* **2022**, *11*, 362. https://doi.org/10.3390/act11120362

Academic Editor: Ramin Sedaghati

Received: 25 October 2022
Accepted: 1 December 2022
Published: 3 December 2022

Publisher's Note: MDPI stays neutral with regard to jurisdictional claims in published maps and institutional affiliations.

1. Introduction

A magnetorheological damper (MRD), as a kind of damping control device with an adjustable damping force, has the advantages of a simple structure, large output, continuously adjustable damping force, and a good control effect [1–3], so it can effectively improve the seismic performance of the structure [4]. As one of the important factors affecting the effect of the structural vibration control [5], the mechanical properties of the MRD should be fully studied by the experiments [6,7].

Over the past twenty years, the variation in the mechanical properties of the MR damper with the current, velocity, displacement amplitude, and the frequency has been comprehensively studied. Xu et al. [8] designed and manufactured a five-stage coil shear valve mode MRD, whose damping force and energy dissipation effect increased significantly with the increase in the current, and the maximum output could reach 260 kN. Zemp et al. [9] designed and manufactured a long-axis MRD with a working course of up to 1 m, and conducted corresponding mechanical property tests, with a maximum output of up to 280 kN. Zhang et al. [10] designed and manufactured a five-stage coil shear valve MRD with an output up to 478 kN and conducted tests on its mechanical properties. Jiang et al. [11] designed a novel MRD with selectable performance parameters to improve the environmental adaptability of the vibration systems equipped with such a damper and to provide a new idea for the design of an MRD.

For the field of the vibration control of the engineering structure, using an MRD to control the structure is one of the important contents. Raju et al. [12] proved that the scissors jack-MR damper can effectively reduce the displacement response of the steel frame bending moment frame model's structure and improve the seismic performance of the structure. Cruze et al. [13] proved that an MRD can reduce the seismic response of reinforced concrete (RC) frame structures under moderate and strong earthquakes. Rakshita et al. [14] conducted a hybrid simulation analysis of the seismic response in a single-story frame with MRDs by using the OpenSees software, and the test results show that the MRD can effectively control the displacement and acceleration response of the frame structures. Chae et al. [15] conducted a real-time hybrid simulation test on the three-story steel frame structure with MRDs and confirmed that the MRD can better control the displacement and acceleration response of the frame structure. Aggumus et al. [16] investigated the effect of the MRD on reducing the vibration amplitude of structures though a test; the results show that the MRD can effectively reduce the vibration of a six-story steel structure, and the MRD arrangement in which the one end is connected to the ground can reduce the vibration amplitudes. The experimental research on the structure with MRD can truly reflect the damping effect of the structure, with accurate results and a high reliability, but the experimental research cycle is long, the cost is high, and the test conditions are limited.

Establishing the calculation model of building structures with MRD and carrying out a numerical simulation is an economical and effective means to study its seismic performance. At present, the RC building structure with an MRD is mainly simplified into an elastic story model [17,18], elastic–plastic story model [19], and plane elastic–plastic beam–column element model [20]. The layer model and plane model can reflect the impact of the MRD on the seismic performance of the structure to a certain extent, and the calculation efficiency is high, but the layer model and plane model cannot reflect the multi-dimensional seismic performance of a spatial structure. Hence, Zhao et al. [21] established the spatial elastic beam system model of the RC frame's structure with an MRD by using the MATLAB programming method, and studied the multi-dimensional damping control effect of the MRD on the structure, but the model did not consider the stiffness degradation caused by structural damage under a strong earthquake. Xu et al. [22] simplified the RC frame with an MRD as a spatial elastic–plastic beam system model, but the model ignored the influence of the slab, and the research results were not closely related to the elastic–plastic response of the structure. Xu et al. [23,24] analyzed the seismic performance of the nonlinear steel frame structure with an MRD by using LS-DYNA software. In the aspect of multi-dimensional vibration reduction [25–27] and the establishment of the dampers' calculation model [28–30], many scholars have done relevant research in recent years, but not many have conducted research related to the MRD. Recently, You et al. [31] established the model of an L-shaped frame structure with MRDs, and studied its multi-dimensional seismic performance and torsional vibration characteristics, but they did not study the effect of the MRD on the hysteretic behavior and yield mechanism of the structures under seismic excitation. To sum up, although many innovative simulation studies have been carried out on the seismic response of frame structures with an MRD, there are still few studies on the multi-dimensional elastic–plastic calculation model and the multi-dimensional elastic–plastic seismic performance of an RC frame structures with an MRD; especially, the influence of the MRD on the cracking, yield, hysteretic behavior, and structural yield mechanism of concrete beams and columns in spatial structures is not completely clear.

In this paper, the position matrix of the MRD was derived at first. Then, a multi-dimensional elastic–plastic calculation model of the MRDs frame structure was established based on the three-fold linear stiffness retrograde model. Finally, a multi-dimensional elastic–plastic model program of a five-story RC frame with an MRD was developed using MATLAB software, and the dynamic time history analysis was carried out under a strong earthquake. The purposes of the study are summarized as (1) to provide a numerical analysis method for the multi-dimensional elastoplastic dynamic time history analysis of the MRD frame structure; (2) to study the multi-dimensional elastic–plastic seismic performance and the damping effect of the MRD frame structure; and (3) to study the effects of the MRD on the cracking, yield, hysteretic behavior of the beam and column members, and the structural yield mechanism.

2. Elastic–Plastic Calculation Model of the RC Frame Structure

2.1. Three-Fold Line Model Considering Stiffness Degradation

For RC building structures, the three-fold linear stiffness degradation model can reflect the stress process of RC beam and column members [32]. The model considers that: (1) when reloading after the previous cycle, the reduction in the stiffness is related to the loading history, and the loading path points to the maximum deformation point of all the previous cycles; (2) the unloading stiffness after the yielding is equal to the secant stiffness at the yield point. Therefore, the three-fold linear stiffness degradation model is selected as the restoring force model of the elastic–plastic region of the beam in this paper, as shown in Figure 1.

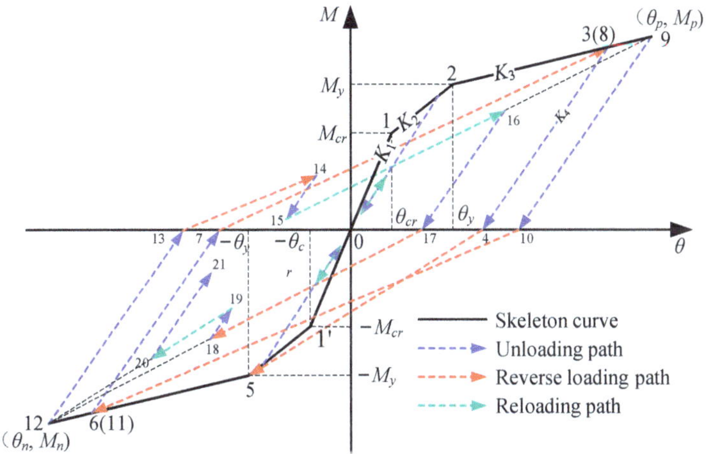

Figure 1. Three-fold linear stiffness retrograde model of RC beam–column element.

The model shown in Figure 1 consists of a skeleton curve and hysteretic path. The skeleton curve is divided into the elastic segment, elastic–plastic segment, and plastic segment. The hysteretic path of each segment is as follows:

(1) Hysteresis path of elastic segment

Segments 0→1 and 0→1′ are the elastic segment of the skeleton curve, and the initial stiffness is K_1. Point 0 is the starting point, and points 1 and 1′ are the cracking points. For the elastic stage, the maximum deformation point of all the previous cycles does not exceed the cracking point; the loading and unloading path will go straight up or down with the initial stiffness K_1.

(2) Hysteresis path of elastic–plastic segment

Segments 1→2 and 1′→5 are the elastic–plastic segment of the skeleton curve, and the stiffness is K_2. Points 2 and 5 are the yield points. For the elastic–plastic stage, the maximum deformation point of all the previous cycles does not exceed the yield point and is not lower than the cracking point; the loading path will go straight up with the stiffness K_2 to the yielding point, and the unloading path will go straight down to the initial point 0. After the unloading path crosses the initial point 0, the model enters the reverse loading phase, and the reverse loading path points to the maximum deformation point of all the previous cycles. If reloading occurs before the unloading path reaches the initial point 0, the reloading path will follow the unloading path.

(3) Hysteresis path of plastic segment

Segments 2→3 and 5→6 are the plastic sections of the skeleton curve, and the stiffness is K_3. For the plastic stage, the maximum deformation point of all the previous cycles is not lower than the yield point, the loading path will go straight up with stiffness K_3, and for the unloading path, for example, the 3→4 segment or 6→7 segment, will go straight down with the secant stiffness K_4 at the yield point. After the unloading path crosses the residual deformation point (such as points 4 and 7), the model enters the reverse loading phase, and the reverse loading path points to the maximum deformation point of all previous cycles, such as 4→5 segment or 7→8 segment. If reloading occurs before the unloading path reaches the residual deformation point, the reloading path will point to the maximum deformation point of all previous cycles, such as the 15→16 segment or 19→20 segment.

The cracking moment M_{cr}, the angular displacement corresponding to the cracking moment θ_{cr}, the yield moment M_y, the angular displacement corresponding to the yield moment θ_y, and the yield point cut stiffness reduction coefficient α_y of each RC beam and column are determined by the Chinese *code for design of concrete structures* (GB50010-2010) and the structural reinforcement diagram.

2.2. Elastic–Plastic Stiffness Matrix of the Variable Stiffness Space Beam Element

The three-segment variable stiffness beam element consists of two types of regions: the linear elastic region in the middle and the fixed length elastic–plastic region at both ends, as shown in Figure 2. Assuming that the RC beam and column is a straight member with an equal section, the section deformation of the member meets the plane section assumption, the member only has a bending failure, and the plastic hinge only appears in the elastic–plastic area at both ends, ignoring the influence of the shear deformation of the member. The length l_p of the elastic–plastic at both ends of the member can be given as follows [33]:

$$l_p = 0.014d_b f_y + 0.12l \tag{1}$$

where d_b is the diameter of the tensile longitudinal reinforcement, f_y is the yield strength of the reinforcement, and l is the length of the member.

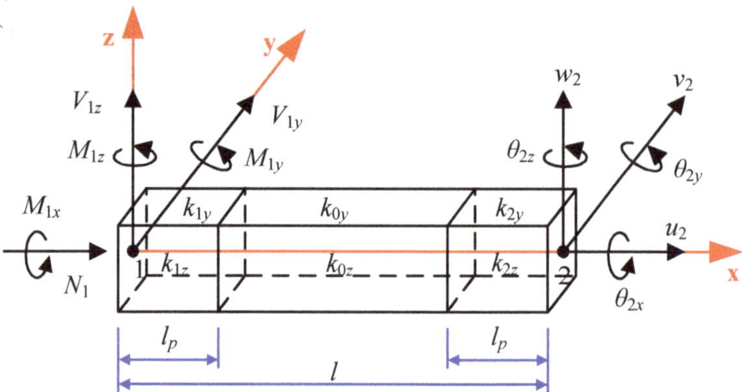

Figure 2. Three-segment variable stiffness model of the RC beam and column member.

The elastic–plastic stiffness matrix of the space beam element is given as follows [34]:

$$k_e = \begin{bmatrix}
k_1 & 0 & 0 & 0 & 0 & 0 & -k_1 & 0 & 0 & 0 & 0 & 0 \\
0 & k_2 & 0 & 0 & 0 & k_4 & 0 & -k_2 & 0 & 0 & 0 & k_6 \\
0 & 0 & k_3 & 0 & -k_5 & 0 & 0 & 0 & -k_3 & 0 & -k_7 & 0 \\
0 & 0 & 0 & k_{14} & 0 & 0 & 0 & 0 & 0 & -k_{14} & 0 & 0 \\
0 & 0 & -k_5 & 0 & k_8 & 0 & 0 & 0 & k_5 & 0 & k_{10} & 0 \\
0 & k_4 & 0 & 0 & 0 & k_9 & 0 & -k_4 & 0 & 0 & 0 & k_{11} \\
-k_1 & 0 & 0 & 0 & 0 & 0 & k_1 & 0 & 0 & 0 & 0 & 0 \\
0 & -k_2 & 0 & 0 & 0 & -k_4 & 0 & k_2 & 0 & 0 & 0 & -k_6 \\
0 & 0 & -k_3 & 0 & k_5 & 0 & 0 & 0 & k_3 & 0 & k_7 & 0 \\
0 & 0 & 0 & -k_{14} & 0 & 0 & 0 & 0 & 0 & k_{14} & 0 & 0 \\
0 & 0 & -k_7 & 0 & k_{10} & 0 & 0 & 0 & k_7 & 0 & k_{12} & 0 \\
0 & k_6 & 0 & 0 & 0 & k_{11} & 0 & -k_6 & 0 & 0 & 0 & k_{13}
\end{bmatrix} \tag{2}$$

The elements within the elastic–plastic stiffness matrix are shown below:

$$k_1 = \frac{EA}{l}, k_2 = \frac{2(a_{1z} + a_{2z} + b_{1z})b_{2z}}{l^2}, k_3 = \frac{2(a_{1y} + a_{2y} + b_{1y})b_{2y}}{l^2}, k_4 = \frac{(2a_{2z} + b_{1z})b_{2z}}{l^2},$$

$$k_5 = -\frac{(2a_{2y} + b_{1y})b_{2y}}{l^2}, k_6 = \frac{(2a_{1z} + b_{1z})b_{2z}}{l^2}, k_7 = -\frac{(2a_{1y} + b_{1y})b_{2y}}{l^2}, k_8 = 2a_{2y}b_{2y}, k_9 = 2a_{2z}b_{2z}$$

$$k_{10} = b_{1y}b_{2y}, k_{11} = b_{1z}b_{2z}, k_{12} = 2a_{1y}b_{2y}, k_{13} = 2a_{1z}b_{2z}, k_{14} = \frac{GJ}{l}$$

in which:

$$a_{1z} = p_{2z}q^3 - p_{1z}(1-q)^3 + p_{1z} + 1, a_{2z} = p_{1z}q^3 - p_{2z}(1-q)^3 + p_{2z} + 1$$

$$a_{1y} = p_{2y}q^3 - p_{1y}(1-q)^3 + p_{1y} + 1, a_{2y} = p_{1y}q^3 - p_{2y}(1-q)^3 + p_{2y} + 1$$

$$b_{1z} = (p_{2z} + p_{1z})q^2(3 - 2q) + 1, b_{2z} = \frac{6k_{0z}}{4a_{1z}a_{2z}l - b_{1z}^2 l}, b_{1y} = (p_{2y} + p_{1y})q^2(3 - 2q) + 1,$$

$$b_{2y} = \frac{6k_{0y}}{4a_{1y}a_{2y}l - b_{1y}^2 l}, p_{1z} = \frac{k_{0z}}{k_{1z}} - 1, p_{2z} = \frac{k_{0z}}{k_{2z}} - 1, p_{1y} = \frac{k_{0y}}{k_{1y}} - 1, p_{2y} = \frac{k_{0y}}{k_{2y}} - 1, q = \frac{l_p}{l}$$

E is the elastic modulus of the concrete, A is the cross-sectional area of the beam and column member, G is the shear modulus, J is the torsional inertia moment, and k_{iz} and k_{iy} (i = 0, 1, 2) are the cross-sectional bending stiffness in Figure 1, as determined by the M-θ relation in Figure 1. There are six force components and six displacement vectors at each end of the beam and column member, including three axial forces (N, V_y, and V_z) along the x-axis, y-axis, and z-axis directions and three bending moments (M_x, M_y, and M_z) around x-axis, y-axis, and z-axis directions. The displacement vector consists of three axial displacements (u, v, and w) along the x, y, and z axes and three angular displacements (θx, θy, and θz) around the x, y, and z axes, as seen in Figure 2.

The mass matrix of the beam and column member and the spatial shell element used in the RC frame's structure are consistent with the literature [21] and will not be repeated here.

3. Differential Equation of Motion of the RC Structure with MRD

3.1. Test Results of MRD

The MRD to be used in the RC frame's structure was made by ourselves and is the same as the MRD used in reference [31]; its mechanical property tests were carried out with the current of 0 A to 0.28 A at the interval of 0.14 A. The schematic diagram of the test loading equipment is shown schematically in Figure 3. As is shown in Figure 3, the test device was mainly composed of a fatigue testing machine and regulated DC power supply. Before the test, the position of the vertical centerline of the loading head of the actuator shall be consistent with that of the piston rod of the MRD, and then the loading head of the actuator and the damper should be connected through a spherical bowl seat, high strength bolt, and spherical cover plate. The other end of the damper was connected with the base through the three steel pipes and the high strength bolt, and the base is fixed by the anchor screw. The test data are automatically recorded on the computer console through the external digital acquisition device.

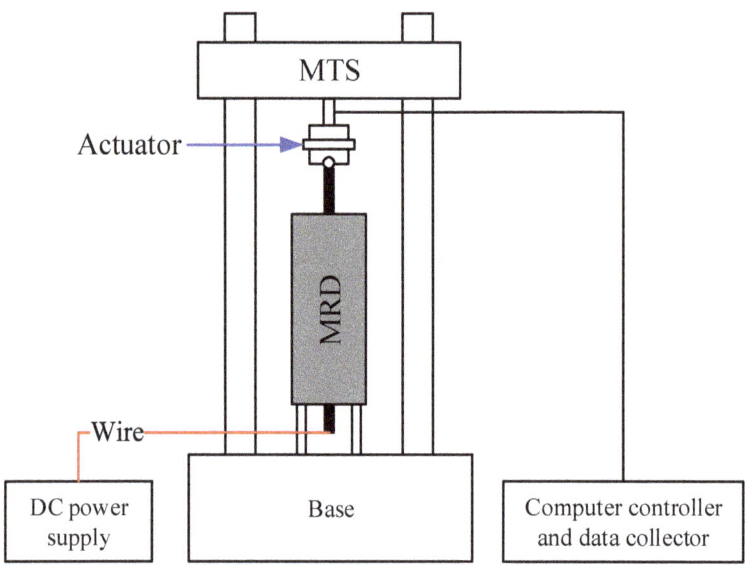

Figure 3. Schematic diagram of test loading equipment.

The mechanical performance fluctuation with displacement is depicted in Figure 4. It can be seen from Figure 4 that the mechanical performance fluctuation of the MRD with the current. When the current was small, the damping force provided by the MRD was also small. As the current increased, so did the damping force. It can be seen from Figure 4a, when the current was 0 A, that the maximum damping force of the MRD was around 5.3 kN; when the current reached 0.14 A and 0.28 A, the maximum damping forces of the MRD are around 60 kN and 120 kN, respectively, as shown in Figure 4b,c.

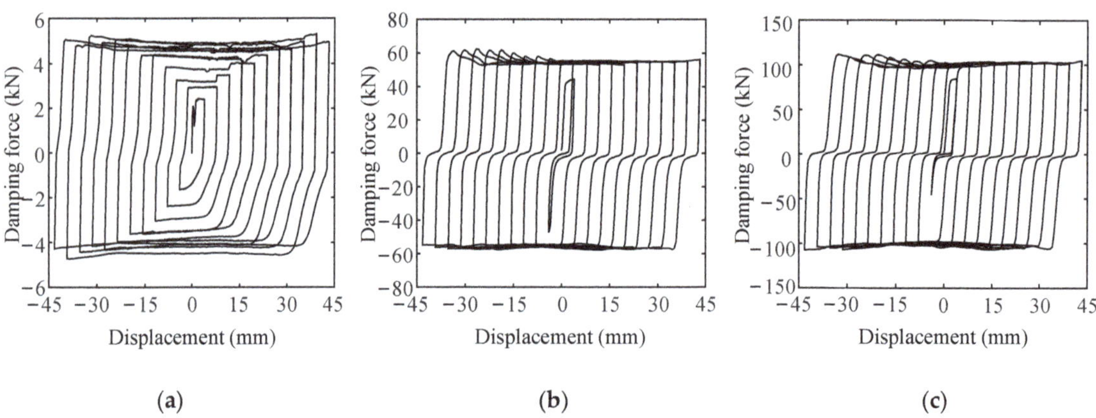

Figure 4. Damping force-displacement hysteresis loops of MRD at different current. (**a**) Current 0A; (**b**) current 0.14A; (**c**) current 0.28A.

3.2. Equilibrium Equation of the RC Structure

Under the action of the earthquake, the set of second-order nonlinear ordinary differential equations of an RC frame structure with MRDs can be expressed as follows [10]

$$M\ddot{u} + C\dot{u} + Ku = -MI\ddot{x}_g - HF \tag{3}$$

where M, C, and K are the mass matrix, the damping matrix, and the stiffness matrix of the frame structure, respectively; u is the displacement column vector of the controlled structure; $\ddot{x}_g$ is the seismic acceleration; I is the unit column vector; F is the control force column vector of the MRD that can be efficiently simulated by adopting analytical hysteresis models [35–37]; and H is the location matrix of the MRD.

C is the Rayleigh damping matrix as Equation (4) and α_1 and α_2 are the two constants [38].

$$C = \alpha_1 M + \alpha_2 K \tag{4}$$

where α_1 and α_2 are the two constants and can be expressed as

$$\alpha_1 = \frac{2\omega_1\omega_2(\zeta_1\omega_2 - \zeta_2\omega_1)}{\omega_2^2 - \omega_1^2}, \alpha_2 = \frac{2(\zeta_2\omega_2 - \zeta_1\omega_1)}{\omega_2^2 - \omega_1^2} \tag{5}$$

where ω_1 and ω_2 are the first and second order natural vibration frequencies of the structure (circular frequency), respectively, and ζ_1 and ζ_2 are the damping ratios of the first and second order frequencies of the structure, respectively.

Using the Wilson-θ method, the dynamic responses of the structure with and without the MRD can be calculated step by step. However, if one wants to carry out nonlinear dynamic analyses without performing iterations and by maintaining a very high level of accuracy, more recent methods can be adopted [39,40].

3.3. Semi-Active Control Algorithm

At present, the optimal control force imposed on RC structures by the MRD is mainly solved by the linear quadratic regulator (LQR) control algorithm [41]. As can be seen from Section 3.1, the damping force provided by our homemade MRD ranges from 3.1 to 120 kN. Because the MRD of the real-time control can effectively trace the required damping force, except the force beyond the range of damping force [42–44]. Thus, the same semi-active control strategy as in reference [34] is adopted in this paper.

3.4. MRD Location Matrix

The schematic installation diagram of the MRD in the structure is shown in Figure 5. The relationships between the damping force of the MRD and the force on the corresponding node of the structure are as follows:

$$\begin{cases} F_{jr} = F_{kr} = F_i \\ F_{pr} = F_{qx} = -F_i \\ F_{jz} = -F_{kz} = -2F_i h/d \end{cases} \tag{6}$$

The location matrix H is used to distribute the control force vector F of the MRD to the corresponding nodes of the structure, whose dimension is n × m, in which n is the amount of the structural degrees of freedom (DOFs), and m is the amount of herringbone supports. There are six DOFs at each node of the space frame structure, including three axial displacements along the x, y, and z directions, and three angular displacements around the x, y, and z axes, which are consistent with the literature [21]. Therefore, for the MRDs on the ith chevron support shown in Figure 5, the elements which correspond to nodes j and k in the location matrix H are: $H(6j - 5, i) = H(6k - 5, i) = 1$ and $H(6j - 3, i) = -H(6k - 3, i) = -2h/d$, and the elements which correspond to nodes p and q in the location matrix H are: $H(6p - 5, i) = H(6q - 5, i) = -1$. While the others without a damping force are all equal to zero. For example, the location matrix $\boldsymbol{H}$ for the MRDs in Figure 5 can be expressed as:

$$
\boldsymbol{H} = \begin{array}{c}
\begin{bmatrix}
\cdots & 1 & \cdots \\
\cdots & 0 & \cdots \\
\cdots & -2h/d & \cdots \\
\cdots & \mathbf{0} & \cdots \\
\cdots & 1 & \cdots \\
\cdots & 0 & \cdots \\
\cdots & -2h/d & \cdots \\
\cdots & \mathbf{0} & \cdots \\
\cdots & 1 & \cdots \\
\cdots & \mathbf{0} & \cdots \\
\cdots & 1 & \cdots
\end{bmatrix}
\begin{array}{l}
6j-5 \\
6j-4 \\
6j-3 \\
\cdots \\
6k-5 \\
6k-4 \\
6k-3 \\
\cdots \\
6p-5 \\
\cdots \\
6q-5
\end{array}
\end{array}
\tag{7}
$$

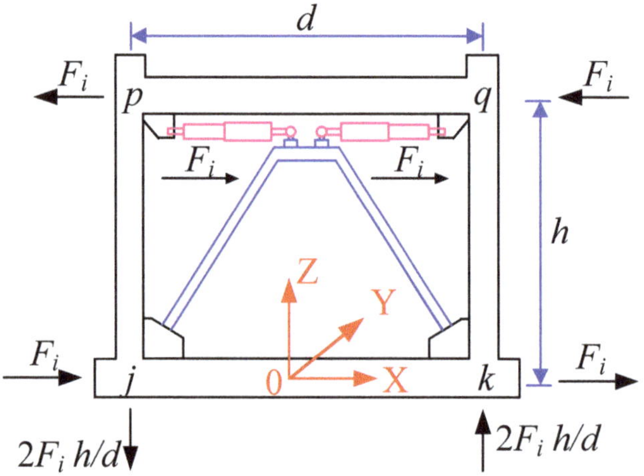

Figure 5. The schematic diagram of damping force distribution.

4. Case Analysis

4.1. The RC Frame Structure with and without MRD

This paper takes the five-story RC frame structure as an example, and the floor height of the structure is 3.6 m. The concrete strength grade is C35, and the thickness of each floor of the structure is 0.12 m. The seismic fortification intensity of the structure is eight (0.2 g), the site category is a class III area, and the design seismic group is the first group. The three-dimensional size of the RC frame's structure with and without MRDs is shown in Figure 6. The maximum inter-story displacement of the frame structure under the action of a strong earthquake occurs at the second floor [21,34], therefore, the MRDs were arranged on the second floor to exert the effect of an MRD energy dissipation and shock absorption as much as possible. The cross-sectional dimensions of the columns are all 0.5 m $\times$ 0.5 m; the cross-sectional dimensions of the X and Y directional beam are all 0.25 m $\times$ 0.6 m and 0.25 m $\times$ 0.5 m, respectively. The section reinforcement of the RC beams and columns are shown in Figure 7, in which the longitudinal reinforcement of the beams and columns is the HRB400 deformed reinforcement; the stirrup is the HRB335 ordinary round reinforcement.

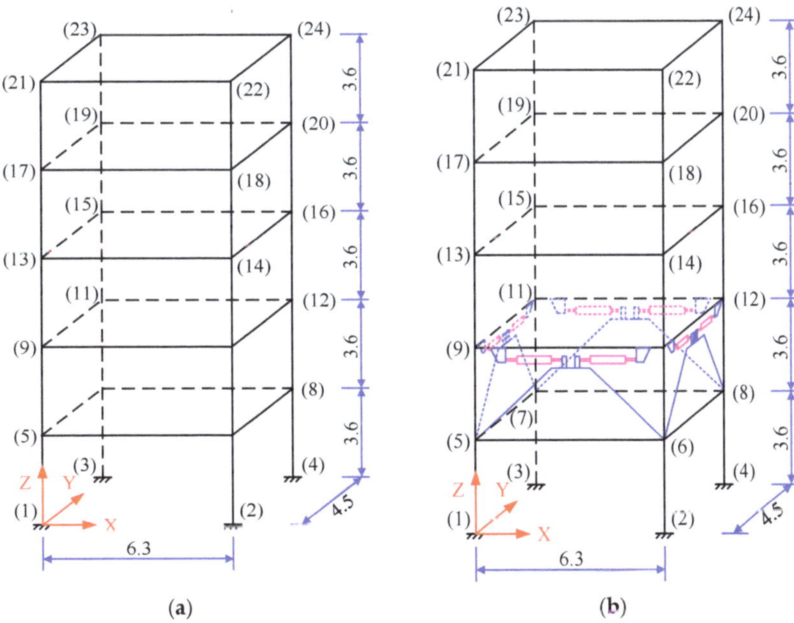

Figure 6. Three-dimensional diagram of the RC frame structure with and without an MRD (unit: m). (**a**) Without MRDs; (**b**) with MRDs.

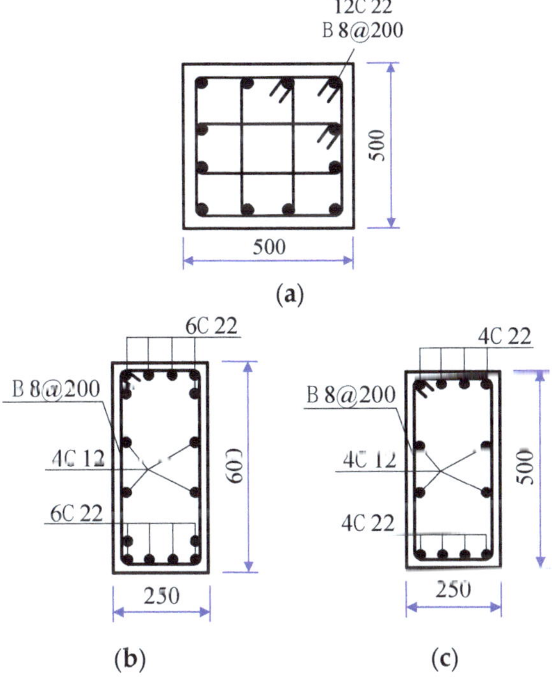

Figure 7. Reinforcement of the beams and columns (unit: mm). (**a**) Column; (**b**) X directional beam; (**c**) Y directional beam.

4.2. Introduction of Dynamic Time History Analysis Process

Based on the conditions and methods mentioned above, this paper used MATLAB to program the elastic–plastic time history analysis program of the RC frame's structure with and without an MRD. The program calculation flowchart is shown in Figure 8. The following assumptions are adopted in programming: (1) ignoring the influence of the second moment caused by a structural deformation; (2) ignoring the influence of the tangential relative displacement at both ends of the member on the bending moment and the rotation angle of the member; (3) ignoring the effect of the diagonal bracing on the mass of the structure; and (4) assume that the lower end of the bottom column is consolidated.

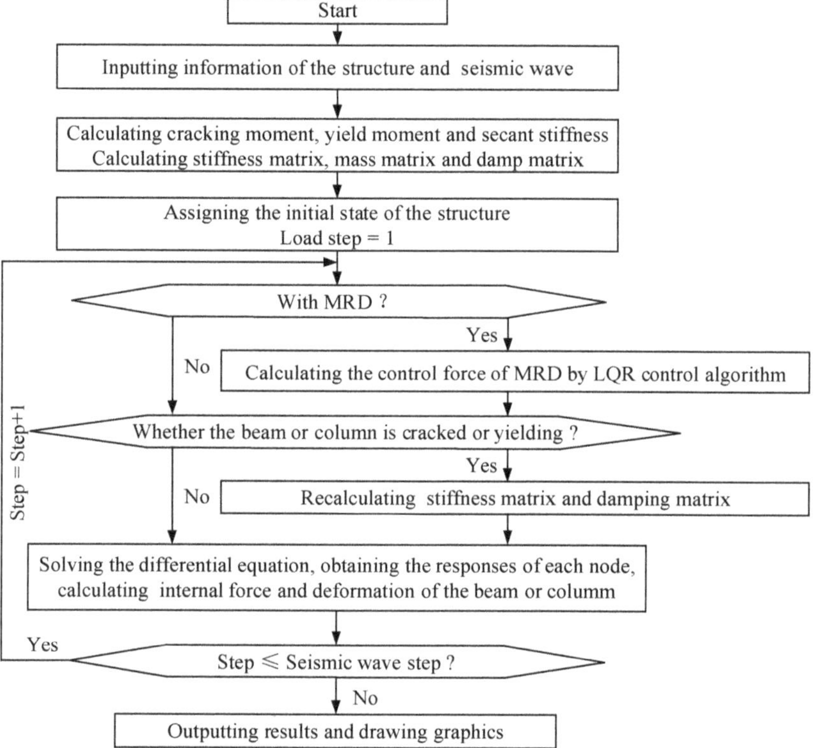

Figure 8. Calculation flowchart of elastic–plastic time history analysis program.

4.3. Verification of Model Validity

Due to the disadvantages of the long period and high cost of experimental simulation, a numerical simulation method was adopted in this paper to verify the effectiveness of an elastic–plastic time history analysis program by using MATLAB. By using ANSYS software, the finite element model of the structure was built, in which the BEAM188 element and SHELL181 element were used to simulate the beam, column, and floor slab, respectively. All the beams and columns in the ANSYS model were divided into four units, and all the floor slabs were divided into 4 × 4 units. In addition, the geometry, density, and material property in the ANSYS model were the same as that in the MATLAB program. Then, the modal analysis was conducted, and the calculated natural frequencies were listed in Table 1, in which the maximum error was only 1.59%. It shows the validity and accuracy of the elastic–plastic history analysis program for the RC frame's structure.

Table 1. Comparison of structural natural frequencies.

Natural Frequencies	1st	2nd	3rd	4th	5th	6th	7th	8th	9th	10th
ANSYS	0.759	0.823	0.896	2.371	2.559	2.777	4.322	4.441	4.940	6.352
MATLAB	0.749	0.810	0.881	2.348	2.524	2.749	4.255	4.487	4.877	6.315
Relative error (%)	1.29	1.59	0.95	0.97	1.36	1.00	1.54	−1.03	1.28	0.58

4.4. Analysis of Structural Damping Results

The elastic–plastic analysis of the seismic response of the RC frame's structure was carried out by using the MATLAB self-made program. According to the Chinese *code for design of concrete structures* (GB50010-2010) and the assumptions of the site type and design seismic grouping in Section 4.1, two seismic waves, the El-Centro wave (NS component) and Tangshan wave (NS component), were selected. The seismic duration was 20 s, the peak acceleration of the two seismic waves was set to 400 cm/s^2, and the ratio of the two-way seismic wave peak acceleration was X:Y = 1:0.85. The damping ratios ζ_1 and ζ_2 in Equation (5) were set to 0.05 according to the Chinese *code for design of concrete structures* (GB50010-2010). The Wilson-θ method was used to solve the differential equations of the structure and took θ − 1.40 The weight matrix coefficients $\alpha = 20$ and $\beta = 7 \times 10^{-6}$ in the LQR control algorithm were determined by a trial calculation.

4.4.1. Comparative Analysis of Multi-Dimensional Damping Results

The displacement time history results of the node 24 of the RC frame's structure with and without an MRD under the action of the El-Centro wave and Tangshan wave are shown in Figure 9. As can be seen from Figure 9, the X and Y directions' displacement time history curves of the node 24 of the structure without an MRD are offset, the reason is that some members in the structure enter the yield stage and have a residual deformation. After setting the MRD in the structure, the residual deformation is greatly reduced, and the offset of the displacement time history curve is reduced.

Figure 9a,b shows that under the action of the El-Centro wave and Tangshan wave, the horizontal bidirectional displacement response of the node 24 in the structure with an MRD was reduced compared with the structure without an MRD. Figure 9a shows that, under the action of an El-Centro wave, the X and Y directional maximum displacements of the top node in the RC structure were 76.66 mm and 74.44 mm, respectively, after setting the MRD in the structure; the X and Y directional maximum displacements of the top node in the RC structure were 52.42 mm and 50.79 mm, which were reduced by 31.63% and 31.76%, respectively. Under the action of the Tangshan wave shown in Figure 9b, the X and Y directional maximum displacements of the node 24 in the RC structure were 109.40 mm and 92.53 mm, respectively. After setting the MRD in the structure, the X and Y directional maximum displacements of the node 24 in the RC structure were 52.65 mm and 55.91 mm, which were reduced by 51.87% and 39.59%, respectively.

The acceleration time history results of the node 24 of the RC frame's structure without an MRD and with an MRD under the action of the El-Centro wave and Tangshan wave are shown in Figure 10. As can be seen from Figure 10, the X and Y directional acceleration time-history curves of the node 24 do not deviate. In addition, under the action of the El-Centro wave and Tangshan wave, the horizontal bidirectional acceleration response of the node 24 of the structure with an MRD is reduced compared with the structure without an MRD, but the effect is not as obvious as the MRD in reducing the displacement response of the structure.

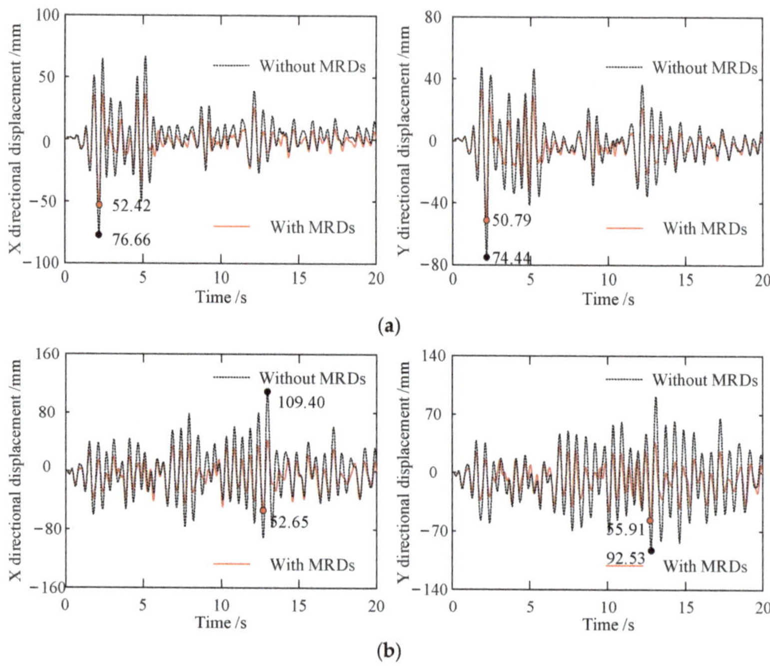

Figure 9. Comparison of horizontal bidirectional total displacement response of node 24 of RC frame structure. (**a**) Under the action of El-Centro wave; (**b**) under the action of Tangshan wave.

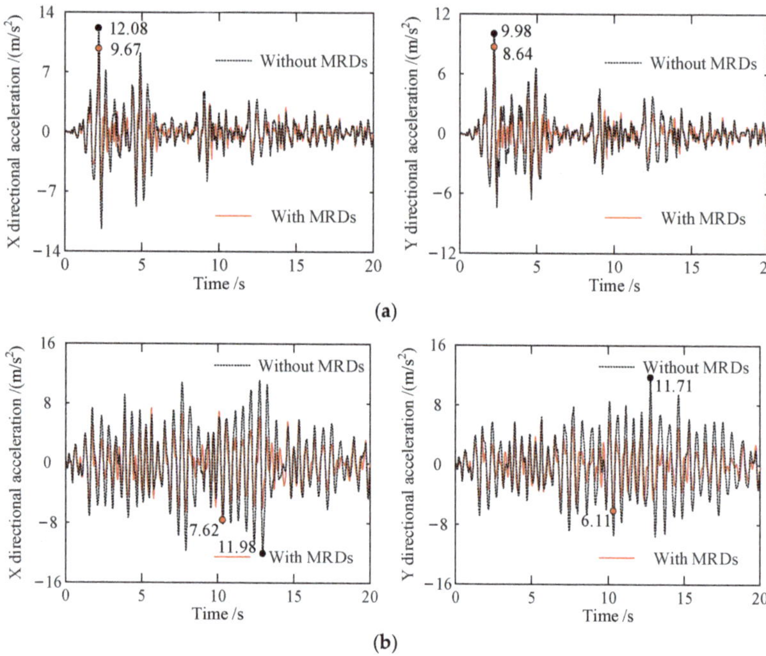

Figure 10. Comparison of horizontal bidirectional total acceleration response of node 24 of RC frame structure. (**a**) Under the action of El-Centro wave; (**b**) under the action of Tangshan wave.

Figure 10a shows that under the action of the El-Centro wave, the X and Y directional maximum accelerations of the top node in the RC structure were 12.08 m/s^2 and 9.98 m/s^2, respectively. After setting the MRD in the structure, the X and Y directional maximum accelerations of the top node in the RC structure were 9.67 m/s^2 and 8.64 m/s^2, which were reduced by 19.96% and 13.50%, respectively. Under the action of the Tangshan wave shown in Figure 10b, the X and Y directional maximum accelerations of the node 24 in the RC structure were 11.98 m/s^2 and 11.71 m/s^2, respectively. After setting the MRD in the structure, the X and Y directional maximum accelerations of the node 24 in the RC structure were 7.62 m/s^2 and 6.11 m/s^2, which were reduced by 36.67% and 47.86%, respectively.

Adding the MRD to the structure can be equivalent to increasing the stiffness and damping of the structure. It can be obtained from Equation (1) that the increase in the stiffness and damping of the structure is conducive to reducing the displacement response of the structure under an earthquake action, as shown in Figure 9. Although increasing the damping of the structure can reduce the acceleration response of the structure under an earthquake action, increasing the stiffness will increase the acceleration response of the structure. Therefore, after adding the MRD, the acceleration damping effect of the structure will be lower than the displacement damping effect, as shown in Figure 10.

4.4.2. Comparative Analysis of Maximum Response Results of Each Story

The envelope diagram of the maximum horizontal acceleration, displacement, and story drift ratio of the RC frame structure with and without an MRD under the El Centro and Tangshan waves are shown in Figure 11. As shown in Figure 11a,b, under the action of the El-Centro wave and Tangshan wave, compared with the structure without an MRD, the maximum horizontal displacement, acceleration, and story drift ratio of each story of the structure with an MRD are reduced.

Taking node 16 on the third story as an example, Figure 11a shows that under the action of the El-Centro wave, the X and Y directional maximum displacements of the RC structure were 55.22 mm and 54.04 mm, respectively. After setting the MRD in the structure, the X and Y directional maximum displacements of the RC structure were 37.15 mm and 35.24 mm, which were reduced by 32.73% and 34.80%, respectively. The X and Y directional maximum accelerations of the RC structure were 9.96 m/s^2 and 8.69 m/s^2, respectively. After setting the MRD in the structure, the X and Y directional maximum accelerations of the RC structure were 7.65 m/s^2 and 6.83 m/s^2, which are reduced by 23.22% and 21.35%, respectively. The X and Y directional maximum story drift ratios of the RC structure were 4.46 × 10^{-3} rad and 4.51 × 10^{-3} rad, respectively. After setting the MRD in the structure, the X and Y directional maximum story drift ratios of the RC structure were 3.18 × 10^{-3} rad and 3.15 × 10^{-3} rad, which were reduced by 28.67% and 29.91%, respectively.

As shown in Figure 11b, under the action of the Tangshan wave, the X and Y directional maximum displacements of the RC structure were 78.21 mm and 65.09 mm, respectively. After setting the MRD in the structure, the X and Y directional maximum displacements of the RC structure were 37.78 mm and 39.43 mm, which were reduced by 51.69% and 39.42%, respectively. The X and Y directional maximum accelerations of the RC structure were 8.41 m/s^2 and 6.66 m/s^2, respectively. After setting the MRD in the structure, the X and Y directional maximum accelerations of the RC structure were 4.89 m/s^2 and 4.42 m/s^2, which were reduced by 41.90% and 33.59%, respectively. The X and Y directional maximum story drift ratios of the RC structure were 6.58 × 10^{-3} rad and 5.53 × 10^{-3} rad, respectively. After setting the MRD in the structure, the X and Y directional maximum story drift ratios of the RC structure were 2.92 × 10^{-3} rad and 3.13 × 10^{-3} rad, which were reduced by 55.64% and 43.49%, respectively.

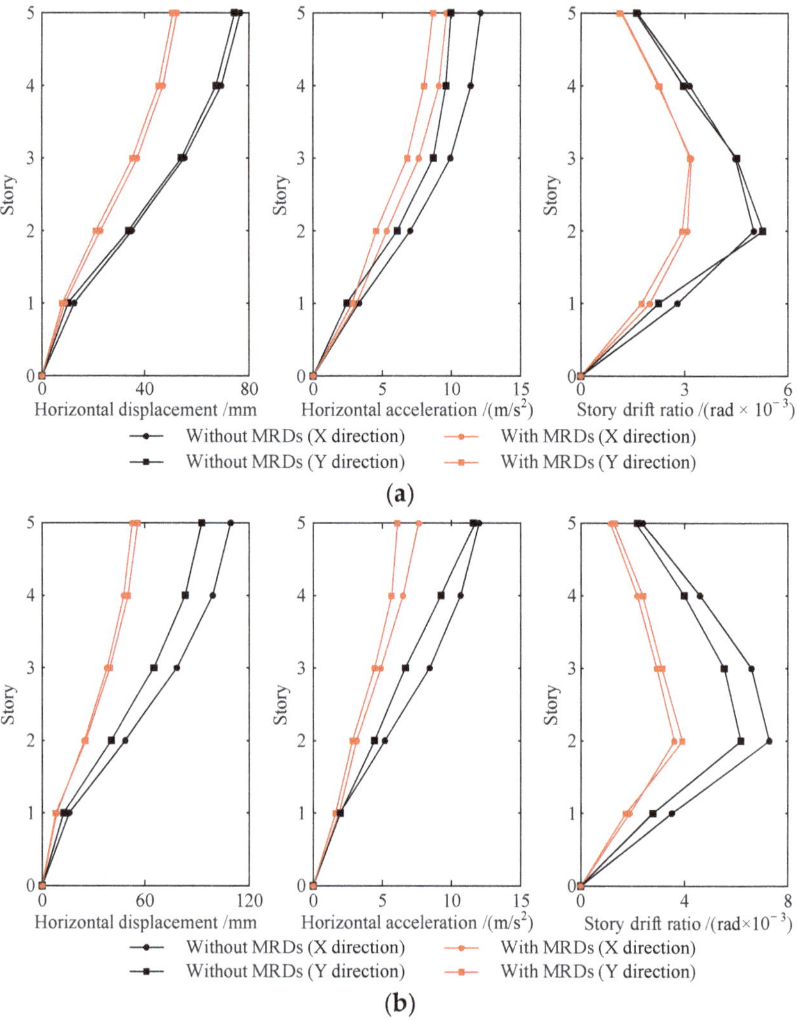

Figure 11. Envelope diagram of maximum horizontal displacement, acceleration and story drift ratio of RC frame structure. (**a**) Under the action of El-Centro wave; (**b**) under the action of Tangshan wave.

As can be seen from Figure 11, the maximum story drift ratio of the RC frame structure with and without an MRD firstly increase and then decrease with the increase in the structural stories, and the maximum values are all less than 1/50. After setting the MRD in the RC frame's structure, the maximum story drift ratio of each story are far less than 1/50, which significantly improves the seismic performance of the RC frame's structure.

4.4.3. Comparative Analysis of Moment Rotation Hysteretic Curve Results

Figures 12 and 13 show the comparison of moment-angle hysteresis curves at the bottom of each column of the RC frame's structure with and without an MRD. As can be seen from Figure 12, under the action of the El-Centro wave, a hysteresis loops appeared in both X and Y directions of the column in the third story of the structure, and the intersection of the hysteresis curves and coordinate axes deviated from the origin, indicating that the column entered the yield stage in X and Y directions, and the maximum residual deformations in X and Y directions were 0.923×10^{-3} rad and 0.914×10^{-3} rad, respectively. After setting the MRD in the structure, the seismic response of the column in the X and Y direction was weakened significantly, the maximum value of the moment rotation curve at the bottom of the column was significantly smaller, the hysteretic loop area was reduced, and the maximum residual deformations in X and Y directions were reduced to 0.205×10^{-3} rad and 0.188×10^{-3} rad, respectively. Similar to the column in the third story, after setting the MRD in the structure, the maximum residual deformations in the X and Y direction of the column in the fourth story decreased from 0.648×10^{-3} rad and 0.672×10^{-3} rad to 0.241×10^{-3} rad and 0.141×10^{-3} rad, respectively.

It can be seen from Figure 13 that under the action of the Tangshan wave, the rotation angles in X and Y directions were large when the column in the third story of the structure was loaded, and a large residual deformation occurred during the unloading, and hysteresis loops appeared in both X and Y directions of the column in the third story of the structure, the intersection of the hysteresis curves and coordinate axes deviated from the origin, which indicates that the column entered the yield stage in the X and Y directions, and the maximum residual deformations in the X and Y directions were 1.628×10^{-3} rad and 2.101×10^{-3} rad, respectively. After setting the MRD in the structure, the seismic response of the column in X and Y direction was weakened significantly, the maximum value of the moment rotation curve at the bottom of the column was significantly smaller, the hysteretic loop area was reduced, and the maximum residual deformations in the X and Y directions were reduced to 0.511×10^{-3} rad and 0.297×10^{-3} rad, respectively. Similar to the column in the third story, after setting the MRD in the structure, the maximum residual deformations in the X and Y direction of the column in fourth story decreased from 1.217×10^{-3} rad and 1.562×10^{-3} rad to 0.264×10^{-3} rad and 0.065×10^{-3} rad, respectively.

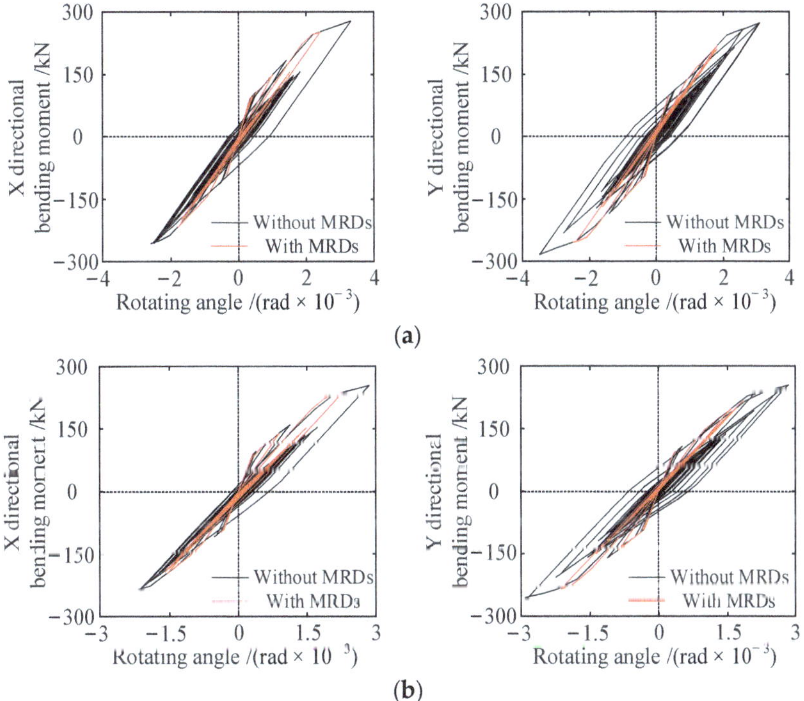

Figure 12. Bending moment-angle hysteresis loops of RC frame structures with and without MRD under El-Centro waves. (**a**) The column in third story; (**b**) the column in fourth story.

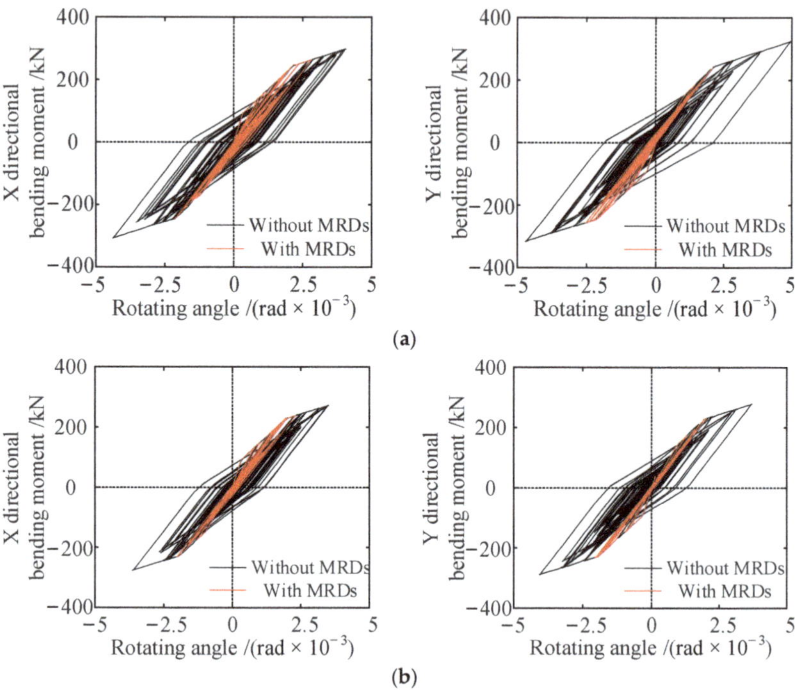

Figure 13. Bending moment-angle hysteresis loops of RC frame structures with and without MRD under Tangshan waves. (**a**) The column in third story; (**b**) the column in fourth story.

Table 2 shows the maximum residual displacement angles in the X and Y directions at the column bottom of each story of the RC frame's structure with and without an MRD under the action of the seismic wave. It can be seen from Table 2 that the MRD can effectively consume the vibration energy transmitted to the structure and reduce the residual deformation at the column bottom of each story of the structure.

Table 2. The maximum residual displacement angle of column bottom of RC frame structure under seismic wave (rad $\times 10^{-3}$).

Story (Node)	El-Centro Wave				Tangshan Wave			
	Without MRD		With MRD		Without MRD		With MRD	
	X-Direction	Y-Direction	X-Direction	Y-Direction	X-Direction	Y-Direction	X-Direction	Y-Direction
1(4)	0	0	0	0	0	0	0	0
2(8)	0.412	0.828	0.014	0.017	1.137	1.686	0.031	0.069
3(12)	0.923	0.914	0.205	0.188	1.628	2.101	0.511	0.297
4(16)	0.648	0.672	0.241	0.141	1.217	1.562	0.264	0.065
5(20)	0.205	0.133	0	0	0.561	0.620	0	0

In addition, whether the X-direction and Y-direction of each member of the RC frame structure with or without an MRD enters the yield stage under the action of two seismic waves can be compared in Figure 14. All the above indicate that the stiffness of the structural components would degrade under the action of seismic waves, but the seismic performance of the RC frame's structure is significantly improved after setting the MRD in the structure.

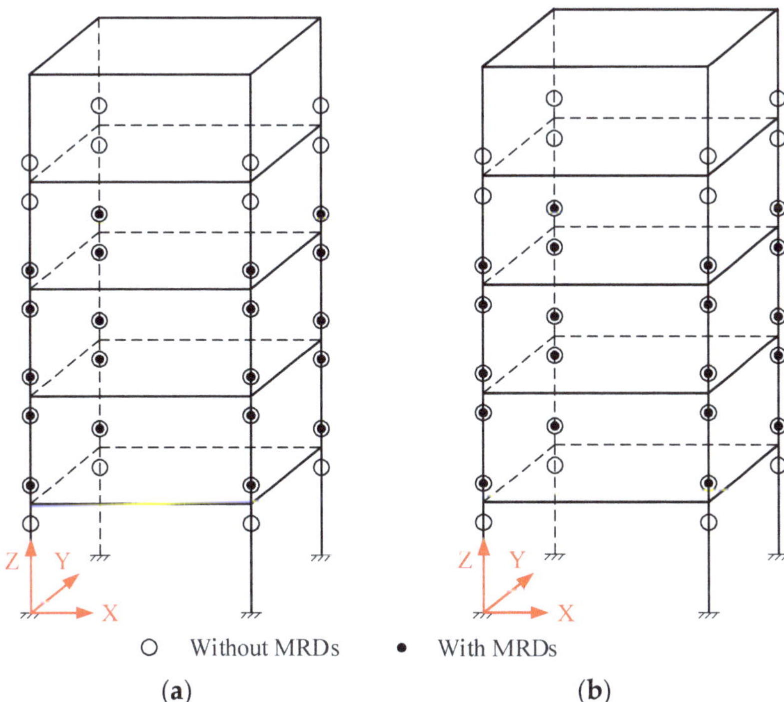

Figure 14. Plastic hinges distribution of RC frame structure with and without MRD under seismic waves. (**a**) X direction; (**b**) Y direction.

4.4.4. Cracking and Yield of Each Member of Structure

Figure 14 shows the plastic hinges distribution of the beams and columns of RC frame's structure with and without an MRD under different seismic waves. Under the action of two seismic waves, the plastic hinge distribution of the RC frame's structure was exactly the same.

It can be seen from Figure 14a,b that under the action of the El-Centro wave and Tangshan wave, most of the column ends of the second to fifth stories of the structure appeared to be a plastic hinge, but the beam ends did not yield. After the MRD was set in the structure, the plastic hinges at the column ends of the structure were reduced. The number of plastic hinges in the X direction and Y direction have been decreased by 37.50%.

Therefore, setting the MRD in the RC frame's structure can effectively inhibit the yield of the structural members, so as to improve the overall seismic performance of the structure. In addition, the multi-dimensional elastic–plastic calculation model and elastic–plastic dynamic time history analysis program established by MATLAB software in this chapter can effectively describe the yield position, sequence, failure degree, and process of each beam and column, so as to judge the plastic deformation concentration position, weak position, and possible failure type of the overall structure.

5. Conclusions

(1) The multi-dimensional elastoplastic calculation model of the MRD frame's structure was established, and the elastoplastic dynamic time history analysis program was developed by using MATLAB software, which could accurately calculate the multi-dimensional elastoplastic response of the structure under a strong earthquake.

(2) After the MRD is set in the frame structure, the maximum horizontal displacement and acceleration of each story decreases. The maximum displacement and acceleration of the top node 24 on the structure in the X direction and Y direction decreased by 51.87%, 39.59%, 36.67%, and 47.86%, respectively; the decrease in the acceleration is not very significant.

(3) For the frame structure with an MRD, the offset of the displacement time history curve of the column in the third story is weakened and the hysteretic loop area of the structural members is significantly reduced. The maximum residual displacement angle in the X and Y directions of the column in the third story decreased from 1.628×10^{-3} rad and 2.101×10^{-3} rad to 0.511×10^{-3} rad and 0.297×10^{-3} rad, indicating that the MRD can effectively consume the vibration energy of the incoming structure and significantly improve the seismic performance of the structure.

(4) The column end of the frame structure without an MRD appears to be more of a plastic hinge, which is the yield mechanism of the column hinge. Compared with the frame structure without the MRD, after setting the MRD in the structure, the number of plastic hinges in the X direction and Y direction were all reduced by 37.50%. Although some structural members still yield, it will not endanger the safety of the whole structure.

Author Contributions: Conceptualization, methodology, investigation, data curation, and writing—review and editing, X.Z. Software, formal analysis, writing—original draft, visualization, and data curation, C.M. Conceptualization, project administration, supervision, and writing—review and editing, J.Z. Software, validation, methodology, data curation, and review and editing, Y.G. Supervision, project administration, resources, and funding acquisition, Y.S. Data curation, writing—original draft, validation, and funding acquisition, J.Y. All authors have read and agreed to the published version of the manuscript.

Funding: This research was funded by the National Natural Science Foundation of China, grant number 51878621; Program for Changjiang Scholars and Innovative Research Team in University of Minister of Education of China, grant number IRT_16R67.

Institutional Review Board Statement: Not applicable.

Informed Consent Statement: Not applicable.

Data Availability Statement: Not applicable.

Acknowledgments: The authors are grateful for the financial support provided by the National Natural Science Foundation of China and the Program for Changjiang Scholars and Innovative Research Team in the University of the Minister of Education of China.

Conflicts of Interest: The authors declare no conflict of interest.

References

1. Zhang, X.C.; Zhang, X.; Zhao, Y.X.; Zhao, J.; Xu, Z. Experimental and numerical studies on a composite MR damper considering magnetic saturation effect. *Eng. Struct.* **2017**, *132*, 576–585. [CrossRef]
2. Xu, Z.D.; Huang, X.H.; Xu, F.H.; Yuan, J. Parameters optimization of vibration isolation and mitigation system for precision platforms using non-dominated sorting genetic algorithm. *Mech. Syst. Signal Process.* **2019**, *128*, 191–201. [CrossRef]
3. Wang, W.X.; Hua, X.G.; Chen, Z.Q.; Wang, X.; Song, G. Modeling, simulation, and validation of a pendulum-pounding tuned mass damper for vibration control. *Struct. Control Health Monit.* **2019**, *26*, e2326. [CrossRef]
4. Sakurai, T.; Morishita, S. Seismic response reduction of a three-storey building by an MR grease damper. *Front. Mech. Eng.* **2017**, *12*, 224–233. [CrossRef]
5. Xu, Z.D.; Jia, D.H.; Zhang, X.C. Performance tests and mathematical model considering magnetic saturation for magnetorheological damper. *J. Intell. Mater. Syst. Struct.* **2012**, *23*, 1331–1349. [CrossRef]
6. Lv, H.Z.; Zhang, S.S.; Sun, Q.; Chen, R.; Zhang, W.J. The dynamic models, control strategies and applications for magnetorheological damping systems: A systematic review. *J. Vib. Eng. Technol.* **2020**, *9*, 131–147. [CrossRef]
7. Parlak, Z.; Engin, T. Time-dependent CFD and quasi-static analysis of magnetorheological fluid dampers with experimental validation. *Int. J. Mech. Sci.* **2012**, *64*, 22–31. [CrossRef]
8. Xu, Z.D.; Sha, L.F.; Zhang, X.C.; Ye, H.-H. Design, performance test and analysis on magnetorheological damper for earthquake mitigation. *Struct. Control Health Monit.* **2013**, *20*, 956–970. [CrossRef]

9. Zemp, R.; De la Llera, J.C.; Weber, F. Experimental analysis of large capacity MR dampers with short- and long-stroke. *Smart Mater. Struct.* **2014**, *23*, 125028. [CrossRef]
10. Zhang, X.C.; Xu, Z.D. Testing and modeling of a CLEMR damper and its application in structural vibration reduction. *Nonlinear Dynamics* **2012**, *70*, 1575–1588. [CrossRef]
11. Jiang, R.L.; Rui, X.T.; Zhu, W.; Yang, F.; Zhang, Y.; Gu, J. Design of multi-channel bypass magnetorheological damper with three working modes. *Int. J. Mech. Mater. Des.* **2022**, *18*, 155–167. [CrossRef]
12. Raju, K.R.; Jame, A.; Gopalakrishnan, N.; Muthumani, K.; Iyer, N.R. Experimental studies on seismic performance of three-storey steel moment resisting frame model with scissor-jack-magnetorheological damper energy dissipation systems. *Struct. Control Health Monit.* **2014**, *21*, 741–755. [CrossRef]
13. Cruze, D.; Gladston, H.; Farsangi, E.N.; Banerjee, A.; Loganathan, S.; Solomon, S.M. Seismic performance evaluation of a recently developed magnetorheological damper: Experimental investigation. *Pract. Period. Struct. Des. Constr.* **2021**, *26*, 04020061. [CrossRef]
14. Rakshita, R.; Daniel, C.; Hemalatha, G.; Sarala, L.; Tensing, D.; Manoharan, S.S. Studies on modeling and control of RCC frame with MR damper. *Smart Technol. Sustain. Dev.* **2021**, *78*, 223–234.
15. Chae, Y.; Ricles, J.M.; Sause, R. Large-scale real-time hybrid simulation of a three-story steel frame building with magneto-rheological dampers. *Earthq. Eng. Struct. Dyn.* **2014**, *43*, 1915–1933. [CrossRef]
16. Aggumus, H.; Cetin, S. Experimental investigation of semiactive robust control for structures with magnetorheological dampers. *J. Low Freq. Noise Vib. Act. Control* **2018**, *37*, 216–234. [CrossRef]
17. Eltahawy, W.; Ryan, K.; Cesmeci, S.; Gordaninejad, F. Displacement/velocity-based control of a liquid spring—MR damper for vertical isolation. *Struct. Control Health Monit.* **2019**, *26*, e2363. [CrossRef]
18. Bhaiya, V.; Shrimali, M.K.; Bharti, S.D.; Datta, T.K. Modified semi-active control with MR dampers for partially observed systems. *Eng. Struct.* **2019**, *191*, 129–147. [CrossRef]
19. Li, L.Y.; Liang, H.Z. Semiactive control of structural nonlinear vibration considering the MR damper model. *J. Aerosp. Eng.* **2018**, *31*, 04018095. [CrossRef]
20. Zafarani, M.M.; Halabian, A.M. Supervisory adaptive nonlinear control for seismic alleviation of inelastic asymmetric buildings equipped with MR dampers. *Eng. Struct.* **2018**, *176*, 849–858. [CrossRef]
21. Zhao, J.; Li, K.; Zhang, X.C.; Sun, Y.; Xu, Z. Multidimensional vibration reduction control of the frame structure with magnetorheological damper. *Struct. Control Health Monit.* **2020**, *27*, e2572. [CrossRef]
22. Xu, F.H.; Xu, Z.D.; Zhang, X.C. Study on the space frame structures incorporated with magnetorheological dampers. *Smart Struct. Syst.* **2017**, *19*, 279–288. [CrossRef]
23. Xu, L.H.; Li, Z.X.; Lv, Y. Nonlinear seismic damage control of steel frame-steel plate shear wall structures using MR dampers. *Earthq. Struct.* **2014**, *7*, 937–953. [CrossRef]
24. Xu, L.H.; Xie, X.S.; Li, Z.X. Seismic performances of magnetorheological flag-shaped damping braced frame structures. *Smart Mater. Struct.* **2020**, *29*, 075032. [CrossRef]
25. Dai, J.; Xu, Z.D.; Gai, P.P.; Hu, Z.-W. Optimal design of tuned mass damper inerter with a Maxwell element for mitigating the vortex-induced vibration in bridges. *Mech. Syst. Signal Process.* **2021**, *148*, 107180. [CrossRef]
26. Xu, Z.D.; Yang, Y.; Miao, A.N. Dynamic analysis and parameter optimization of pipelines with multidimensional vibration isolation and mitigation device. *J. Pipeline Syst. Eng. Pract.* **2021**, *12*, 04020058. [CrossRef]
27. Lu, H.; Xu, Z.D.; Iseley, T.; Matthews, J.C. Novel data-driven framework for predicting residual strength of corroded pipelines. *J. Pipeline Syst. Eng. Pract.* **2021**, *12*, 04021045. [CrossRef]
28. Ge, T.; Xu, Z.D.; Yuan, F.G. Predictive model of dynamic mechanical properties of VE damper based on acrylic rubber–graphene oxide composites considering aging damage. *J. Aerosp. Eng.* **2022**, *35*, 04021132. [CrossRef]
29. Sun, B.; Liu, X.; Xu, Z.D. A Multiscale Bridging Material Parameter and Damage Inversion Algorithm from Macroscale to Mesoscale Based on Ant Colony Optimization. *J. Eng. Mech.* **2022**, *148*, 04021150. [CrossRef]
30. Li, H.; Xu, Z.D.; Gomez, D.; Gai, P.; Wang, F.; Dyke, S.J. A modified fractional order derivative zener model for rubber like devices for structural control. *J. Eng. Mech.* **2022**, *148*, 04021119. [CrossRef]
31. You, J.T.; Yang, Y.; Fan, Y.F.; Zhang, X. Seismic response study of L-shaped frame structure with magnetorheological dampers. *Appl. Sci.* **2022**, *12*, 5976. [CrossRef]
32. Xu, Z.D.; Guo, Y.Q. Fuzzy control method for earthquake mitigation structures with magnetorheological dampers. *J. Intell. Mater. Syst. Struct.* **2006**, *17*, 871–881. [CrossRef]
33. Panagiotakos, T.B.; Fardis, M.N. Deformations of reinforced concrete members at yielding and ultimate. *ACI Struct. J.* **2001**, *98*, 135–148.
34. Rao, S.S. *The Finite Element Method in Engineering*, 6th ed.; Butterworth-Heinemann: Oxford, UK, 2017.
35. Vaiana, N.; Rosati, L. Classification and unified phenomenological modeling of complex uniaxial rate-independent hysteretic responses. *Mech. Syst. Signal Process.* **2022**, *182*, 1132. [CrossRef]
36. Vaiana, N.; Sessa, S.; Marmo, F.; Rosati, L. A class of uniaxial phenomenological models for simulating hysteretic phenomena in rate-independent mechanical systems and materials. *Nonlinear Dyn.* **2018**, *93*, 1647–1669. [CrossRef]
37. Vaiana, N.; Sessa, S.; Rosati, L. A generalized class of uniaxial rate-independent models for simulating asymmetric mechanical hysteresis phenomena. *Mech. Syst. Signal Process.* **2021**, *146*, 2162. [CrossRef]

38. Yang, H.X.; Shen, T.; Yao, J.G. Study on Earthquake Response Analysis of Reinforced Concrete Masonry Structure. *Adv. Mater. Res.* **2011**, *1336*, 244–248. [CrossRef]
39. Vaiana, N.; Sessa, S.; Marmo, F.; Rosati, L. Nonlinear dynamic analysis of hysteretic mechanical systems by combining a novel rate-independent model and an explicit time integration method. *Nonlinear Dyn.* **2019**, *98*, 2879–2901. [CrossRef]
40. Vaiana, N.; Sessa, S.; Paradiso, M.; Marmo, F.; Rosati, L. An Efficient Computational Strategy for Nonlinear Time History Analysis of Seismically Base-Isolated Structures. In Proceedings of the AIMETA 2019 XXIV Conference, Rome, Italy, 15–19 September 2019; Springer: Cham, Switzerland, 2020; pp. 1340–1353, Lecture Notes in Mechanical Engineering.
41. Motra, G.B.; Mallik, W.; Chandiramani, N.K. Semi-active vibration control of connected buildings using magnetorheological dampers. *J. Intell. Mater. Syst. Struct.* **2011**, *22*, 1811–1827. [CrossRef]
42. Ubaidillah, H.K.; Kadir, F.A.A. Modelling, characterisation and force tracking control of a magnetorheological damper under harmonic excitation. *Int. J. Model. Identif. Control* **2011**, *13*, 9–21. [CrossRef]
43. Russo, R.; Terzo, M. Modelling, parameter identification, and control of a shear mode magnetorheological device. *Proc. Inst. Mech. Eng. Part I-J. Syst. Control. Eng.* **2011**, *225*, 549–562. [CrossRef]
44. Weber, F. Robust force tracking control scheme for MR dampers. *Struct. Control Health Monit.* **2015**, *22*, 1373–1395. [CrossRef]

Article

Numerical Simulation and Torsional Vibration Mitigation of Spatial Eccentric Structures with Multiple Magnetorheological Dampers

Yang Yang [1] and Ying-Qing Guo [2,*]

1   China-Pakistan Belt and Road Joint Laboratory on Smart Disaster Prevention of Major Infrastructures, Southeast University, Nanjing 210096, China
2   College of Mechanical and Electronic Engineering, Nanjing Forestry University, Nanjing 210037, China
*   Correspondence: gyingqing@njfu.edu.cn; Tel.: +86-13951835602

Abstract: Eccentric structures will have torsional vibrations subjected to earthquakes, which can accelerate the damage of structures, and even become the main cause of building collapse. Semi-active control systems equipped with multiple magnetorheological (MR) dampers have been widely applied in structural vibration control. In this study, numerical models of spatial eccentric structures with multiple MR dampers were established, and time history analysis was conducted to mitigate torsional vibrations of eccentric structures. Firstly, a full-scale spatial eccentric structure model with both plan asymmetry and vertical irregularity was established in OpenSEES, and the accuracy of the structure model was verified by comparisons with model results from SAP2000. Then, the mathematical model of MR dampers was introduced to the structure model using the 'Truss' element and self-defined material in OpenSEES, and damping forces obtained from the MR damper model were compared with experimental data. Finally, modal analysis and nonlinear time history analysis of the eccentric structure model equipped with multiple MR dampers subjected to different seismic excitations were performed. Comparisons between the seismic responses of the uncontrolled structure and the structure with multiple MR dampers were carried out to demonstrate the effectiveness of the MR control system. Numerical results show that the control system with multiple MR dampers can significantly attenuate the torsional vibrations of eccentric structures, and thus possess significant engineering application prospects.

Keywords: eccentric structure; torsional vibration; magnetorheological damper; OpenSEES; modal analysis; time history analysis

Citation: Yang, Y.; Guo, Y.-Q. Numerical Simulation and Torsional Vibration Mitigation of Spatial Eccentric Structures with Multiple Magnetorheological Dampers. *Actuators* **2022**, *11*, 235. https://doi.org/10.3390/act11080235

Academic Editor: Hongli Ji

Received: 14 July 2022
Accepted: 15 August 2022
Published: 16 August 2022

Publisher's Note: MDPI stays neutral with regard to jurisdictional claims in published maps and institutional affiliations.

1. Introduction

In order to meet the needs of functional diversity and urban planning of modern architecture, the structures often present asymmetric layout and irregular facade. The structures whose center of mass and rigidity are not coincident can be collectively referred to as eccentric structures [1]. Under the action of earthquake, the inertia force passes through the center of mass, while the restoring force of the lateral resisting members passes through the center of rigidity. Therefore, the eccentric structures not only vibrate horizontally, but also have torsional motion around the center of rigidity, forming the coupled translation-torsion vibration [2,3]. Meanwhile, the actual earthquake excitations contain multiple components, not only translational components in different directions, but also torsional components, which can lead to the torsional responses of structures. Theoretical research and earthquake damage investigations indicate that the torsional responses can concentrate the deformation in some columns and amplify the acceleration at certain floors, which will make the structure susceptible to further damage, especially for eccentric structures vulnerable to seismic excitations and wind loadings, and even become the main factor leading to the collapse of buildings in some cases [4,5]. Therefore, it is of great practical significance to mitigate the torsional vibration of eccentric structures under earthquake action.

Numerous studies have manifested that the traditional seismic design, such as improving the structural stiffness and ductility or using strong materials, cannot guarantee the safety of the structure under future dynamic loads, and cannot meet the economic requirements [6]. On this basis, the concept of structural vibration control was first proposed in the 1970s, which uses the control system attached to the structure to exert a group of control forces actively or passively in order to mitigate the structural response [7–11]. Among various vibration control methods, the semi-active control has shown obvious superiority in that it remarkably outperforms passive control and requires much less external energy than active control [12–16]. As a typical subset of semi-active control systems, MR dampers have the characteristics of high controllability, low energy consumption, fast response, mechanical simplicity, and reliable damping effect, and thus have been widely studied and applied in automobile suspension, cable and civil structures, and aerospace engineering [17–21].

In recent years, a wide range of analytical and experimental studies on structural torsional vibration control using MR dampers have been carried out. Yoshida et al. [22,23] proposed a MR control system to reduce the coupled translation-torsion motions in asymmetric buildings based on a clipped-optimal control algorithm, and this method was numerically assessed by two full scale irregular building models and experimentally verified by a two-story building with an asymmetric stiffness distribution. The results showed that the MR control system can significantly reduce the torsional coupled responses of irregular buildings. Li et al. [24] adopted a multi-state control strategy to mitigate the coupled translation and torsion responses of a three-story reinforced concrete frame–shear wall eccentric structure by three MR dampers, and the shaking table test results showed that MR dampers are effective for torsional seismic response control. Shook et al. [25] experimentally investigated the application of four MR dampers for the torsional response control of a 3-story, 9 m torsion-benchmark building using the fuzzy logic controller optimized by genetic algorithm. Bharti et al. [26] verified the effectiveness of MR damper-based control systems for torsional response mitigation through a numerical idealized one-story one-bay plan asymmetric building model and two MR dampers based on the Lyapunov stability theory. Hu et al. [27] adopted two pairs of eccentrically placed MR dampers to control the vibration of a 10-story irregular steel frame building by the clipped-optimal strategy using LQR algorithm. The numerical results demonstrated that the MR dampers can effectively reduce the inter-story drifts as well as roof displacements and accelerations of irregular structures. Zafarani et al. [28] proposed a coupled fuzzy logical control algorithm to simultaneously control two MR dampers, and its effectiveness was verified numerically by simulating nonlinear seismic response of a one-way asymmetric inelastic single-story structure model through time history analysis. Zafarani and Halabian [29,30] proposed an adaptive model-based strategy to mitigate the inelastic torsional responses of one-story and multi-story plan-asymmetric structures with MR dampers, where the changes of the system can be considered in determining the control force of MR dampers. Al-Fahdawi et al. [31] used multiple MR dampers to connect two full-scale coupled buildings for the vibration control of structural responses under bi-directional earthquakes, and the MR dampers were controlled by the adaptive neuro-fuzzy and simple adaptive control methods.

However, previous research efforts on the torsional vibration control using MR dampers have obvious limitations. In the numerical analysis of MR damped structures, simplified plane structure models with idealized linear beam-column elements have been established in most studies, which cannot capture the torsional vibration characteristics of spatial eccentric structures. It is difficult to simulate the nonlinear characteristics of MR dampers in time history analysis of MR damped structures using common finite element software. What is more, in most cases for structural vibration control using multiple MR dampers, the implemented MR dampers in the control system have obvious disadvantages. The control strategies for control systems with multiple MR dampers proposed by existing research are all 'single-input and single-output' control modes, that is, a separate controller is set for each MR damper, without the cooperation and interaction between different dampers [24,25,27]. Such decentralized control strategies require a large number of controllers, which greatly increases the cost of the control system, and easily leads to control imbalance and poor stability. Additionally, for the structures with both torsional and translational vibration, the suitable positions of MR dampers for these two types of vibrations are different, but the existing research does not distinguish the two types of dampers well [26–29]. Therefore, for control systems with multiples dampers, it is of practical significance to study the control strategy and device placement for control systems with multiple MR dampers.

In this study, numerical simulation of spatial eccentric structures with multiple MR dampers were established, and modal analysis and time history analysis were conducted to reveal the effectiveness of MR control system in mitigating torsional vibrations of eccentric structures. Firstly, a self-programed full-scale spatial structure model with both plan asymmetry and vertical irregularity was established in OpenSEES to exhibit the torsional vibrations. Then, the mathematical model of MR dampers was introduced to the structure model based on the new material development function of OpenSEES, and the damping forces obtained from the MR damper model were compared with performance tests data. Finally, to evaluate the effectiveness of the MR control system with multiple MR dampers, modal analysis and nonlinear time history analysis of the numerical MR damped structure subjected to seismic excitations were carried out, and these numerical results were compared with seismic performances of the uncontrolled structure

2. Numerical Modeling of Spatial Eccentric Structures

Full-scale spatial structures have numerous degrees of freedom, and the inelastic coupled translation-torsion vibrations under strong earthquakes are complex, which places high demands on the non-linear analysis and solution capabilities of finite element software. Meanwhile, in order to accurately evaluate the control effect of MR damping systems, it is necessary to introduce the mechanical model of MR dampers and the real-time control strategy to the time–history analysis of structures. Therefore, the finite element software is required to have flexible programmability and secondary development capabilities.

In this study, the OpenSEES (the Open System for Earthquake Engineering Simulation, version 3.3.0, the Pacific Earthquake Engineering Research Center, CA, USA) software developed by Berkeley was implemented for structural analysis. OpenSEES is an object-oriented, open-source software framework which can calculate the response of structures under earthquake excitations [32]. The most prominent advantage of OpenSEES is its open-source feature, which can develop and share new materials, new elements, or new algorithms through C++ language, providing a software platform for the introduction of nonlinear mechanical properties of MR dampers and real-time control algorithms to the numerical time–history analysis of MR damped structures.

2.1. Spatial Structure Modeling

For spatial structures with irregular plane and elevation, its vibration form is complex, and the simplified models are difficult to describe the coupled translational and torsional vibration characteristics. It is necessary to establish a full-scale three-dimensional spatial model. Herein, a typical full-scale spatial eccentric structure was selected as the numerical example to stimulate the coupled translation-torsion vibration responses under earthquake actions. In practical engineering, the eccentricity of real building structures is often caused by the irregularity in both horizontal and vertical directions. Therefore, a full scale ten-story reinforced concrete (RC) frame building with both plan asymmetry and vertical irregularity was modeled for this numerical study. The height of the first floor is 4.5 m and the upper floors are all 3.3 m high. The structure has three bays, where the span of the middle bay is 3 m and the side bay is 6.6 m. The cross section of the columns is 650 mm by 650 mm, the dimension of the beams at the side bay is 300 mm by 600 mm, and the dimension of the beams at the middle bay is 300 mm by 350 mm.

The structure model has both plan asymmetry due to the 'L-shaped' floor plan and vertical irregularity due to setbacks above the sixth floor, whose typical floor plans can be seen in Figure 1. In order to truly reflect the spatial torsional vibrations of eccentric structures, the numerical structure in this study was modeled as a three-dimensional beam-column element system [33,34], considering the elastic–plastic deformation of beam and column members under strong earthquakes, as can be seen in Figure 2. The lateral load resistance system of the 'L-shaped' structure is strong in one direction but weak in the other, and due to the eccentricity of stiffness in the lateral load resistance system, the building is prone to torsional vibrations on the vertical axis.

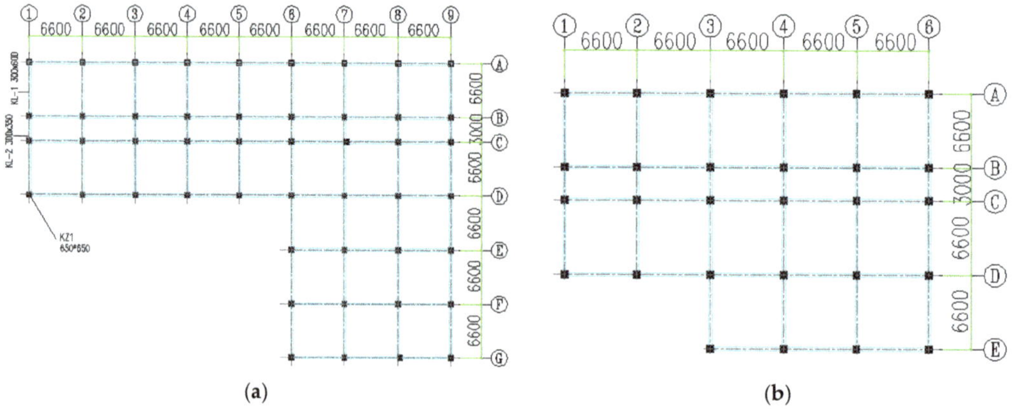

(a) **(b)**

Figure 1. Typical floor plans of the ten-story L-shaped structure model. (**a**) 1st–6th floor (**b**) 7th–10th floor.

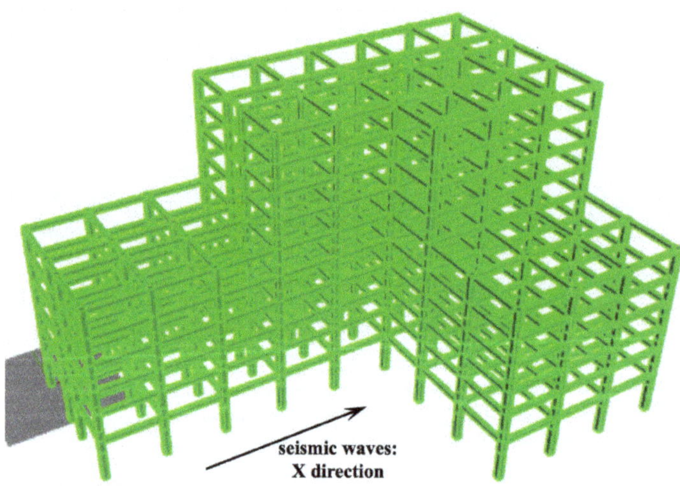

Figure 2. Three-dimensional view of the full-scale spatial eccentric structure model.

In the numerical modeling through OpenSEES, the fiber model is used to simulate the member sections, in which the beam, column, and other member sections in the structure are discretized into several small elements, and each small element adopts the uniaxial constitutive relation of the corresponding material. The skeleton curve of the concrete constitutive model adopts the Kent–Scott–Park model, and the stress–strain relationship of the steel bar is described by the Menegotto–Pinto model [35,36]. In OpenSEES, the 'concrete 02' and 'steel 02' command are used to construct the concrete and steel material respectively. All the degrees of freedom of the bottom nodes of the structure are fixed, simulating the fixed connection between the real structure and the ground. In order to fully simulate the multi-directional translational and torsional vibration of the structure, the translational and torsional degrees of freedom (UX, UY, UZ, RX, RY, RZ) of the remaining nodes in the three directions are free.

2.2. Structure Model Verification

In order to verify the validity and accuracy of the self-programmed structure model in OpenSEES, comparisons of modal analysis results and dynamic time history responses calculated from OpenSEES (version 3.3.0) and SAP2000 (version 23.0) were carried out. The structure model set up in SAP2000 with node numbers can be seen in Figure 3. The El-Centro N-S component ground motions were used in the time history analysis, and the amplitude is scaled to 70 cm/s^2. Seismic waves were applied unidirectionally in the X direction of the structure. The damping ratio of all modes was set to 5%.

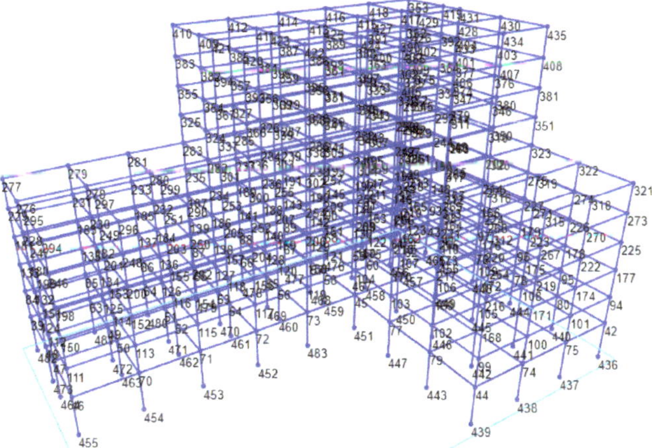

Figure 3. The spatial structure model with node numbers in SAP2000.

The comparisons of modal analysis results are provided in Table 1, and comparisons of the acceleration and displacement time history responses of the top node (node 435) are plotted in Figure 4. The vibration type of each mode of the structure can be determined according to the participation mass ratio. When the participation mass ratio in the RZ direction of the structure is much larger than the sum of the participation mass ratios in the UX and UY directions, it means that the structure exhibits obvious torsional vibration; conversely, it means that the structure vibrates in the X-direction or Y-direction. As can be seen in Table 1 and Figure 4, the differences of the modal analysis results obtained from OpenSEES and SAP2000 are limited to 15%, and the acceleration time history responses of the top node in OpenSEES and SAP2000 are close, which indicates that the structure model established in OpenSEES is effective and accurate.

Table 1. Comparisons of modal analysis results from OpenSEES and SAP2000.

Mode	Frequency (Hz)		Error (%)	Participating Mass Ratios			Vibration Type
	OpenSEES	SAP2000		UX	UY	RZ	
1	1.0943	1.2106	9.61	0.00041	0.72	0.03633	Translation vibration-Y
2	1.2549	1.3427	6.54	0.76	0.00102	0.00640	Translation vibration-X
3	1.2706	1.3957	8.96	0.02175	0.00122	0.53	Torsion vibration-Z
4	1.6530	1.8435	10.33	0.00269	0.00172	0.01864	Torsion vibration-Z
5	1.8223	1.9544	6.76	0.00273	0.07293	0.11	Coupled translation-torsion vibration
6	1.9566	2.1730	9.96	0.00775	0.02567	0.11	Torsion vibration-Z
7	2.2289	2.2265	0.11	0.02872	2.267×10^{-5}	0.01696	Coupled translation-torsion vibration
8	2.7088	2.6666	1.58	0.00555	0.00053	0.00443	Coupled translation-torsion vibration

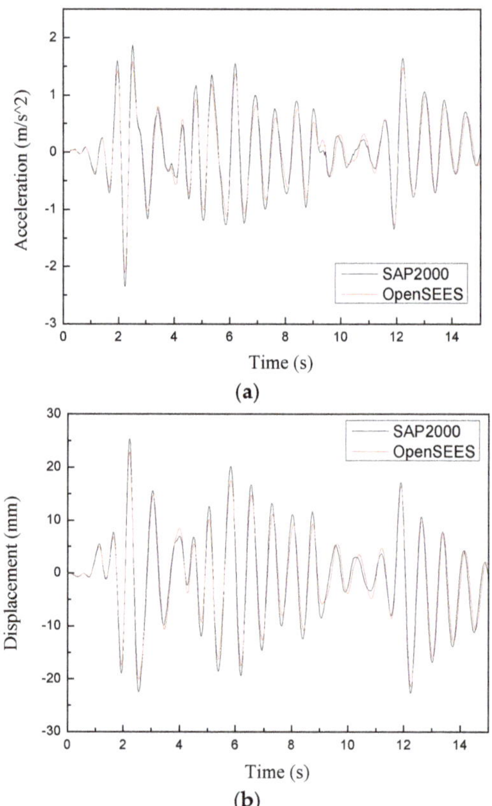

Figure 4. Response comparisons of the structure model in OpenSEES and SAP2000. (**a**) Acceleration, (**b**) Displacement.

What is more, it can be revealed from Table 1 that the vibration types of the third and higher structure modes are torsional vibration or coupled translation-torsional vibration, and the ratio of the first torsional period to the first translational period of the structure is 0.86, close to the upper limit value 0.9 in the Chinese seismic design code [37], which further proves that the established numerical model is a typical spatial eccentric structure with weak torsional resistance. Therefore, it is necessary to suppress the torsional irregularity of the eccentric structure using vibration control devices.

3. MR Damper Modeling

3.1. Performance Tests of MR Dampers

The vibration control devices implemented to the established structure model for vibration mitigation are MR dampers. Utilizing the reversible fluid–solid conversion characteristics of MR fluid under the action of a magnetic field, the output of MR dampers can be adjusted in real time according to the external excitations and the vibration response of structures, by changing the excitation current. Due to the intrinsically nonlinear characteristics of MR dampers, accurate and efficient mathematical models to properly describe their behavior are crucial for the design of semi-active control systems and the responses predication of MR damped structures.

In order to provide experimental data basis for parameter identification of MR dampers, performance tests of a single-coil MR damper were carried out [38], as can be seen in Figure 5. The MR fluid used in this MR damper was the highly stabilized MR fluids based on MWCNTs/GO composites coated ferromagnetic particles prepared in our previous studies [39], and the basic properties of the MR fluid are listed in Table 2.

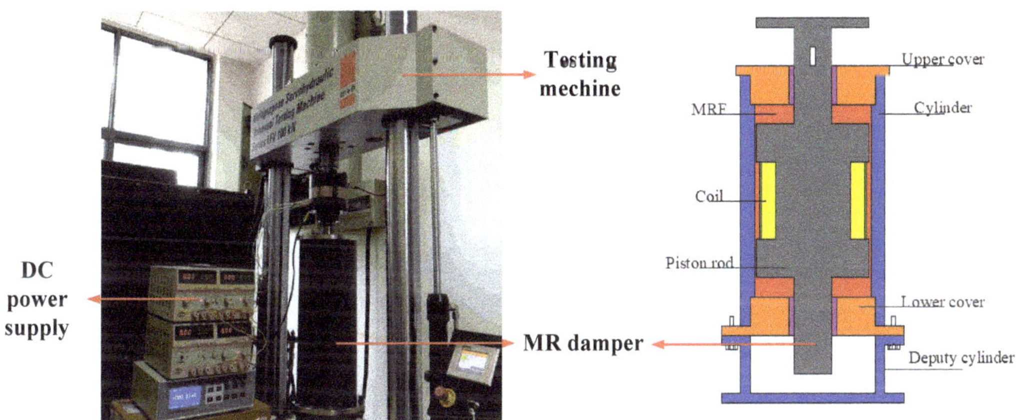

Figure 5. Schematic diagram and performance tests of a single-coil MR damper.

Table 2. Basic properties of MR fluid.

Particle Size (µm)	Coating Thickness (µm)	Volume Fraction (%)	Sedimentation Rate (%)	Yield Shear Stress τ_y (kPa)	Viscosity η (Pa.s)
1.5	0.015	40	3.5 (120 days)	7.5–10 (0.3 T)	2 2.5 (25 mm/s)

The magnetic characteristics of MR dampers have significant influence on the mechanical properties, so it is necessary to study the magnetic field distribution of this MR damper. The single-coil MR damper is axisymmetric, so it can be simplified as a two-dimensional model with ANSYS, with the axial direction of the damper as the symmetric axis. Figure 6 shows the magnetic induction lines and the magnetic induction intensity distributing of the single-coil MR damper. As can be seen from this figure, the magnetic induction lines all go perpendicularly through the damping gap, and the magnetic induction intensities in the damping gap is uniformly distributed, with nearly 200 mT, when the excitation current is 2.0 A.

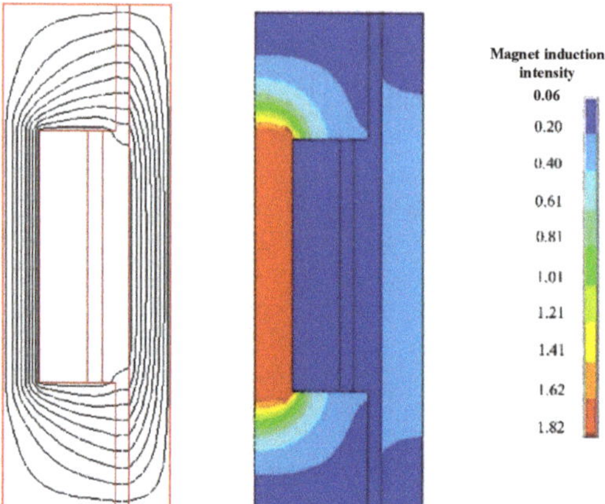

Figure 6. The vector diagram and the nephogram of the magnetic induction intensities of a single-coil MR damper with the excitation current of 2.0 A.

The force–displacement and force–velocity curves of the MR damper were plot'ted in Figure 7a,b, respectively. In Figure 7, the excitation frequency was 0.1 Hz, and the displacement amplitude was 5 mm. As can be seen from this figure, the force–displacement and force–velocity curves of MR dampers exhibit significant nonlinear hysteresis characteristics. In Figure 7a, when the excitation current gradually increases from 0 A to 1.5 A, the damping force increases greatly, but when the excitation current increases from 1.5 A to 2.0 A, the damping force does not change much, with only a small increment. It can be inferred that the saturation current of this MR damper is 1.5 A. Additionally, when the excitation current increases from 0 A to 1.5 A, the damping force increases from 2 kN to 8 kN, which shows that the damper has good adjustable performance. It can be seen from Figure 7b that the damping force of the MR damper first increases rapidly and then tends to be flat with the increase of velocity. In the high-velocity region, the damping force of the MR damper is basically unchanged, because this period is the process of the damper piston moving rapidly from one end to the other end, and the damping force is stable around the peak value during this process. When the piston moves to one end and is ready to turn, the damping force will drop rapidly.

Based on the performance tests results and the phenomenological models for MR dampers [40], a microstructure-based sigmoid model for describing the mechanical properties of MR dampers affected by MR fluid microstructure, magnetic saturation, excitation properties, and current was proposed in our earlier publications [41]. This model does not need to solve the differential equations, with only three undetermined parameters, and thus is suitable for applications in the practical vibration control of MR damped structures. In this study, the damping force of MR dampers f_d was calculated using this model, as can be seen in Equation (1):

$$f_d = \left[\left(2.07 + \frac{12Q\eta}{12Q\eta + 0.4wh^2\tau_y} \right) \cdot \frac{\tau_y \cdot L_t \cdot A_p}{h} \right] \cdot \frac{1 - e^{-k(\dot{x}+\dot{x}_h)}}{1 + e^{-k(\dot{x}+\dot{x}_h)}} + C_b\dot{x} \tag{1}$$

in which k, $\dot{x}_h$, and C_b are the parameters of the sigmoid model related to the current, excitation frequency, and displacement amplitude, and can be identified based on the performance test results; $\dot{x}$ is the velocity of the piston rod; Q is the volumetric flow rate; η is the viscosity of MR fluid with no magnetic field; w is the mean circumference of the damper's annular damping gap; h is the gap between the outer cylinder and the piston; τ_y is the yield shear stress of MR fluid; A_p is the cross-sectional area of the piston head,

$A_p = (D^2 - d^2)/4$; D is the piston diameter; d is the piston rod diameter; L_t is the working length of the damping gap.

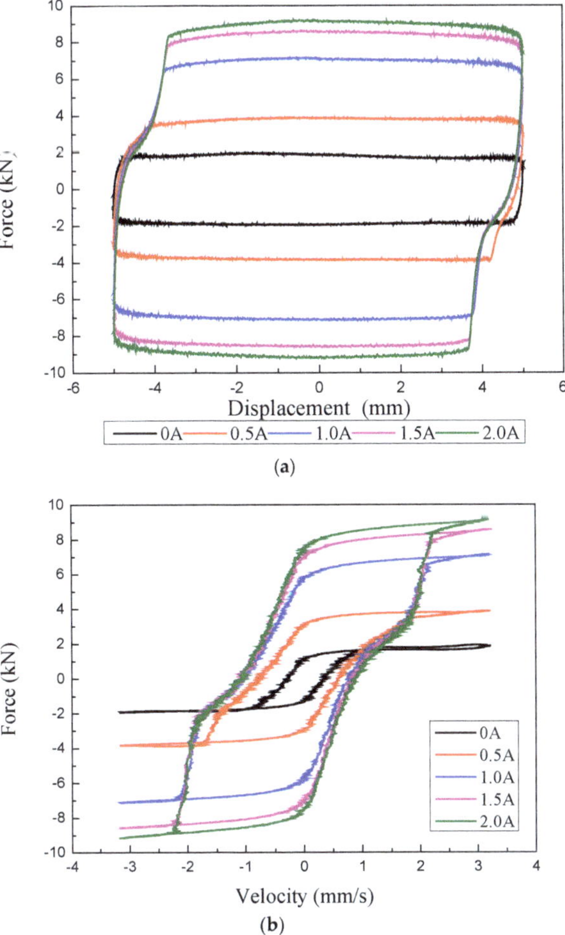

(a)

(b)

Figure 7. Force–displacement and force–velocity curves. (**a**) Force–displacement curves. (**b**) Force–velocity curves.

The parameters to be identified in this model include the curve slope k, the crossing velocity $\dot{x}_h$, and the damping coefficient C_b. The curve slope k is the slope of the hysteresis curve when the velocity is equal to zero, the cross velocity $\dot{x}_h$ is obtained by the intercept of the force–velocity curves on the abscissa axis, and the damping coefficient C_b is the slope of the high velocity region of the hysteresis curve, as can be seen in Figure 8. All three parameters are related to the excitation current and maximum velocity $\dot{x}_{max}$. In this study, parametric identification was performed in Origin software, and 'nonlinear surface fitting' was used to fit the relationship between the three parameters with excitation current and maximum velocity. The function used in the parametric identification is Poly2D.

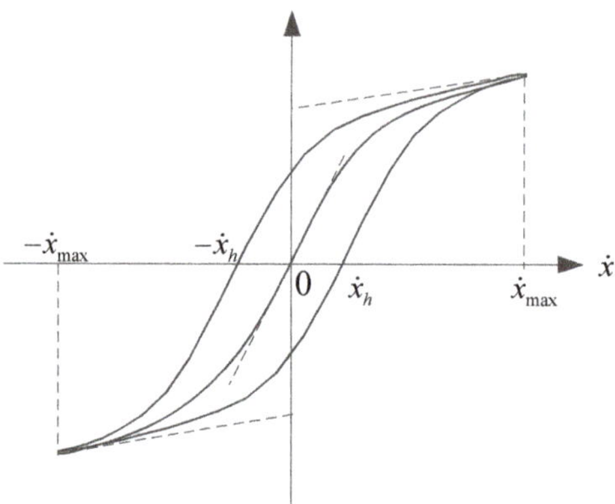

Figure 8. The undetermined parameters of the sigmoid model.

Considering the size of the spatial structure model, the dimensions of the MR damper were selected so that the MR damper can obtain the maximum output of 10 kN and the adjustable coefficient over 3, as listed in Table 3.

Table 3. Dimensions of the MR damper.

Piston Diameter D (mm)	Piston Rod Diameter d (mm)	Damping Gap h (mm)	Working Length L_t (mm)
116	40	2	80

3.2. MR Damper Modeling and Verification

The proposed mathematical model in Equation (1) essentially defines the force–velocity relationship of MR dampers, so this relationship can be simulated in OpenSEES as a new material. The force generated by MR dampers on the structure is only in the axial direction, and the 'Truss' element in OpenSEES is an element that transmits only axial force, thus the 'Truss' element and the self-defined new material can be combined to establish a mathematical model for describing the nonlinear mechanical properties of MR dampers.

The uniaxial material in OpenSEES that describes the stress–strain relationship is adopted to define the new material. In order to convert it into the force–velocity relationship of MR dampers, the 'setTrailStrain' function is used to obtain strain and strain rate (the velocity value $\dot{x}$ in Equation (1)). Then, the material property of the 'Truss' element is defined as the self-defined new material, and the cross-section area of the element is defined as 1. Finally, the damping force of the MR damper can be obtained from the axial force of the element.

By consolidating the element at one point and applying a sinusoidal displacement load to another point, the damping force of the MR damper model at different time can be obtained. In Figure 9a, the sinusoidal displacement loads applied to the element have the excitation frequency of 0.1 Hz, the displacement amplitude of 20 mm, and the current of 0.4 A and 0.8 A, and in Figure 9b, the sinusoidal displacement loads applied to the element have the excitation frequency of 1.0 Hz, the current of 0.4 A, and the displacement amplitude of 10 mm and 20 mm, which are both the same as the loading conditions of the MR damper performance tests, and the parameters of the model in OpenSEES are the same as the dimensions of the MR damper [38], so the accuracy of the MR damper model established in OpenSEES can be verified by comparing the damping forces calculated from OpenSEES with performance test results.

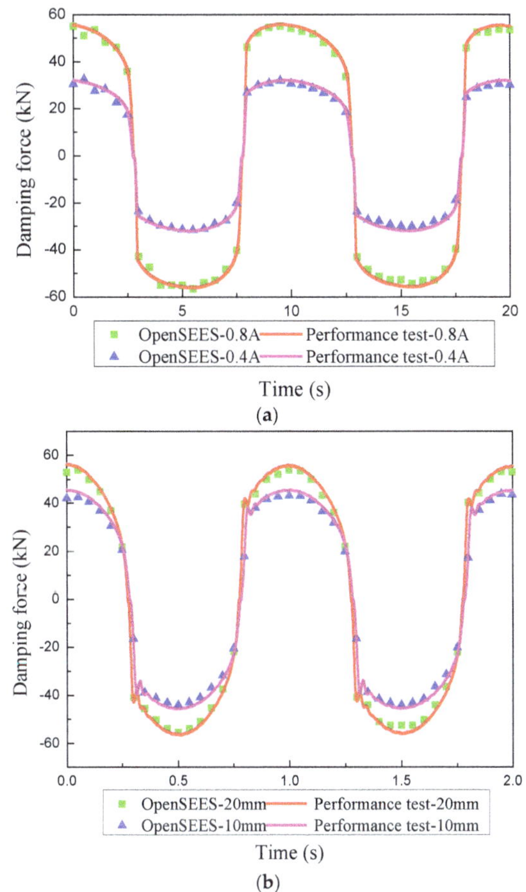

Figure 9. Comparisons between the damping force calculated from OpenSEES and performance test results. (**a**) 0.1 Hz, 20 mm. (**b**) 1.0 Hz, 0.4 A.

Figure 9 compares the damping forces calculated using the developed model in OpenSEES and the performance test results [38], in which the basic dimensions of the MR damper and properties of MR fluid are the same as the model parameters in Tables 2 and 3. As can be seen from these figures, the damping forces of the MR damper calculated by the OpenSEES model are basically consistent with the performance test results, and can reflect the real-time variation trend of the damping force with time. Therefore, it can be concluded that using the 'Truss' element and the self-defined new material in OpenSEES, the developed model can finely describe the nonlinear dynamic properties of MR dampers. Introducing the MR damper model into the established structure model, the numerical MR damped structure model is established, and the dynamic responses of the MR damped structure under earthquake actions can be obtained.

4. Control System with Multiple MR Dampers

The essence of structural vibration control is to minimize the structural vibration response by establishing the appropriate feedback relationship between the control force and the measured structural vibration response as well as external excitation. Therefore, structural vibration control is essentially an optimization process of control parameters.

For eccentric structures with coupled translation-torsion responses, a control system equipped with multiple MR dampers is necessary, which makes the structural vibration control a more complex problem. On the one hand, for effective coupled torsion-translation vibration mitigation, the feedbacks to the control system need the combination of several parameters, including structural vibration response signals (displacement, acceleration, torsion angle) and earthquake excitation signals. On the other hand, in order to realize real-time cooperative control of multiple MR dampers, the control system is required to output different excitation currents to multiple dampers at the same time. Therefore, the control system involving multiple MR dampers is essentially a multi-input, multi-output (MIMO) system.

When multiple MR dampers are arranged in real-life high-rise buildings, it is impractical to equip each damper with a controller, and not every story of the structure has room for control and sensing systems. What is more, if too many controllers are involved in the control system, the stability and robustness of the system are not guaranteed. Fewer controller and sensors may be more applicable, stable, and economical for practical application.

In this study, only two controllers are implemented to the control system with multiple MR dampers, where one controller outputs current to the dampers responsible for translational vibration control ('translational control damper'), while the other controller outputs current to the dampers for torsional vibration control ('torsional control damper'). The two controllers simultaneously output different currents to the two types of MR dampers, considering the cooperative work between the translational control dampers and the torsional control dampers, so as to achieve the optimal control on the coupled translation-torsion vibration of spatial eccentric structures, which is the basic concept of the control strategy proposed in this study.

4.1. Performance Criteria of Eccentric Structures

Firstly, in order to optimize the damping effect of the MR control system, a parameter representing the overall structural response requires to be minimized. For eccentric structures with coupled translation-torsion vibration, inter-story drift ratio and inter-story torsion angle were assigned as the performance criteria, in which the inter-story drift ratio characterizes the translational vibration, and the inter-story torsion angle reflects the torsional vibration.

4.2. Feedbacks to MR Control System

Then, the control system needs appropriate feedback to adjust the excitation current to MR dampers. As the output of MR dampers is directly related to the corresponding velocity, the velocity at the installation position of the MR damper was chosen as the feedback to the control system. In addition, in order to avoid the overcontrol of the structure by the MR control system under small earthquakes, the amplitude of the earthquake excitations was also considered.

4.3. Control Strategy

In general, the equation of motion for a spatial eccentric building equipped with multiple MR dampers subjected to earthquake excitation can be written as:

$$M_s\ddot{x} + C_s\dot{x} + K_s x = -M_s I\ddot{x}_g + DU \tag{2}$$

in which M_s, C_s, and K_s are the mass, damping, and stiffness matrices of the eccentric structure, respectively; x, $\dot{x}$, and $\ddot{x}$ are the displacement, velocity, and acceleration vector of the eccentric structure; $\ddot{x}_g$ is the acceleration vector of the earthquake excitation; D is the location matrix of MR dampers; U is the control force vector of MR dampers; I is the identity matrix.

Equation (2) can be reduced to first-order and written in the state-space form as:

$$\dot{Z} = AZ + BU + E\ddot{x}_g \tag{3}$$

$$Z = \begin{bmatrix} x & \dot{x} \end{bmatrix}^T \tag{4}$$

$$A = \begin{bmatrix} 0 & I \\ -M_s^{-1}K_s & -M_s^{-1}C_s \end{bmatrix}, \ B = \begin{bmatrix} 0 \\ M_s^{-1}D \end{bmatrix}, \ E = \begin{bmatrix} 0 \\ -I \end{bmatrix} \tag{5}$$

in which Z is the state vector; A is the system matrix; B is the distribution matrix of the MR damping force; E is the distribution matrix of the earthquake excitation.

Figure 10 is the work flow diagram of the control system with multiple MR dampers, and the detailed steps to implement the control strategy for the control system with multiple MR dampers can be seen from this figure. The spatially eccentric structure will have coupled translation-torsion vibrations under earthquake excitations. The proposed control strategy will output corresponding command signals to the two controllers in the system according to the structural vibration response and seismic excitation collected by the sensors. The two controllers output different currents to the two types of MR dampers arranged in the structure to ensure the cooperative work between them. Additionally, the vibration signal of MR dampers will also be output to the control strategy, and the control system output by the control strategy can be adjusted in real time. Ultimately, two types of MR dampers apply torsional and translational control forces to the structure, forming a complete closed-loop control of the coupled translation-torsion vibrations of the spatial eccentric structures.

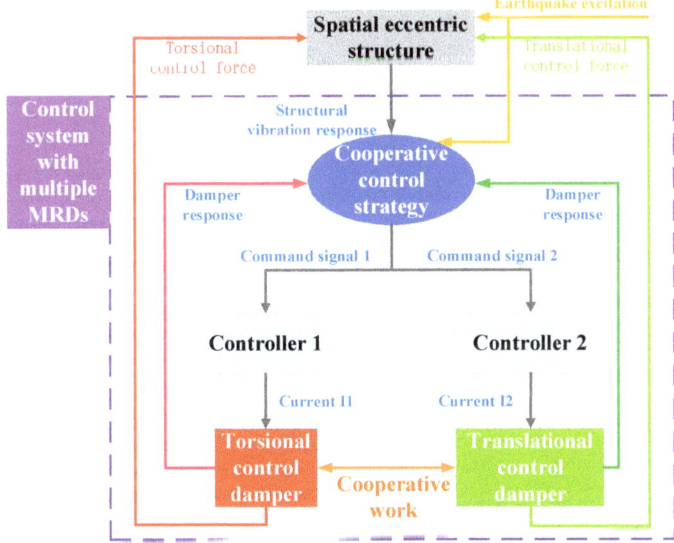

Figure 10. Work flow diagram of the control system with multiple MR dampers.

In Figure 10, there are two problems to be solved for the practical application of the control system with multiple MR dampers. One is how to determine the excitation currents of the two controllers, and the other is how to allocate the two excitation currents to ensure the cooperative work of the two types of MR dampers. Firstly, the excitation current $I1$ is input to the torsional control damper, thus the current value is determined by the real-time response of the torsional control dampers. Similarly, the current value of the excitation current $I2$ is determined by the real time response of the translational control damper. Equations (6) and (7) are the criterion for current determination of the torsional control and translational MR dampers, respectively. Since there are multiple MR dampers, the current values $I1$ and $I2$ obtained from Equations (6) and (7) are within a range. In Equations (6) and (7), I_{max} is the saturation current of the MR dampers, n is the total number of the torsional control dampers, m is the total number of the translational control dampers, I_{1i} is the excitation current for the ith torsional control damper, v_i is the velocity of the ith torsional control damper, I_{2t} is the excitation current for the tth translational control damper, v_t is the velocity of the tth translational control damper, μ is the coefficient describing the relationship between the velocity and the output of MR dampers, which needs to be determined according to the performance test results. In this paper, μ is set as 3×10^6 N·s/m.

$$I_{1i} = \begin{cases} I_{max} \ (\mu \cdot v_i \geq I_{max}) \\ \mu \cdot v_i \ (\mu \cdot v_i < I_{max}) \end{cases} i = 1 \sim n \tag{6}$$

$$I_{2t} = \begin{cases} I_{max} \ (\mu \cdot v_t \geq I_{max}) \\ \mu \cdot v_t \ (\mu \cdot v_t < I_{max}) \end{cases} t = 1 \sim m \tag{7}$$

Secondly, it is necessary to select two excitation current values from the two current ranges obtain from Equations (6) and (7) to complete the cooperative work of the two types of MR dampers. In order to achieve a good control effect on both the translational and torsional vibrations at the same time, inter-story drift ratio and inter-story torsion angle are selected as the evaluation indicators. The objective function J_z is shown in Equation (8), where γ_t and γ_{t0} are the inter-story drift ratio of the top floor of the MR damped structure and the uncontrolled structure, respectively; θ_t and θ_{t0} are the inter-story torsion angle of the top floor of the MR damped structure and the uncontrolled structure, respectively; α and β are the weighting coefficients of inter-story drift ratio and inter-story torsion angle, respectively, which are both taken as 0.5 in this paper.

$$J_z = \alpha \frac{\gamma_t}{\gamma_{t0}} + \beta \frac{\theta_t}{\theta_{t0}} \tag{8}$$

Finally, the genetic algorithm is adopted to optimize the two excitation current values, and the goal is to minimize the objective function, and Equations (6) and (7) are the constraint conditions. Within the range of current $I1$ and $I2$, multiple sets of combined values are selected as the initial population, and new populations are obtained through continuous crossover and mutation, the fitness of each generation is calculated, and the optimal set of current $I1$ and $I2$ is finally obtained. In the optimization process, the optimal excitation currents for controller 1 and controller 2 can be obtained, and the cooperative control of the torsional control dampers and translational control dampers is achieved.

4.4. Placement of Multiple MR Dampers

For the semi-active control system with multiple MR dampers, another important issue is the placement of control devices, which has significant influence on its damping effect. In this study, the placement configuration of MR dampers is determined based on the inter-story drift ratios of the structure. Vertically, the floors where the inter-story drift ratios are significantly larger than other floors are the weak floors of the structure, and dampers should be arranged on these floors. In each floor, MR dampers for torsional vibration mitigation need to be placed at the corners or edges, as far away as possible from the center of mass, while the dampers for translation vibration mitigation are placed near the center of mass.

For the 10-story frame structure model established in this study, the lateral structural deformation is mainly concentrated in the lower floors, so both translational control dampers and torsional control dampers are placed in 1th–6th floors, and the 7th–10th floors only have torsional control dampers. The detailed placement of MR dampers in different floors can be seen in Figure 11. In this figure, the green devices are the translational control dampers, and the red devices are torsional control dampers. As can be seen from Figure 11, a total number of 120 MR dampers are implemented in this structure for vibration control.

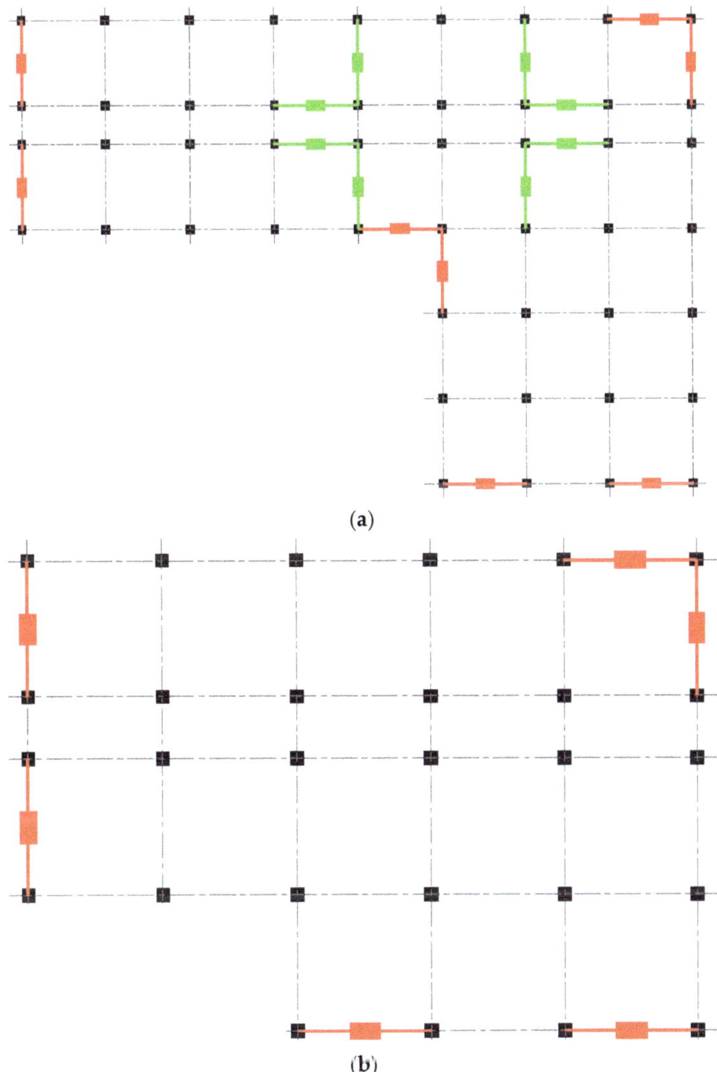

Figure 11. The placement of multiple MR dampers in different floors of the structure. (**a**) 1th–6th floor (**b**) 7th–10th floor.

5. Results and Discussions

The multiple MR dampers in Figure 11 with related controllers, sensors and power sources form a complete semi-active control system, and the proposed control strategy is responsible for calculating corresponding currents for different dampers in the control system to mitigate the coupled translation-torsion vibration of the structure. In order to evaluate the performance of the control system with multiple MR dampers, the simulation results and comparisons with seismic responses of the uncontrolled structure are discussed.

5.1. Modal Analysis

Modal analysis can provide the parameters that reflect the performance of a structure, and can be used to qualitatively access the basic dynamic properties. Therefore, the modal analysis of the MR damped structure is carried out firstly. Table 4 shows the modal analysis results of the MR damped structure. Compared with the model analysis results of the uncontrolled structure in Table 1, the dynamic properties of the eccentric structure are significantly changed due to the application of MR dampers. The uncontrolled eccentric structure exhibits obvious coupled translation-torsion vibration characteristics, while the vibration responses of the MR damped structure are mostly translational (the first three vibration modes are translation vibrations, and there is no coupled translation-torsion vibration), indicating that the control system with multiple MR dampers can remarkably mitigate the torsional vibration of spatial eccentric structures. The ratio of the first torsional period to the first translational period of the structure decreases from 0.86 to 0.53, which proves that the torsional stiffness of the structure has been significantly improved with the control system of multiple MR dampers.

Table 4. Modal analysis results of the MR damped structure.

Mode	Frequency (Hz)	Participating Mass Ratios			Vibration Type
		UX	**UY**	**RZ**	
1	1.3826	0.00041	0.835	0.00271	Translation vibration-Y
2	1.5860	0.7927	0.00082	0.00351	Translation vibration-X
3	1.7265	0.01873	0.00122	0.0053	Translation vibration-X
4	1.9308	0.00269	0.01232	0.00424	Translation vibration-Y
5	2.2733	0.00073	0.07293	0.00033	Translation vibration-Y
6	2.5860	0.00075	0.00316	0.6123	Torsion vibration-Z
7	2.7002	0.00872	1.352×10^{-5}	0.00056	Translation vibration-X
8	2.9523	0.00614	0.00053	0.00043	Translation vibration-X

5.2. Time History Analysis

To quantitatively evaluate the effectiveness of the proposed control strategy for multiple MR dampers, nonlinear dynamic time history responses of the numerical structure due to earthquakes were simulated. The earthquake excitations used in the time history analysis are the unidirectional El Centro and Taft ground motions, and the amplitude is scaled to 400 cm/s^2. Both ground motions were applied in the The duration of the earthquake excitations is set to 15 s, with the step of 0.02 s. The significant parameters selected to assess the control performance of the MR control system in torsional and translational vibration mitigation are node acceleration, node displacement and inter-story torsion angel. The node displacement response of is displacement relative to the ground, while the node acceleration response of is the absolute acceleration.

The node acceleration and displacement responses are obtained from the time history results of the node at the top floor (node 435 in Figure 3). Figure 12 shows the top node acceleration time history responses of the uncontrolled and the structure with multiple MR dampers controlled by the cooperative control strategy. The corresponding displacement responses are plotted in Figure 13. As can be seen from these two figures, with the implementation of the control system with multiple MR dampers, the node acceleration and displacement responses are significantly attenuated. In addition, under different seismic waves, the damping effect of the control system with multiple MR dampers on the structural displacement is obviously better than that of the acceleration. This is because the setting of MR dampers will increase the stiffness of the structure, thereby amplifying the acceleration response of the structure to a certain extent, and ultimately leading to a limited control effect of the acceleration responses.

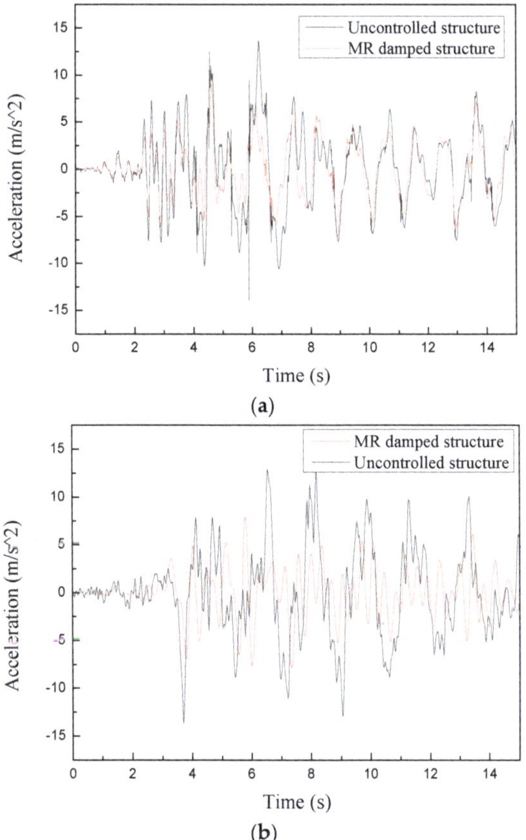

Figure 12. Acceleration time history responses of the uncontrolled and MR damped structure. (**a**) El Centro. (**b**) Taft.

In order to verify the suppression effect of the multiple MR dampers in the torsional irregularity of spatial eccentric structures, the inter-story torsion angles of the uncontrolled and MR damped structures are compared. The inter-story torsion angle can be obtained by subtracting the time–history curves of the rotation angles of the upper and lower floors around the rigid center and then taking the maximum absolute value, and it can intuitively reflect the torsion degree of each story for torsional irregular structures, as can be seen in Figure 14.

The inter-story torsion angles of the uncontrolled and MR damped structure are plotted in Figure 15. After the implementation of the control system with multiple MR dampers controlled by the cooperative control strategy, the inter-story torsion angles of all floors are reduced to 2%, within the code limit, showing that the torsional vibrations of the spatial eccentric structure are obviously mitigated.

What is more, it can be seen from the inter-story torsion angles of the uncontrolled structure that due to the sudden change in vertical stiffness in the 7–10 floors of the structure, the top of the structure exhibits amplification of the torsional vibration response. After the implementation of MR dampers that control the torsional vibration, the torsional vibrations at the top floors are effectively mitigated.

In summary, it can be revealed from these simulation results and comparisons that the semi-active control systems with multiple MR dampers are effective in mitigating both the translational and torsional vibrations of spatial eccentric structures.

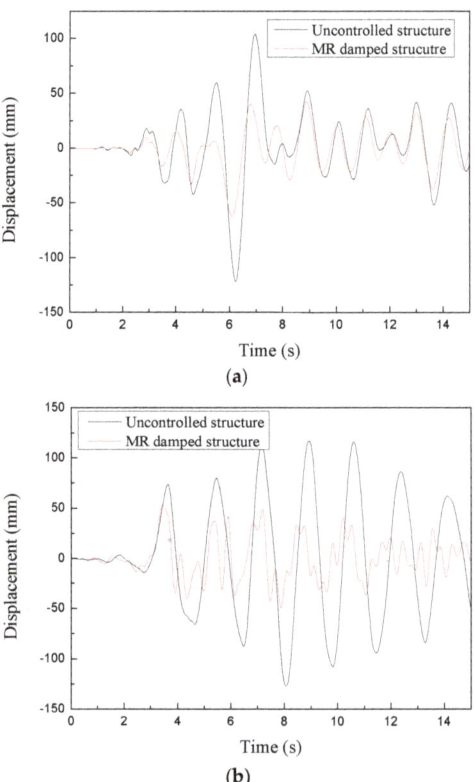

Figure 13. Displacement time history responses of the uncontrolled and MR damped structure. (**a**) El Centro. (**b**) Taft.

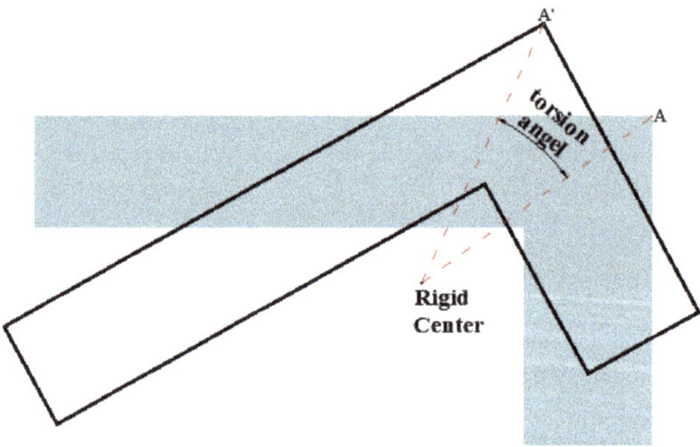

Figure 14. Schematic of the inter-story torsion angle.

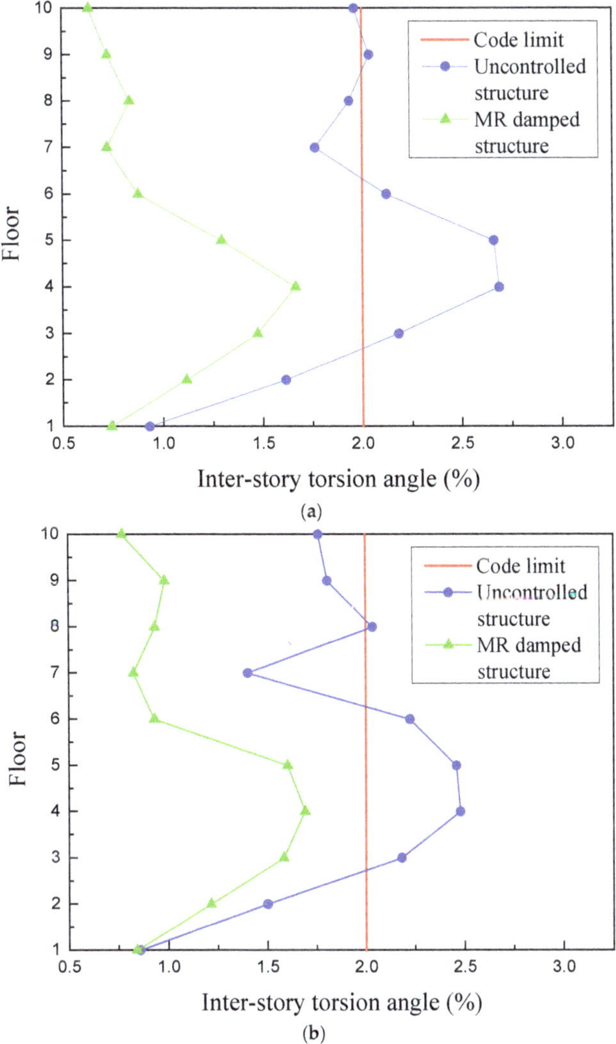

Figure 15. Inter-story torsion angles of the uncontrolled and MR damped structure. (**a**) El Centro. (**b**) Taft.

6. Conclusions

In this study, numerical models of spatial eccentric structures with multiple MR dampers were established in OpenSEES, and numerical analysis was conducted to reveal the effectiveness of the control system with multiple MR dampers. The following are the main conclusions drawn from this study:

(1) The self-programmed structure model in OpenSEES can accurately describe the coupled translation-torsion vibration characteristics of spatial eccentric structures, and the nonlinear mechanical properties of MR dampers can be simulated by combining the 'Truss' element and self-defined new material in OpenSEES.

(2) The semi-active control system with multiple MR dampers using the proposed control strategy is numerically proven to be effective in mitigating both torsional and translational responses of eccentric structures. For translational vibration control, the

acceleration and displacement time history responses have been significantly mitigated. For torsional vibration mitigation, the inter-story torsion angles are limited to 2% after the implementation of multiple MR dampers. The proposed cooperative control strategy for multiple MR dampers only needs two controllers, which is more economical and reliable, and thus has significant engineering application prospects for control systems with multiple MR dampers.

Author Contributions: Conceptualization, Y.Y. and Y.-Q.G.; methodology, Y.Y.; software, Y.Y.; validation, Y.Y. and Y.-Q.G.; formal analysis, Y.Y.; investigation, Y.Y.; resources, Y.Y.; data curation, Y.Y.; writing—original draft preparation, Y.Y.; writing—review and editing, Y.-Q.G.; visualization, Y.-Q.G.; supervision, Y.-Q.G.; project administration, Y.-Q.G.; funding acquisition, Y.-Q.G. All authors have read and agreed to the published version of the manuscript.

Funding: This research was funded by the Program of Chang Jiang Scholars of Ministry of Education, the Tencent Foundation through the XPLORER PRIZE, Ten Thousand Talent Program (Innovation Leading Talents), National Natural Science Foundation of China with Grant No. 51878355 and the program of China Scholarships Council (No. 202006090104) for the visiting at the National University of Singapore.

Institutional Review Board Statement: Not applicable.

Informed Consent Statement: Not applicable.

Data Availability Statement: Not applicable.

Conflicts of Interest: The authors declare no conflict of interest.

References

1. Lin, J.L.; Tsai, K.C. Seismic analysis of two-way asymmetric building systems under bi-directional seismic ground motions. *Earthq. Eng. Struct. Dyn.* **2008**, *37*, 305–328. [CrossRef]
2. De Stefano, M.; Pintucchi, B. A review of research on seismic behaviour of irregular building structures since 2002. *Bull. Earthq. Eng.* **2008**, *6*, 285–308. [CrossRef]
3. Almazan, J.L.; Espinoza, G.; Aguirre, J.J. Torsional balance of asymmetric structures by means of tuned mass dampers. *Eng. Struct.* **2012**, *42*, 308–328. [CrossRef]
4. Xu, Y.L.; Shum, K.M. Multiple-tuned liquid column dampers for torsional vibration control of structures: Theoretical investigation. *Earthq. Eng. Struct. Dyn.* **2003**, *32*, 309–328. [CrossRef]
5. Mevada, S.V.; Jangid, R.S. Seismic response of torsionally coupled building with passive and semi-active stiffness dampers. *Int. J. Adv. Struct. Eng.* **2015**, *7*, 31–48. [CrossRef]
6. Yao, J.T. Concept of structural control. *J. Struct. Div.* **1972**, *98*, 1567–1574. [CrossRef]
7. Yang, J.N. Application of optimal control theory to civil engineering structures. *J. Eng. Mech. Div.* **1975**, *101*, 819–838. [CrossRef]
8. Housner, G.; Bergman, L.A.; Caughey, T.K.; Chassiakos, A.G.; Claus, R.O.; Masri, S.F.; Skelton, R.E.; Soong, T.T.; Spencer, B.F.; Yao, J.T. Structural control: Past, present, and future. *J. Eng. Mech.* **1997**, *123*, 897–971. [CrossRef]
9. Xu, Z.D.; Liao, Y.X.; Ge, T.; Xu, C. Experimental and theoretical study on viscoelastic dampers with different matrix rubbers. *J. Eng. Mech. ASCE* **2016**, *142*, 04016051. [CrossRef]
10. Scerrato, D.; Giorgio, I.; Della Corte, A.; Madeo, A.; Dowling, N.E.; Darve, F. Towards the design of an enriched concrete with enhanced dissipation performances. *Cem. Concr. Res.* **2016**, *84*, 48–61. [CrossRef]
11. Xu, Z.D.; Yang, Y.; Miao, A.N. Dynamic analysis and parameter optimization of pipelines with multidimensional vibration isolation and mitigation device. *J. Pipeline Syst. Eng. Pract.* **2021**, *12*, 04020058. [CrossRef]
12. Yang, G.; Spencer, B.F., Jr.; Jung, H.J.; Carlson, J.D. Dynamic modeling of large-scale magnetorheological damper systems for civil engineering applications. *J. Eng. Mech.* **2004**, *130*, 1107–1114. [CrossRef]
13. Xu, Z.D.; Sha, L.F.; Zhang, X.C.; Ye, H.H. Design, performance test and analysis on MR damper for earthquake mitigation. *Struct. Control Health Monit.* **2013**, *20*, 956–970. [CrossRef]
14. Xu, Z.D.; Shen, Y.P.; Guo, Y.Q. Semi-active control of structures incorporated with magnetorheological dampers using neural networks. *Smart Mater. Struct.* **2003**, *12*, 80–87. [CrossRef]
15. Yang, Y.; Xu, Z.D.; Guo, Y.Q. Seismic performance of magnetorheological damped structures with different MR fluid perfusion densities of the damper. *Smart Mater. Struct.* **2021**, *30*, 065008. [CrossRef]
16. Jansen, L.M.; Dyke, S.J. Semiactive control strategies for MR dampers: Comparative study. *J. Eng. Mech.* **2000**, *126*, 795–803. [CrossRef]
17. Kim, Y.; Langari, R.; Hurlebaus, S. Semiactive nonlinear control of a building with a magnetorheological damper system. *Mech. Syst. Signal Processing* **2009**, *23*, 300–315. [CrossRef]

18. Weber, F. Semi-active vibration absorber based on real-time controlled MR damper. *Mech. Syst. Signal Processing* **2014**, *46*, 272–288. [CrossRef]
19. Xu, Y.W.; Xu, Z.D.; Guo, Y.Q.; Zhou, M.; Zhao, Y.L.; Yang, Y.; Dai, J.; Zhang, J.; Zhu, C.; Ji, B.H.; et al. A programmable pseudo negative stiffness control device and its role in stay cable vibration control. *Mech. Syst. Signal Processing* **2022**, *173*, 109054. [CrossRef]
20. Xu, Z.D.; Jia, D.H.; Zhang, X.C. Performance tests and mathematical model considering magnetic saturation for magnetorheological damper. *J. Intell. Mater. Syst. Struct.* **2012**, *23*, 1331–1349. [CrossRef]
21. Xu, Z.D.; Guo, Y.Q. Fuzzy control method for earthquake mitigation structures with magnetorheological dampers. *J. Intell. Mater. Syst. Struct.* **2006**, *17*, 871–881. [CrossRef]
22. Yoshida, O.; Dyke, S.J.; Giacosa, L.M.; Truman, K.Z. Experimental verification of torsional response control of asymmetric buildings using MR dampers. *Earthq. Eng. Struct. Dyn.* **2003**, *32*, 2085–2105. [CrossRef]
23. Yoshida, O.; Dyke, S.J. Response control of full-scale irregular buildings using magnetorheological dampers. *J. Struct. Eng.* **2005**, *131*, 734–742. [CrossRef]
24. Li, H.N.; Li, X.L. Experiment and analysis of torsional seismic responses for asymmetric structures with semi-active control by MR dampers. *Smart Mater. Struct.* **2009**, *18*, 075007. [CrossRef]
25. Shook, D.A.; Roschke, P.N.; Lin, P.Y.; Loh, C.H. Semi-active control of a torsionally-responsive structure. *Eng. Struct.* **2009**, *31*, 57–68. [CrossRef]
26. Bharti, S.D.; Dumne, S.M.; Shrimali, M.K. Earthquake response of asymmetric building with MR damper. *Earthq. Eng. Eng. Vib.* **2014**, *13*, 305–316. [CrossRef]
27. Hu, Y.W.; Liu, L.F.; Rahimi, S. Seismic Vibration Control of 3D Steel Frames with Irregular Plans Using Eccentrically Placed MR Dampers. *Sustainability* **2017**, *9*, 1255. [CrossRef]
28. Zafarani, M.M.; Halabian, A.M.; Behbahani, S. Optimal coupled and uncoupled fuzzy logic control for magneto-rheological damper-equipped plan-asymmetric structural systems considering structural nonlinearities. *J. Vib. Control* **2018**, *24*, 1364–1390. [CrossRef]
29. Zafarani, M.M.; Halabian, A.M. Supervisory adaptive nonlinear control for seismic alleviation of inelastic asymmetric buildings equipped with MR dampers. *Eng. Struct.* **2018**, *176*, 849–858. [CrossRef]
30. Zafarani, M.M.; Halabian, A.M. A new supervisory adaptive strategy for the control of hysteretic multi-story irregular buildings equipped with MR-dampers. *Eng. Struct.* **2020**, *217*, 110786. [CrossRef]
31. Al-Fahdawi, O.A.S.; Barroso, L.R. Adaptive neuro-fuzzy and simple adaptive control methods for full three-dimensional coupled buildings subjected to bi-directional seismic excitations. *Eng. Struct.* **2021**, *232*, 111798. [CrossRef]
32. Mazzoni, S.; McKenna, F.; Scott, M.H.; Fenves, G.L. Open System for Earthquake Engineering Simulation (OpenSEES) Command Language Manual [EB/OL]. Available online: http://opensees.berkeley.edu/wiki/index.php/Command_Manual (accessed on 10 July 2022).
33. Zhao, J.; Li, K.; Zhang, X.C.; Sun, Y.P.; Xu, Z.D. Multidimensional vibration reduction control of the frame structure with magnetorheological damper. *Struct. Control Health Monit.* **2020**, *27*, e2572. [CrossRef]
34. Chen, X.W.; Lin, Z. *Structural Nonlinear Analysis Program Opensees Theory and Tutorial*; China Architecture & Building Press: Beijing, China, 2014.
35. Scott, B.D.; Park, R.; Priestley, M.J.N. Stress-strain behavior of concrete confined by overlapping hoops at low and high strain rates. *J. Am. Concr. Inst.* **1982**, *79*, 13–27.
36. Taucer, F.; Spacone, E.; Filippou, F.C. *A Fiber Beam-Column Element for Seismic Response Analysis of RC Structures*; EERC Report 91/17; Earthquake Engineering Research Center, University of California: Berkeley, CA, USA, 1991.
37. *Ministry of Housing and Urban-Rural Development of People's Republic of China*; Code for Seismic Design of Buildings (GB50011-2010); China Architecture & Building Press: Beijing, China, 2010.
38. Yang, Y.; Xu, Z.D.; Xu, Y.W.; Guo, Y.Q. Analysis on influence of the magnetorheological fluid microstructure on the mechanical properties of magnetorheological dampers. *Smart Mater. Struct.* **2020**, *29*, 115025. [CrossRef]
39. Xu, Z.D.; Sun, C.L. Single-double chains micromechanical model and experimental verification of MR fluids with MWCNTs/GO composites coated ferromagnetic particles. *J. Intell. Mater. Syst. Struct.* **2021**, *32*, 1523–1536. [CrossRef]
40. Vaiana, N.; Sessa, S.; Marmo, F.; Rosati, L. Nonlinear dynamic analysis of hysteretic mechanical systems by combining a novel rate-independent model and an explicit time integration method. *Nonlinear Dyn.* **2019**, *98*, 2879–2901. [CrossRef]
41. Yang, Y.; Xu, Z.D.; Guo, Y.Q.; Sun, C.L.; Zhang, J. Performance tests and microstructure-based sigmoid model for a three-coil magnetorheological damper. *Struct. Control Health Monit.* **2021**, *28*, e2819. [CrossRef]

Article

Research on PID Position Control of a Hydraulic Servo System Based on Kalman Genetic Optimization

Ying-Qing Guo *, Xiu-Mei Zha, Yao-Yu Shen, Yi-Na Wang and Gang Chen

College of Mechanical and Electronic Engineering, Nanjing Forestry University, Nanjing 210037, China; zhaxiumei@njfu.edu.cn (X.-M.Z.); shenyaoyu96@163.com (Y.-Y.S.); wangyina@njfu.edu.cn (Y.-N.W.); chengang8059@njfu.edu.cn (G.C.)
* Correspondence: gyingqing@njfu.edu.cn; Tel.: +86-139-5183-5602

Abstract: With the wide application of hydraulic servo technology in control systems, the requirement of hydraulic servo position control performance is greater and greater. In order to solve the problems of slow response, poor precision, and weak anti-interference ability in hydraulic servo position controls, a Kalman genetic optimization PID controller is designed. Firstly, aiming at the nonlinear problems such as internal leakage and oil compressibility in the hydraulic servo system, the mathematical model of the hydraulic servo system is established. By analyzing the working characteristics of the servo valve and hydraulic cylinder in the hydraulic servo system, the parameters in the mathematical model are determined. Secondly, a genetic algorithm is used to search the optimal proportional integral differential (PID) controller gain of the hydraulic servo system to realize the accurate control of valve-controlled hydraulic cylinder displacement in the hydraulic servo system. Under the positioning benchmark of step signal and sine wave signal, the PID algorithm and the genetic optimized PID algorithm are compared in the system simulation model established by Simulink. Finally, to solve the amplitude fluctuations caused by the GA optimized PID and reduce the influence of external disturbances, a Kalman filtering algorithm is added to the hydraulic servo system to reduce the amplitude fluctuations and the influence of external disturbances on the system. The simulation results show that the designed Kalman genetic optimization PID controller can be better applied to the position control of the hydraulic servo system. Compared with the traditional PID control algorithm, the PID algorithm optimized by genetic algorithm improves the system's response speed and control accuracy; the Kalman filter is a good solution for the amplitude fluctuations caused by GA-optimized PID that reduces the influence of external disturbances on the hydraulic servo system.

Keywords: hydraulic servo system; valve-controlled hydraulic cylinder; PID control; genetic optimization; Kalman filter

Citation: Guo, Y.-Q.; Zha, X.-M.; Shen, Y.-Y.; Wang, Y.-N.; Chen, G. Research on PID Position Control of a Hydraulic Servo System Based on Kalman Genetic Optimization. *Actuators* **2022**, *11*, 162. https://doi.org/10.3390/act11060162

Academic Editor: Tatiana Minav

Received: 12 May 2022
Accepted: 13 June 2022
Published: 15 June 2022

Publisher's Note: MDPI stays neutral with regard to jurisdictional claims in published maps and institutional affiliations.

1. Introduction

Hydraulic systems are widely used in industrial and mobile applications due to their high power to mass ratio and reliability. As a closed-loop control hydraulic system, the hydraulic servo system has the advantages of a hydraulic system and has the characteristics of fast response and high servo accuracy. Therefore, a hydraulic servo system is widely used in automation direction. In recent years, research on hydraulic servo systems has focused on trajectory tracking, state estimation, fault diagnosis, and parameter identification to achieve high-performance control [1]. One of the most desirable requirements is high precision position control with fast response, especially for valve-operated hydraulic cylinders, the actuators of hydraulic servo systems.

The accurate modeling of the controlled object is essential for the position control of the hydraulic servo system. Establishing an accurate mathematical model of the hydraulic servo system is very complicated due to the dead zone in the flow region, the static friction of the fluid, the compressibility and internal leakage of the fluid, the complex flow pressure characteristics of the control valve, and other factors that lead to the problem of a highly nonlinear and largely uncertain hydraulic servo system. In the last few years, some important work has been carried out by related researchers in the modeling of hydraulic servo systems. Xing et al. [2] analyzed the functions of servo valves and hydraulic cylinders and established the transfer function of an automatic depth-controlled electrohydraulic system using a first-principles approach. Mete et al. [3] considered the effects of compressibility, friction, servo valve internal leakage, actuator leakage, and inertia on the electro-hydraulic servo system and established a mathematical model of the main components of the internal leakage electro-hydraulic servo system. Zheng et al. [4] considered the problems of dead zones, saturation nonlinearity, time lag, and time variability in servo-hydraulic presses. Yao et al. [5] used the known information of the electro-hydraulic servo system to build an adaptive inverse model of the system. For the system's nonlinear parameters, online estimation was used to adapt the dynamic inverse model in real-time. Ye et al. [1] considered nonlinear factors such as dead zone, saturation, discharge coefficient, and friction during modeling a valve-controlled cylinder system for a hydraulic excavator. Li et al. [6] constructed adaptive parts of the hydraulic actuator model with parameter uncertainty to handle it, and considered the residuals of parameter adaptation and the unmodeled dynamics part with a robust part. Shen et al. [7] analyzed the basic principles of the new hydraulic transformer, established a mathematical model of the system, and simplified the state-space equations of the system by making appropriate assumptions about the system. Knohl and Unbehauen [8] studied the problem of large dead zones of valves in electro-hydraulic servo systems by linearizing the cylinder and load forces as a second-order system and integrator while connecting the dead zone part and the linear part in a series to describe the hydraulic system. Bao et al. [9] implemented a multi-pump, multi-actuator hydraulic system modeling based on the dynamic analysis of hydraulic systems. Zhang et al. [10] trained a BP neural network model to obtain a nonlinear relationship between motor speed and cylinder two-chamber pressure as input and cylinder speed as output and constructed a soft measurement model for the position of the direct-drive hydraulic system by integrating the calculated speed of the network. Nguyen et al. [11] investigated the motion dynamics of actuators under torque in a mechanical system considering parameter uncertainties, unmodeled uncertainties, and perturbations. They realized the position model of the electro-hydraulic servo system with detailed analysis of servo valve and hydraulic system models. Zhang et al. [12] considered the clearance problem between spool and sleeve in the construction of the mathematical model of the nonlinear position servo control system of magnetically coupled rodless hydraulic cylinders, established the mass flow relationship of each valve port by analyzing the proportional control valve structure and through experimental tests, and used the friction model after experimental tests for the friction model of the valve-controlled cylinder, so as to realize the establishment of the system dynamics equations. These studies provide some methods for solving nonlinear problems in modeling hydraulic servo systems.

In addition, to achieve good position control, it is necessary to study the control methods used in the hydraulic servo system, such as PID, fuzzy algorithm, neural network optimization, etc. PID control has been commonly used in hydraulic systems with the advantages of simple structure, easy implementation, and mature theoretical analysis. The critical step in PID control is the effective adjustment of three adjustable gains: proportional gain, integral gain, and differential gain. However, PID control has linear characteristics and cannot meet the requirements of nonlinear systems alone. In recent decades, many intelligent optimization techniques have been used to regulate PID gains to solve the control problems of nonlinear and complex systems. Chang [13] used an artificial bee colony (ABC) algorithm to search for PID control parameters to enhance the control performance of a continuously stirred kettle reactor. He [14] proposed an improved artificial bee colony algorithm to optimize the gain of the PID controller and improve the control performance of the strip deviation control system. Hao et al. [15] used the particle swarm optimization

(PSO) algorithm to optimize the PID parameters and improved PSO in inertia weights, learning coefficients, and elite variances to improve trajectory tracking accuracy in electro-hydraulic position servo systems. Zheng et al. [4] introduced a fuzzy PID control method to improve the overall performance of the electro-hydraulic position servo system to establish fuzzy inference rules capable of adaptively adjusting the PID parameters in terms of the error and error variation of the system. Wang et al. [16] introduced a fuzzy controller based on the particle swarm algorithm to rectify the parameters of the PID. Shutnan et al. [17] used the clone selection algorithm to rectify the PID parameters for the path tracking problem of a robotic manipulator. Guo et al. [18] combined PID control and generalized predictive control to exploit the advantages of each and improve the system performance. Odili et al. [19] applied the African Buffalo Optimization algorithm to rectify the PID controller parameters for effective control of the voltage regulator (AVR). Bingul et al. [20] used the cuckoo search algorithm to optimize the design of the PID controller parameters in an AVR system. Loucif [21] used a novel optimization algorithm of the Whale Optimization Algorithm to determine the optimal parameters of the PID controller to achieve better trajectory tracking of the robot operator. Xue et al. [22] proposed an advanced flaring (AFW) algorithm based on the adaptive principle and bimodal Gaussian function and established a PID parameter rectification model combined with AFW. Gao et al. [23] proposed an artificial fish swarm algorithm, which has a good optimization effect on the PID of the motion servo control system. Liu et al. [24] proposed a method for intelligent tuning of PID parameters based on iterative learning control, enabling the PID parameters of the AFM to be self-tuned according to the shape of the sample. Li [25] proposed a PID control strategy based on bacterial foraging optimization to improve the performance of variable air volume-air conditioning systems. Liu et al. [26] proposed an improved fruit fly optimization algorithm that combines a PID control strategy with a cloud model algorithm to improve the response performance of a magnetorheological liquid brake proportional-integral differential controller.

Genetic algorithms (GA) are inspired by natural selection and have stochastic optimization properties to optimize any problem without prior knowledge. It also provides faster convergence to near-optimal solutions [27]. At the same time, compared to traditional optimization algorithms that obtain optimal solutions by a single initial value iteration, genetic algorithms can evolve by searching for multiple solutions in the space simultaneously, thus reducing the risk of falling into a local optimum [28–30]. Therefore, the study of the adaptive adjustment of PID parameters using a genetic algorithm has certain reliability and development.

This paper implements modeling of the system by considering nonlinear factors such as internal leakage and oil compressibility of the system. To address the problems of slow response and low accuracy in the position control of valve-controlled hydraulic servo systems, the PID control is used as the basis to improve the performance of the system. In this work, a GA algorithm is used to search the PID parameters to improve the control performance of the system. To solve the amplitude fluctuations caused by the GA algorithm optimized PID controller in the position response of the hydraulic servo system and to reduce the influence of external disturbances on the system, a Kalman filter is added to the designed GA-optimized PID controller to improve the anti-interference capability of the system.

2. Valve-Controlled Hydraulic Servo System Model

2.1. Valved Hydraulic Servo System

As shown in Figure 1, the valve-controlled hydraulic servo system has the following operating principles. The system consists of eight parts: hydraulic cylinder, displacement sensor, servo valve, gear pump, oil cylinder, servo motor, servo driver, and controller. The servo driver starts the servo motor and controls the speed of the motor as well as forward and reverse rotation. The servo driver starts the servo motor and controls the speed of the motor and forward and reverse rotation, and the servo motor drives the gear pump to rotate. When the servo motor is turning, the gear pump is turning, the cylinder side is the suction chamber, the gear teeth in the suction chamber are disengaged one after another, the suction tube sucks in oil from the cylinder under atmospheric pressure, and the system side is the pressure chamber, the pressure chamber is engaged one after another, and the oil sucked in by the suction tube is squeezed. The oil is delivered to the system side. When the servo motor reverses, the gear pump reverses. The cylinder side is the pressure chamber, the system side is the suction chamber, and the oil is delivered to the cylinder from the system side. The controller controls the opening of the servo valve through the voltage amount. The opening of the servo valve determines the amount of oil in and out of the hydraulic cylinder. The servo valve opening is large, and the oil entering the hydraulic cylinder from the system side of the oil output from the hydraulic cylinder to the system side is more. Thus, the displacement of the hydraulic cylinder movement is large, and vice versa, the displacement is small. The displacement of the movement is fed back to the controller through the displacement sensor at the top of the hydraulic cylinder, forming a closed-loop control loop.

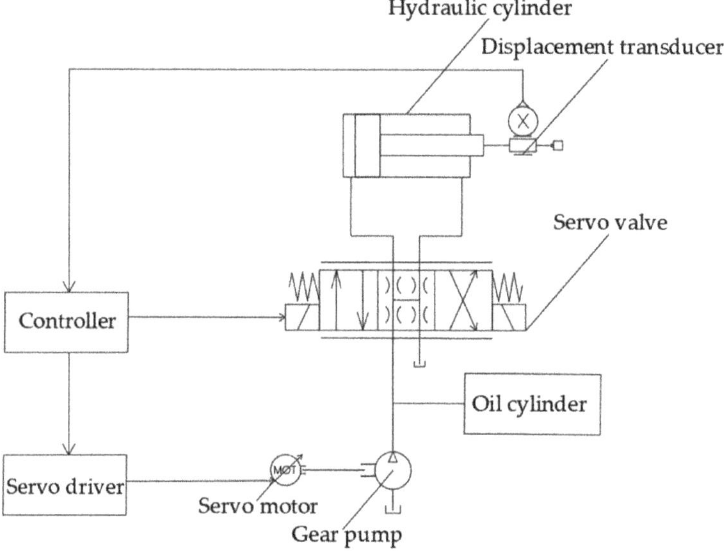

Figure 1. Schematic diagram of valve-controlled hydraulic servo system.

2.2. Mathematical Modeling of Valve-Controlled Hydraulic Servo System

2.2.1. Hydraulic Cylinder Modeling

The principle of the valve-controlled hydraulic cylinder is shown in Figure 2. Among them, p_s is the system oil supply pressure; p_0 is the return pressure of the system; q_1 is the inflow flow rate of the hydraulic cylinder inlet chamber; q_2 is the outflow flow rate of the hydraulic cylinder return chamber; and x_v is the input displacement of the slide valve spool. p_1 and p_2 are the pressure of the oil inlet chamber and the oil return chamber, respectively; C_{ep} is the external leakage coefficient of the hydraulic cylinder; $C_{ep}p_1$ and $C_{ep}p_2$ are the external leakage flow at the sealed piston rod; C_{ip} is the internal leakage coefficient of the hydraulic cylinder; and $C_{ip}(p_1 - p_2)$ is the internal leakage flow at the sealed piston rod, where $p_1 - p_2$ is the load pressure, equivalent to p_L. V_1 and V_2 are the volume of the oil inlet chamber and the oil return chamber, respectively. A_p is the effective area of the hydraulic cylinder piston; m_t is the sum of the converted mass of the piston end load and the piston mass; F_L is the force of any accidental load force on the piston; and x_p is the displacement of the hydraulic cylinder piston.

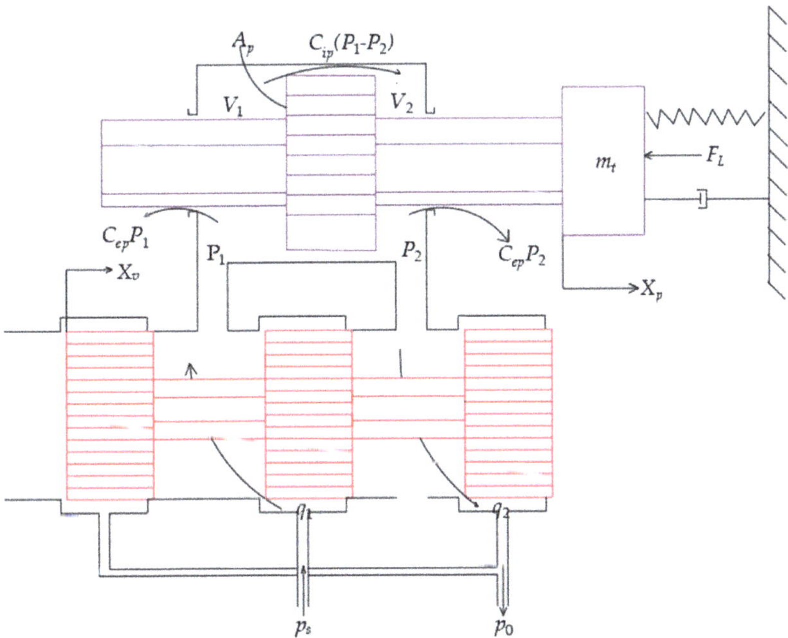

Figure 2. Valve-controlled hydraulic cylinder schematic.

A slide valve is studied with four symmetrical throttling windows, with constant supply pressure and zero return pressure [31]. The linearized flow equation of the valve is as follows.

$$q_L = K_q x_v - K_e p_L \tag{1}$$

where q_L is the system load flow rate; K_q is the flow gain of the slide valve; and K_e is the flow pressure coefficient of the slide valve. Due to leaks and the effect of fluid compressibility, the flow into and out of the hydraulic cylinder is unequal. To simplify the analysis process, q_L is defined as

$$q_L = \frac{q_1 + q_2}{2} \tag{2}$$

According to the inflow flow rate q_1, outflow flow rate q_2 can be obtained [32].

$$q_L = A_p \frac{dx_p}{dt} + C_{tp} p_L + \frac{V_t}{4\beta_e} \frac{dp_L}{dt} \tag{3}$$

where C_{tp} is the total leakage coefficient of the hydraulic cylinder; $C_{tp} = C_{ip} + C_{ep}/2$; V_t is the total compression volume of the working chamber of the hydraulic cylinder; and $V_t = V_1 + V_2$; β_e is the effective volume modulus of elasticity. The load characteristics will impact the dynamic characteristics of the hydraulic power element, and the load force generally contains damping force, inertia force, elastic force, etc. The equilibrium equation between the output force of the hydraulic cylinder and the load force is shown in Equation (4) [31].

$$A_p p_L = m_t \frac{d^2 x_p}{dt^2} + B_p \frac{dx_p}{dt} + K x_p + F_L \tag{4}$$

where B_p is the viscous damping factor corresponding to the piston and the load and K is the stiffness corresponding to the load spring. Equations (1), (3), and (4) describe the dynamic characteristics of the valve-controlled hydraulic cylinder. In order to facilitate the analysis of the system characteristics, the form of the transfer function is used instead of a differential Equation to describe the dynamic characteristics of the valve-controlled hydraulic cylinder. The Laplace transformation of Equations (1), (3), and (4) yields Equation (5).

$$\begin{cases} Q_L = K_q X_v - K_c P_L \\ Q_L = A_p X_p s + C_{tp} P_L + \frac{V_t}{4\beta_e} P_L s \\ A_p P_L = m_t X_p s^2 + B_p X_p s + K X_p + F_L \end{cases} \tag{5}$$

Since the input of the system is X_v, F_L, and the output is X_p, to obtain the relationship between the input and output, the intermediate variables Q_L, P_L in Equation (5) are eliminated to obtain Equation (6).

$$X_p = \frac{\frac{K_q}{A_p} X_v - \frac{K_{ce}}{A_p^2}\left(1 + \frac{V_t}{4\beta_e A_p^2} s\right) F_L}{\frac{V_t m_t}{4\beta_e A_p^2} s^3 + \left(\frac{m_t K_{ce}}{A_p^2} + \frac{B_p V_t}{4\beta_e A_p^2}\right) s^2 + \left(1 + \frac{B_p K_{ce}}{A_p^2} + \frac{K V_t}{4\beta_e A_p^2}\right) s + \frac{K K_{ce}}{A_p^2}} \tag{6}$$

where K_{ce} is the total flow pressure coefficient, and $K_{ce} = K_c + C_{tp}$. In the hydraulic servo system, generally, $B_P \ll A_p^2 / K_{ce}$; therefore, Equation (6) can be simplified as follows.

$$X_p = \frac{\frac{K_q}{A_p} X_v - \frac{K_{ce}}{A_p^2}\left(1 + \frac{V_t}{4\beta_e A_p^2} s\right) F_L}{\frac{V_t m_t}{4\beta_e A_p^2} s^3 + \left(\frac{m_t K_{ce}}{A_p^2} + \frac{B_p V_t}{4\beta_e A_p^2}\right) s^2 + \left(1 + \frac{K V_t}{4\beta_e A_p^2}\right) s + \frac{K K_{ce}}{A_p^2}} \tag{7}$$

When the load stiffness K is 0.

$$X_p = \frac{\frac{K_q}{A_p} X_v - \frac{K_{ce}}{A_p^2}\left(1 + \frac{V_t}{4\beta_e A_p^2} s\right) F_L}{\frac{V_t m_t}{4\beta_e A_p^2} s^3 + \left(\frac{m_t K_{ce}}{A_p^2} + \frac{B_p V_t}{4\beta_e A_p^2}\right) s^2 + s} \tag{8}$$

Then, when under external no-load, F_L equals to 0, the transfer function between the displacement X_p of the piston of the hydraulic cylinder and the input displacement X_v of the spool is

$$\frac{X_p}{X_v} = \frac{\frac{K_q}{A_p}}{\frac{V_t m_t}{4\beta_e A_p^2} s^3 + \left(\frac{m_t K_{ce}}{A_p^2} + \frac{B_p V_t}{4\beta_e A_p^2}\right) s^2 + s} \tag{9}$$

Let

$$\frac{2\beta_e A_p^2}{V_t} = K_h \tag{10}$$

$$\sqrt{\frac{K_h}{m_t}} = \omega_h \tag{11}$$

$$\frac{K_{ce}}{2A_p}\sqrt{\frac{2\beta_e m_t}{V_t}} + \frac{B_p}{2A_p}\sqrt{\frac{V_t}{2\beta_e m_t}} = \zeta_h \tag{12}$$

where K_h is the spring stiffness of the hydraulic pressure; ω_h is the hydraulic inherent frequency; and ζ_h is the damping ratio of the hydraulic pressure.

According to Equations (10)–(12), Equation (9) can be expressed as

$$\frac{X_p}{X_v} = \frac{\frac{K_q}{A_p}}{\frac{s^3}{\omega_h^2} + \frac{2\zeta_h}{\omega_h} s^2 + s} \tag{13}$$

The no-load flow characteristic Q is the characteristic flow corresponding to the load pressure drop $p_L = 0$, which is known from (1).

$$Q = K_q X_v \tag{14}$$

Equation (15) is obtained from Equations (13) and (14).

$$\frac{X_p}{Q} = \frac{\frac{1}{A_p}}{\frac{s^3}{\omega_h{}^2} + \frac{2\zeta_h}{\omega_h}s^2 + s} \tag{15}$$

2.2.2. Servo Valve Modeling

The input signal of the electro-hydraulic servo valve is the current signal. The output signal is the no-load flow signal of the servo valve. Its essence is a nonlinear element. The form of its transfer function used depends on the magnitude of the hydraulic intrinsic frequency of the power element. When the servo valve bandwidth is close to the hydraulic intrinsic frequency, the servo valve can be approximated as a second-order oscillatory system.

$$\frac{Q}{I} = \frac{K_f}{\frac{s^2}{\omega_f{}^2} + \frac{2\zeta_f}{\omega_f}s + 1} \tag{16}$$

where K_{sf} is the flow gain of the servo valve; ω_f is the intrinsic frequency of the servo valve; and ζ_f is the damping ratio of the servo valve.

2.2.3. Mathematical Modeling of the Valve-Controlled Hydraulic Servo System

According to the parameters of the hydraulic cylinder and the servo valve, see Table 1, the specific servo valve and the hydraulic cylinder model can be determined.

Table 1. Hydraulic cylinder and servo valve parameters.

Parameters	Numerical Value
Work trip	220 mm
Hydraulic cylinder bore	101 mm
Piston rod bore	63 mm
Valve rated current	40 mA
Total compression volume (V_t)	0.0010758 m^3
Load Quality (m_t)	630 kg
Effective piston area (A_p)	0.00489 m^2
Servo valve flow gain (K_{sf})	0.021 m/A
Effective bulk modulus of elasticity (β_e)	7×10^8 N/m^2
Oil supply pressure (p_s)	2.1 Mpa
Total flow pressure coefficient (K_{ce})	6.83×10^{-12} m $\cdot$ (N $\cdot$ s)$^{-1}$
Hydraulic natural frequency (ω_h)	314 rad/s
Hydraulic damping ratio (ζ_h)	0.2
Natural frequency of servo valve (ω_f)	753.6 rad/s
Damping ratio of servo valve (ζ_f)	0.7

The displacement command signal input by the system is a voltage signal, and the input signal of the servo valve is a current signal, so it is necessary to add a servo amplifier between the system input and the servo valve. After comparing and amplifying the input displacement command signal (voltage signal) and the displacement feedback signal (voltage signal) of the system, a control current proportional to the deviation voltage is output to the servo valve. The amplification factor of the servo amplifier is:

$$K_a = \frac{I}{\Delta U} \tag{17}$$

where ΔU is the deviation between the input displacement signal U and the feedback displacement signal U_c. The displacement command signal input by the system takes a voltage signal of ± 10 V as the input, and the input current of the electro-hydraulic servo valve is $0 \sim \pm 40$ mA. It can be determined that K_a is equal to 0.004.

The output signal of the valve-controlled hydraulic servo system is a displacement signal, and the feedback signal is a displacement signal. In order to compare it with the input signal, a feedback amplifier is added to convert the displacement signal into a voltage signal. The amplification factor of the feedback amplifier is:

$$K_{fk} = \frac{U_c}{X_p} \tag{18}$$

The movable working stroke of the hydraulic cylinder is 220 mm, and the input voltage of the system is ± 10 V, from which K_{fk} can be determined to be equal to 90.9.

According to the mathematical model of each part of the hydraulic servo system, the control flow chart of the hydraulic servo valve control system is established as shown in Figure 3.

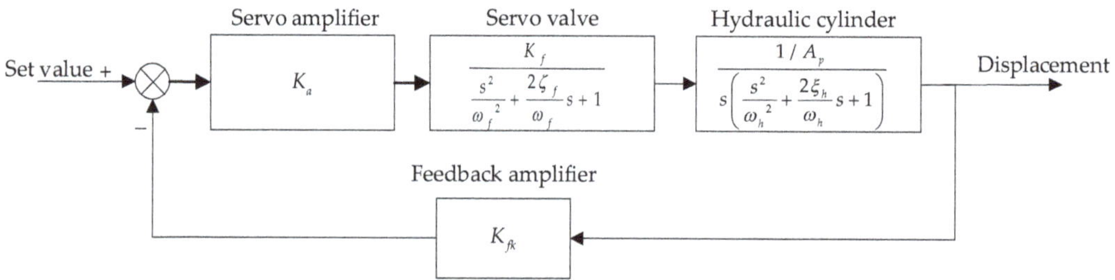

Figure 3. Control flow chart of valve-controlled hydraulic cylinder.

3. Control Algorithm of Hydraulic Servo System

3.1. PID Control Algorithm

In the closed-loop control system, the PID controller is widely used in the industry. It has the advantages of good stability, convenient adjustment, and high reliability [6,7]. Using the PID controller for position control in the hydraulic servo system, as shown in Figure 4, can increase the accuracy and response speed of the hydraulic servo position control to some extent.

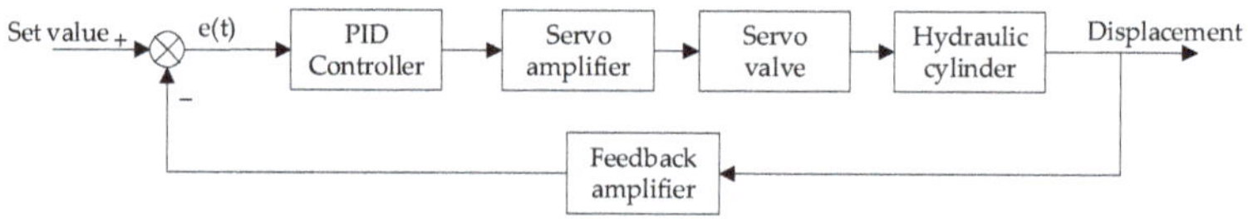

Figure 4. Hydraulic servo system PID control.

PID control consists of three units: proportional, integral, and differential. PID control adapts to different system requirements by adjusting the gain value of these three units. The PID controller can be expressed as the following mathematical formula [33]:

$$u(t) = K_p e(t) + K_i \int_0^t e(\tau)d\tau + K_d \frac{d}{dt}e(t) \tag{19}$$

where $u(t)$ is defined as the output function of the controller at time t; K_p is the proportional coefficient; K_i is the integral coefficient; K_d is the differential coefficient; e is the error (e

equals the set value minus the feedback value); t is the current time; and τ is the integral variable. Its expression in the continuous domain s is:

$$u(s) = (K_p + \frac{K_i}{s} + K_d s)E(s) \tag{20}$$

The performance of the PID controller depends on the appropriateness of the selection of the PID gain parameter. Although the gain of the three units of the PID is easy to adjust, different system transfer functions correspond to different PID gain parameters. In the process of gain adjustment, the three units will conflict, resulting in the PID controller not achieving a better performance. Therefore, the gain parameters corresponding to the three units of the PID need to be traded off to obtain the best overall control results. So the key to the PID controller is to adjust the adjustable parameters.

3.2. Optimization of the PID Algorithm by Genetic Algorithm

At present, many optimization methods are used to adjust the PID controller parameters, such as PSO, ABC, and so on. ABC can avoid falling into local optimal stagnation in optimization, but it will have the problem of slow convergence; the PSO algorithm converges fast, but easily falls into the local optimal solution and is unstable [34]. Compared with the above intelligent algorithms, the GA can obtain the global optimal solution according to the principle of natural evolution and converge quickly. The PID controller optimized by GA is used in the hydraulic servo system, as shown in Figure 5, so that the output response can better track the target.

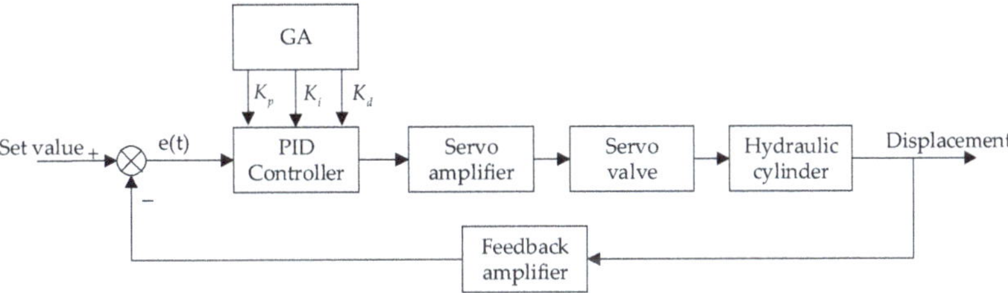

Figure 5. PID Controller optimized by genetic algorithm.

GA represents an evolutionary process similar to the Darwinian model. This process begins with the creation of an initial random group of individuals. These individuals, the points in the state space, adopt the value of the criterion to be optimized. It goes through the stages of selection, hybridization, and mutation from one generation to another to produce a new population for the next generation. The evaluation of the fitness function makes it possible to provide the optimal solution of the generation under consideration until the unique optimal solution is imposed by the convergence condition of the iterative system [35]. In order to overcome the shortcomings of the PID algorithm, the GA with crossover and mutation operation is used to adjust and optimize the three parameters of PID. GA can flexibly construct a heuristic process and corresponding objective function and can randomly and quickly search the optimal solution and find the global optimal solution in the process of a parameter search. The specific process is as follows:

1. Code and population initialization.

Coding is the process of converting the three parameters, K_p , K_i, and K_d, in the PID into chromosomes that genetic algorithms can manipulate. The content of the coding is to correspond the genotypes of individuals in biology to potentially feasible solutions. In this paper, the real number coding method is adopted, and the integer in the given range 0 to 9 is used to represent the parameters. The Ziegler-Nichols method is used to determine the optimization interval of PID parameters. The sequence of N, randomly generated

$[K_p, K_i, K_d]$, is used as the initial cluster, while the maximum evolutionary generation is set considering the complexity of the calculation.

2. Calculate the value of individual fitness function:

The absolute value time-integrated performance index of the system error is used as the minimum objective function value of the parameter, and the optimal index J for parameter selection is

$$J = \int_0^\infty c_1 |e(t)| dt + c_4 t_s \tag{21}$$

Among them, J is the optimal index, $|e(t)|$ is the absolute value of error, and t_s is the system adjustment time. A penalty factor is added to the optimal index J. In order to avoid the appearance of overshoot and prevent the response time from being slow due to the long rise time, the amount of overshoot is taken as one of the optimal indexes when overshoot occurs, and the rise time is taken as one of the optimal indexes when the response is slow, at which time the optimal index J is

$$J = \int_0^\infty (c_1 |e(t)| + c_2 pos + c_3 t_r) dt + c_4 t_s \tag{22}$$

Among them, pos is the system overshoot, t_r is the rise time of the system and c_1, c_2, c_3, and c_4 are the weights corresponding to $|e(t)|$, pos, t_r and t_s. The choice of weights affects the corresponding performance. Adjusting the relative sizes of c_1, c_2, c_3, and c_4 can indicate the importance given to $|e(t)|$, pos, t_r, and t_s. In the program simulation for test debugging, c_1 is set to 0.5, the value of c_2 is set to 0.4, c_3 is set to 0.9, and c_4 is set to 0.4. The fitness function is a tool to determine the genotypic performance of an individual. It can be seen as the driving force for individuals to evolve to salient features. To translate the designed optimal metrics into an adaptation function, the fitness function is designed as

$$F = \frac{1}{J} = \frac{1}{\int_0^\infty (c_1 |e(t)| + c_2 pos + c_3 t_r) dt + c_4 t_s} \tag{23}$$

3. Select operation:

The core idea of the selection operation is to calculate each individual's fitness in the population. Individuals with higher adaptive values have a higher probability of being saved to the next generation, whereas individuals with lower adaptive values are more likely to be eliminated. Using the ranking selection strategy for selection operations, the individuals with larger adaptation values will be selected into the next generation, which can avoid premature convergence to a certain extent.

4. Crossover and variation:

Crossover operation and mutation operation are the main methods for the genetic algorithm to generate new gene individuals. In this paper, the crossover probability P_c is set to 0.8 and the variation probability P_m is set to 0.1. For the variant operation, the random number function is used to determine the location of individual gene mutation. Then, the mutation probability P_m is used to reverse the binary code to produce new individuals for mutation. For crossover operations, the method of uniform crossover is adopted. For two successfully paired individuals A and B, each gene on their locus can be exchanged with the same probability P_c, thus forming new individuals A_1 and B_1.

5. Decode:

Decoding is converting the integer value within 0 to 9 into the actual value of the parameter. When the global optimal value of the parameter selected by the genetic algorithm is the integer value K, the corresponding actual value after decoding is shown in the Formula (24).

$$K_p = K \cdot \frac{\max(K_p) - \min(K_p)}{9 - 0} + \min(Kp) \tag{24}$$

where max (K_p) is the maximum value in the parameter K_p optimization interval, and min (Kp) is the minimum value in the parameter K_p optimization interval. The decoding mode of the K_i, K_d is similar to that of the formula (24).

On this basis, the workflow of optimizing PID parameters based on a genetic algorithm is given, as shown in Figure 6.

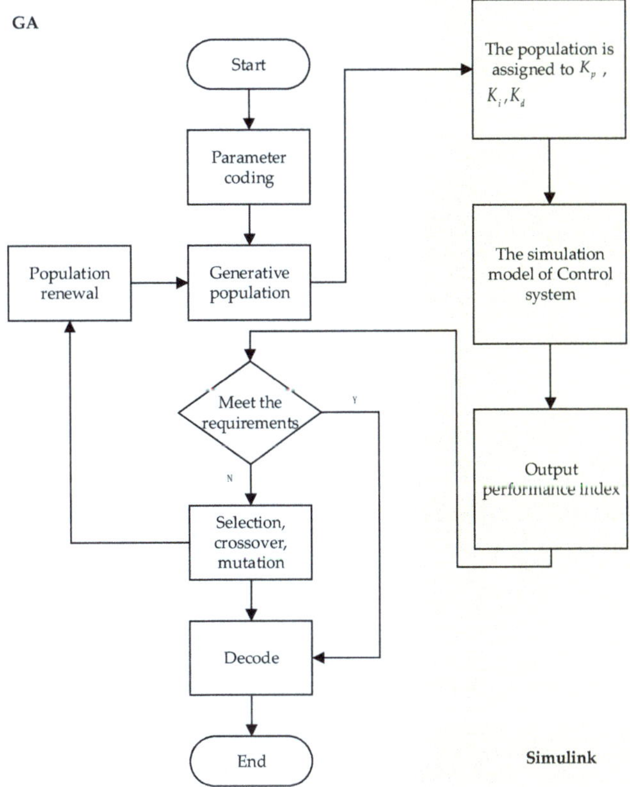

Figure 6. Schematic diagram of GA-tuning PID parameters.

3.3. Kalman Filter

The Kalman filter is an algorithm for estimating unknown variables (states) of linear dynamic systems according to noise and measurements linearly related to the system state. The Kalman filter tracking algorithm mainly consists of two parts: prediction and update [36–40]:

The forecast section is shown in (25):

$$\begin{cases} \hat{x}'_k = A\hat{x}_{k-1} + Bu_k \\ P'_k = AP_{k-1}A^T + Q \end{cases} \tag{25}$$

where $\hat{x}'_k$ is the prior state estimation of the system state x_k; $\hat{x}_{k-1}$ is the a posteriori state estimation of the previous system state x_{k-1}; u_k is the external control quantity of the system; P'_k is the covariance of the prior estimation error (the difference between x_k and $\hat{x}'_k$); P_{k-1} is the covariance of the prior estimation error (the difference between x_{k-1} and $\hat{x}_{k-1}$); A is the transition matrix; A^T is the transpose matrix of A; B is the control matrix; and Q is the covariance of the system noise ω_k.

The update section is shown in (26):

$$\begin{cases} K_k = P'_k H^T (H P'_k H^T + R)^{-1} \\ \hat{x}_k = \hat{x}'_k + K_k(z_k - H\hat{x}') \\ P_k = (1 - K_k H)P'_k \end{cases} \tag{26}$$

where K_k is the Kalman gain; H is the observation matrix; H^T is the transpose of H; $\hat{x}_k$ is the posterior state estimation of system state x_k; z_k is the system observation; P_k is the covariance of the posterior estimation error (the difference between x_k and $\hat{x}_k$); and R is the covariance of the observation noise v_k.

The combination of the Kalman filter and the GA-optimized controller is applied to the position control of the hydraulic servo system. As shown in Figure 7, it can reduce the impact of external interference (such as periodic vibration of servo valve, etc.) on the system.

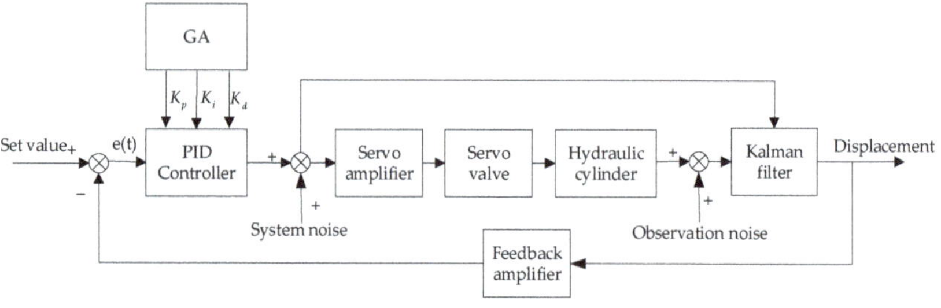

Figure 7. PID position control of hydraulic servo system based on Kalman genetic optimization.

4. Simulation and Analysis

The simulation model of the hydraulic servo system is built in MATLAB/Simulink, and the PID and GA optimized PID control of the system is carried out without interference, as shown in Figure 8. As shown in Figure 9a,b, the step signal and sinusoidal signal are used as the input of the system in turn, and the step response and sinusoidal response under PID control and GA-optimized PID control are compared and analyzed. The system simulation results are shown in Figure 9c,d. The corresponding outputs of the PID control and the GA-optimized PID controller are shown in Figure 10a,b for the case of step and sinusoidal signal inputs, respectively. The selections of the three parameters K_p, K_i, and K_d of the normal PID and GA-optimized PID are shown in Table 2.

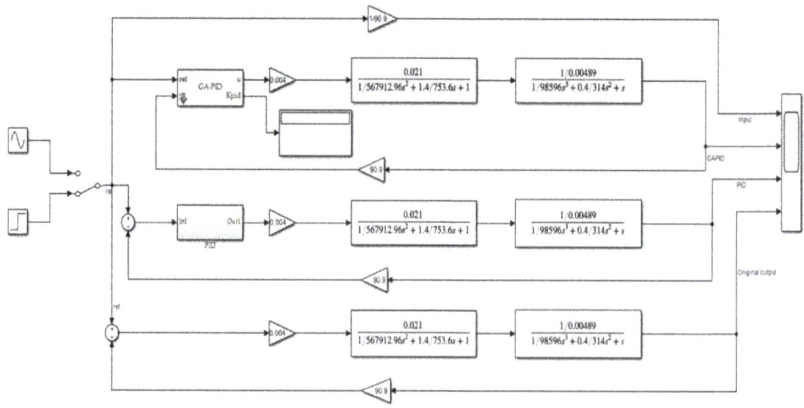

Figure 8. Optimization of PID simulation diagram by GA.

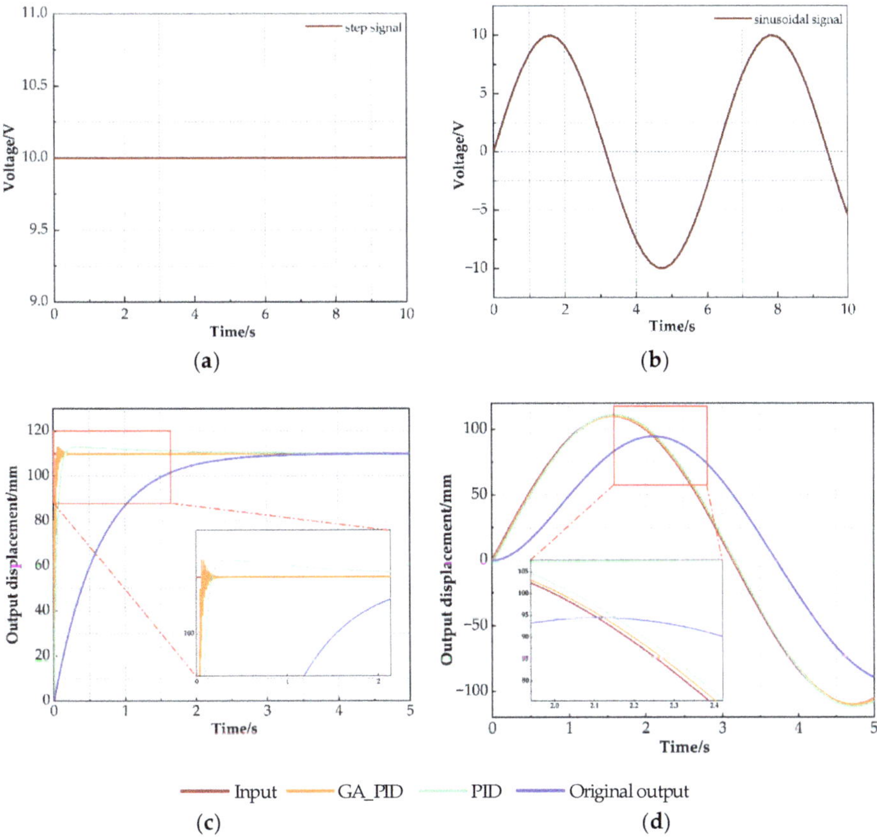

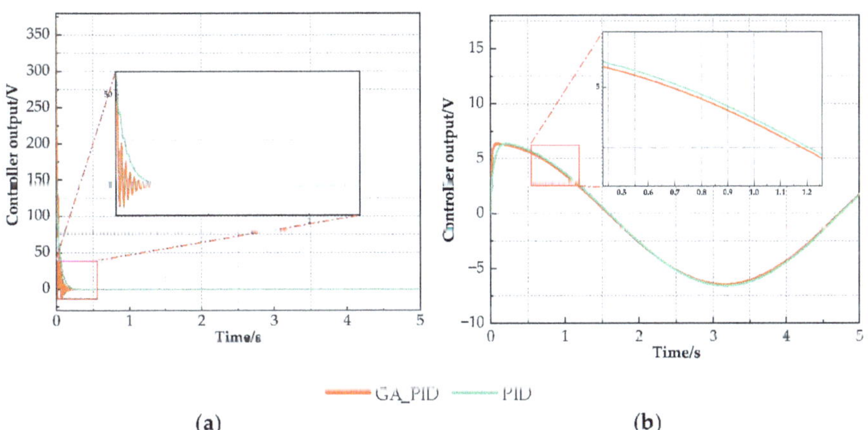

Figure 9. Simulation results of the system without noise. (**a**) Step signal input; (**b**) Sinusoidal signal; (**c**) System simulation results under step signal input; (**d**) System simulation results under sinusoidal signal input.

Figure 10. Output of PID and GA-optimized PID controllers. (**a**) Controller output under step signal input; (**b**) Controller output under sinusoidal signal input.

Table 2. Parameter values of ordinary PID and GA-optimized PID.

Parameters	PID	GA-PID
K_p	20	39.72
K_i	1	0.9914
K_d	0.1	0.033

As can be seen from Figure 9c, the maximum response of ordinary PID is 0.1132, and the overshoot is 2.9%. The maximum response of PID optimized by GA is 0.1129, and the overshoot is 2.7%. The overshoot of the optimized PID of GA is 0.2% lower than that of ordinary PID. The stable time of the ordinary PID is 1164 s, the adjustment time of PID optimized by GA is 0.063 s, and the adjustment time after optimization is 1101 s better than that before optimization. Taking the time when the output response rises from 10% of the target value to 90% of the target value as the rise time, the rise time of the ordinary PID is 0.0845 s, the rise time of the GA optimized PID is 0.0309 s. The peak time of ordinary PID is 0.248 s. The peak time of PID optimized by GA is 0.061 s. According to the rise time and peak time, under the step signal input, the response speed of the ordinary PID is obviously slower than that of the GA-optimized PID.

As can be seen from Figure 9d, when the control algorithm does not optimize the input sine signal, there is a large amplitude attenuation and phase lag; under the control of PID optimized by PID and GA, the corresponding output displacement of the valve-controlled hydraulic servo system under the sinusoidal signal input has been improved to a certain extent. From the local magnification diagram of the system sinusoidal signal corresponding to the output displacement in Figure 9d, it can be clearly seen that the output response of the sinusoidal signal of the system under the PID control optimized by GA is closer to the input signal of the system in terms of phase and amplitude than that of the ordinary PID. Therefore, the tracking performance of the GA optimized PID is better than that of the ordinary PID. That is, the control performance of the GA optimized is better.

As can be seen from Figure 10a, for the error between the input and output of the system, the PID is close to zero after 0.261 s of regulation, while the system takes 0.227 s with the GA optimized PID regulation, a reduction of 0.034 s over the PID's regulation time. This shows that the GA-optimized PID has a faster adjustment to error and better tracking performance of the input signal. As can be seen from Figure 10b, the error between the input and output of the system is corrected after both controllers and the error remains stable, whereas the error is zero when passing the mid-point, preventing the accumulation of errors, but the GA-optimized PID correction is faster than the unoptimized PID.

As can be seen from Figure 9c,d, although the output response of the hydraulic servo system after the genetic algorithm-optimized PID controller is somewhat better than the normal PID, the genetic algorithm-optimized PID causes large amplitude fluctuations, which are not allowed in the hydraulic servo system. For the fluctuations caused by the genetic algorithm-optimized PID and the possible external disturbances during the operation of the hydraulic servo system, Kalman filtering is introduced in this paper to deal with them. System noise and observation noise are added to the hydraulic servo system to simulate the hydraulic servo system subjected to external disturbance. A simulation model of the hydraulic servo system is built in Simulink, as shown in Figure 11, to analyze and compare the system immunity under PID, GA-optimized PID control, and GA-optimized PID control after the introduction of the Kalman filter.

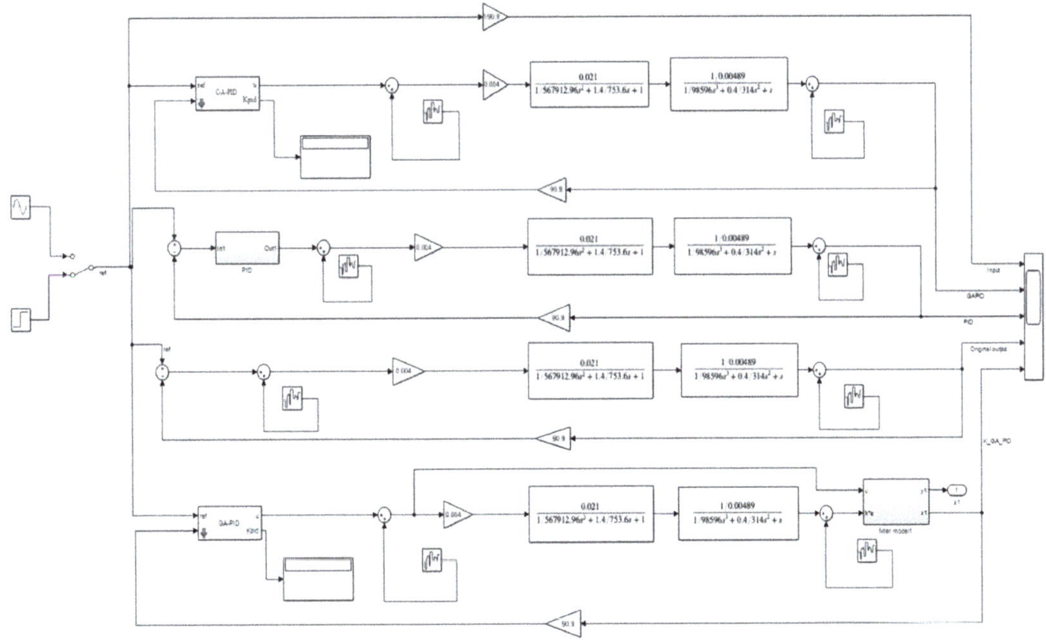

Figure 11. System Simulation Diagram of Kalman filter-optimized genetic PID.

The system noise and observation noise are set to 0.002 energy white noise, as shown in Figure 12. At the same time, the covariance Q and R of system noise ω_k and observation noise v_k are set to 1, and the step signal and sinusoidal signal are taken as the input of the system. The step response and sinusoidal response, as shown in Figure 9a,b, under PID control, GA-optimized PID control, and Kalman- and GA-optimized PID control are compared and analyzed. The system simulation results are shown in Figure 13a,b. The corresponding outputs of the PID control and the GA-optimized PID controller are shown in Figure 13a,b for the case of step and sinusoidal signal inputs, respectively.

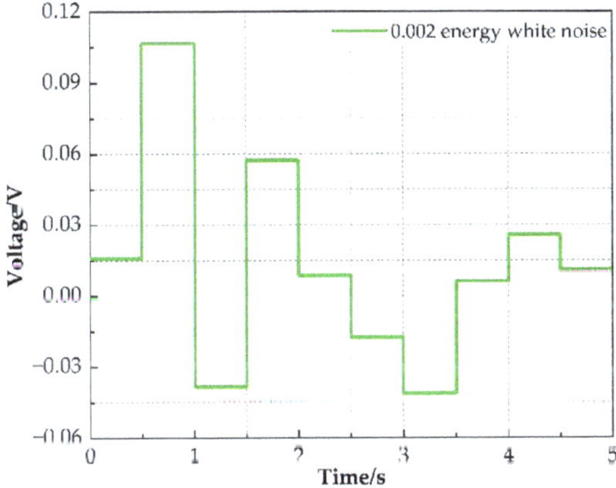

Figure 12. 0.002 energy white noise.

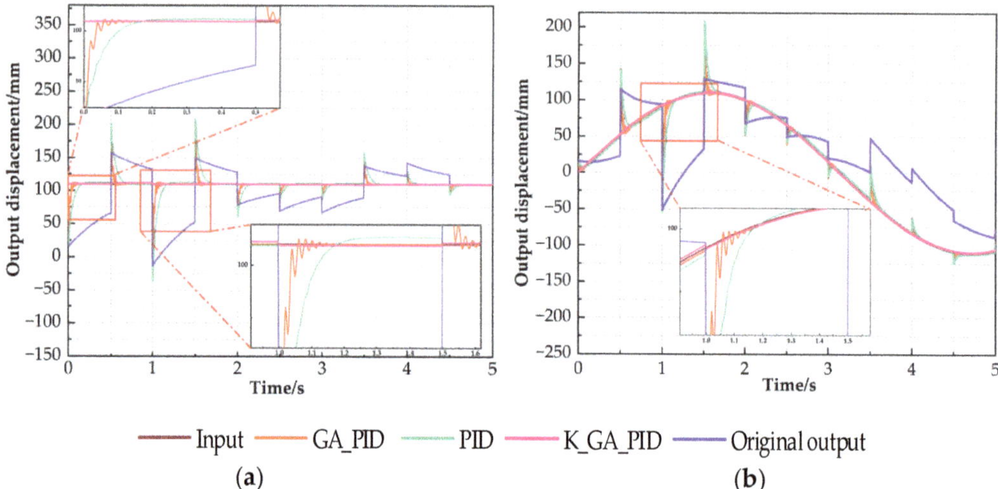

Figure 13. System simulation results under 0.002 energy white noise. (**a**) System simulation results under step signal input; (**b**) System simulation results under sinusoidal signal input.

From Figure 13, it can be seen that under the interference of system noise and observation noise with an energy of 0.002, it takes 0.172 s for PID to restore equilibrium and 0.099 s for PID optimized by GA. As can be seen from Figure 14, the PID takes 0.26 s to eliminate the effect of disturbances on the system error when subjected to external disturbances, compared to 0.22 s after GA optimization. At the same time, GA_PID with the Kalman filter module needs almost no time to adjust. It can be seen that the anti-jamming ability of PID is the worst, the anti-jamming ability of PID optimized by GA is better than that of PID, and the anti-jamming performance of GA_PID after Kalman filter is the best, which not only solves the amplitude disturbance of the system caused by the GA optimized PID, but also largely reduces the effect of external disturbance on the system.

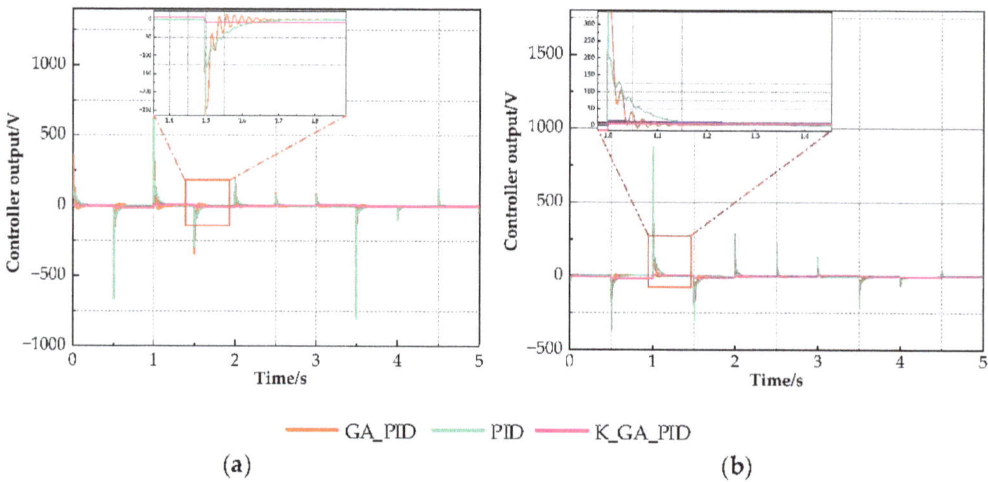

Figure 14. Output of PID and GA-optimized PID controllers under external disturbances. (**a**) Controller output under step signal input; (**b**) Controller output under sinusoidal signal input.

5. Conclusions

Aiming at the nonlinearity and uncertainty of the hydraulic servo system, the mathematical model of the valve-controlled hydraulic servo system is constructed. In order to improve the performance of the position control of the hydraulic servo system, a set of PID controllers based on Kalman genetic optimization is designed, and the controller is applied to the established hydraulic servo system for simulation analysis. The results show that the controller designed in this paper can improve the position tracking of the performance of the hydraulic servo system and the anti-interference ability. The GA algorithm is used on the PID controller to realize the tuning of the PID's three parameters and solve the problem that it is difficult to adjust the PID parameters. The introduction of the Kalman filter on the GA-optimized PID controller solves the amplitude fluctuations in the initial stage caused by the GA-optimized PID while reducing the effect of external disturbances on the system and enhancing the system's anti-interference capability.

Author Contributions: Conceptualization, Y.-Q.G. and X.-M.Z.; methodology, X.-M.Z.; software, X.-M.Z.; validation, Y.-Y.S., Y.-N.W. and G.C.; formal analysis, X.-M.Z.; investigation, Y.-N.W.; resources, Y.-Q.G.; data curation, X.-M.Z.; writing—original draft preparation, X.-M.Z.; writing—review and editing, Y.-Q.G.; visualization, X.-M.Z.; supervision, Y.-Q.G.; project administration, Y.-Q.G.; funding acquisition, Y.-Q.G. All authors have read and agreed to the published version of the manuscript.

Funding: This work was supported by the National Key R&D Programs of China under Grant 2019YFE0121900, the National Natural Science Foundation of China under Grant 51878355, and the National Natural Science Foundation of China Key Projects under Grant 52130807.

Institutional Review Board Statement: Not applicable.

Informed Consent Statement: Not applicable.

Data Availability Statement: Not applicable.

Conflicts of Interest: The authors declare no conflict of interest.

References

1. Ye, Y.; Yin, C.-B.; Gong, Y.; Zhou, J.-j. Position control of nonlinear hydraulic system using an improved PSO based PID controller. *Mech. Syst. Signal Proc.* **2017**, *83*, 241–259. [CrossRef]
2. Xing, Z.-Y.; Gao, Q.; Jia, L.-M.; Wu, Y.-Y. Modelling and Identification of Electrohydraulic System and Its Application. *IFAC Proc. Vol.* **2008**, *41*, 6446–6451. [CrossRef]
3. Kalyoncu, M.; Haydim, M. Mathematical modelling and fuzzy logic based position control of an electrohydraulic servosystem with internal leakage. *Mechatronics* **2009**, *19*, 847–858. [CrossRef]
4. Zheng, J.-M.; Zhao, S.-D.; Wei, S.-G. Application of self-tuning fuzzy PID controller for a SRM direct drive volume control hydraulic press. *Control Eng. Pract.* **2009**, *17*, 1398–1404. [CrossRef]
5. Yao, J.; Jiao, Z. Electrohydraulic Positioning Servo Control Based on Its Adaptive Inverse Model. In Proceedings of the 6th IEEE Conference on Industrial Electronics and Applications (ICIEA), Beijing, China, 21–23 June 2011; pp. 1698–1701. [CrossRef]
6. Li, X.; Yao, J.; Zhou, C. Output feedback adaptive robust control of hydraulic actuator with friction and model uncertainty compensation. *J. Frankl. Inst.* **2017**, *354*, 5328–5349. [CrossRef]
7. Shen, W.; Wang, J.; Huang, H.; He, J. Fuzzy sliding mode control with state estimation for velocity control system of hydraulic cylinder using a new hydraulic transformer. *Eur. J. Control* **2018**, *18*, 101–114. [CrossRef]
8. Knohl, T.; Unbehauen, H. Adaptive position control of electrohydraulic servo systems using ANN. *Mechatronics* **2000**, *10*, 127–143. [CrossRef]
9. Bao, R.; Wang, Q.; Wang, T. Energy-Saving Trajectory Tracking Control of a Multi-Pump Multi-Actuator Hydraulic System. *IEEE Access* **2020**, *8*, 179156–179166. [CrossRef]
10. Zhang, S.; Chen, T.; Minav, T.; Cao, X.; Wu, A.; Liu, Y.; Zhang, X. Position Soft-Sensing of Direct-Driven Hydraulic System Based on Back Propagation Neural Network. *Actuators* **2021**, *10*, 322. [CrossRef]
11. Nguyen, M.; Dao, H.; Ahn, K. Active Disturbance Rejection Control for Position Tracking of Electro-Hydraulic Servo Systems under Modeling Uncertainty and External Load. *Actuators* **2021**, *10*, 20. [CrossRef]
12. Zhang, Y.; Li, K.; Cai, M.; Wei, F.; Gong, S.; Li, S.; Lv, B. Establishment and Experimental Verification of a Nonlinear Position Servo System Model for a Magnetically Coupled Rodless Cylinder. *Actuators* **2022**, *11*, 50. [CrossRef]
13. Chang, W.-D. Nonlinear CSTR control system design using an artificial bee colony algorithm. *Simul. Model. Pract. Theory* **2013**, *31*, 1–9. [CrossRef]

14. Wang, H.; Du, H.; Cui, Q.; Song, H. Artificial bee colony algorithm based PID controller for steel stripe deviation control system. *Sci. Prog.* **2022**, *105*, 1–22. [CrossRef] [PubMed]
15. Feng, H.; Ma, W.; Yin, C.; Cao, D. Trajectory control of electro-hydraulic position servo system using improved PSO-PID controller. *Autom. Constr.* **2021**, *127*, 103722. [CrossRef]
16. Wang, Z.; Wang, Q.; He, D.; Liu, Q.; Zhu, X.; Guo, J. An Improved Particle Swarm Optimization Algorithm Based on Fuzzy PID Control. In Proceedings of the 4th International Conference on Information Science and Control Engineering (ICISCE), Changsha, China, 21–23 July 2017; pp. 835–839. [CrossRef]
17. Shutnan, W.A.; Abdalla, T.Y. Artificial Immune system based Optimal Fractional Order PID Control Scheme for Path Tracking of Robot manipulator. In Proceedings of the International Conference on Advances in Sustainable Engineering and its Application (ICASEA), Wasit, Iraq, 14–15 March 2018. [CrossRef]
18. Guo, W.; Wang, W.; Qiu, X. An Improved Generalized Predictive Control Algorithm Based on PID. In Proceedings of the International Conference on Intelligent Computation Technology and Automation, Changsha, China, 20–22 October 2008. [CrossRef]
19. Odili, J.B.; Kahar, M.N.M.; Noraziah, A. Parameters-tuning of PID controller for automatic voltage regulators using the African buffalo optimization. *PLoS ONE* **2017**, *12*, e0175901. [CrossRef]
20. Bingul, Z.; Karahan, O. A novel performance criterion approach to optimum design of PID controller using cuckoo search algorithm for AVR system. *J. Frankl. Inst.* **2018**, *355*, 5534–5559. [CrossRef]
21. Loucif, F.; Kechida, S.; Sebbagh, A. Whale optimizer algorithm to tune PID controller for the trajectory tracking control of robot manipulator. *J. Braz. Soc. Mech. Sci. Eng.* **2019**, *42*, 1. [CrossRef]
22. Xue, J.-J.; Wang, Y.; Li, H.; Meng, X.-F.; Xiao, J.-Y. Advanced Fireworks Algorithm and Its Application Research in PID Parameters Tuning. *Math. Probl. Eng.* **2016**, *2016*, 2534632. [CrossRef]
23. Gao, X.; Shi, D.; Mu, Z. PID Optimization of Motion Servo Control System Based on Improved Artificial Fish Swarm Algorithm. In Proceedings of the Chinese Automation Congress (CAC), Xian, China, 30 November–2 December 2018. [CrossRef]
24. Liu, H.; Li, Y.; Zhang, Y.; Chen, Y.; Song, Z.; Wang, Z.; Zhang, S.; Qian, J. Intelligent tuning method of PID parameters based on iterative learning control for atomic force microscopy. *Micron* **2018**, *104*, 26–36. [CrossRef]
25. Chen, L. The Optimization of PID Control Strategy in VAV System Based on Bacterial Foraging Algorithm. In Proceedings of the 19th International Conference on Network-Based Information Systems (NBiS), Ostrava, Czech Republic, 7–9 September 2016. [CrossRef]
26. Liu, X.; Shi, Y.; Xu, J. Parameters Tuning Approach for Proportion Integration Differentiation Controller of Magnetorheological Fluids Brake Based on Improved Fruit Fly Optimization Algorithm. *Symmetry* **2017**, *9*, 109. [CrossRef]
27. Veerasamy, G.; Kannan, R.; Siddharthan, R.; Muralidharan, G.; Sivanandam, V.; Amirtharajan, R. Integration of genetic algorithm tuned adaptive fading memory Kalman filter with model predictive controller for active fault-tolerant control of cement kiln under sensor faults with inaccurate noise covariance. *Math. Comput. Simul.* **2021**, *191*, 256–277. [CrossRef]
28. Xu, Z.-D.; Guo, Y.-F.; Wang, S.-A.; Huang, X.-H. Optimization analysis on parameters of multi-dimensional earthquake isolation and mitigation device based on genetic algorithm. *Nonlinear Dyn.* **2013**, *72*, 757–765. [CrossRef]
29. Ouyang, Z.L.; Zou, Z.J. Nonparametric modeling of ship maneuvering motion based on Gaussian process regression optimized by genetic algorithm. *Ocean. Eng.* **2021**, *238*, 109699. [CrossRef]
30. Xu, Z.-D.; Huang, X.-H.; Xu, F.-H.; Yuan, J. Parameters optimization of vibration isolation and mitigation system for precision platforms using non-dominated sorting genetic algorithm. *Mech. Syst. Signal Process.* **2019**, *128*, 191–201. [CrossRef]
31. Chen, W.; Wei, Q.; Zhang, Y. Research on anti-interference of ROV based on particle swarm optimization fuzzy PID. In Proceedings of the Chinese Automation Congress (CAC), Shanghai, China, 6–8 November 2020. [CrossRef]
32. Chao, C.-T.; Sutarna, N.; Chiou, J.-S.; Wang, C.-J. An Optimal Fuzzy PID Controller Design Based on Conventional PID Control and Nonlinear Factors. *Appl. Sci.* **2019**, *9*, 1224. [CrossRef]
33. Wu, T.-Y.; Jiang, Y.-Z.; Su, Y.-Z.; Yeh, W.-C. Using Simplified Swarm Optimization on Multiloop Fuzzy PID Controller Tuning Design for Flow and Temperature Control System. *Appl. Sci.* **2020**, *10*, 8472. [CrossRef]
34. Xu, F.-X.; Liu, X.-H.; Chen, W.; Zhou, C.; Cao, B.-W. Fractional order PID control for steer-by-wire system of emergency rescue vehicle based on genetic algorithm. *J. Central South Univ.* **2019**, *26*, 2340–2353. [CrossRef]
35. Agrebi, H.; Benhadj, N.; Chaieb, M.; Sher, F.; Amami, R.; Neji, R.; Mansfield, N. Integrated Optimal Design of Permanent Magnet Synchronous Generator for Smart Wind Turbine Using Genetic Algorithm. *Energies* **2021**, *14*, 4642. [CrossRef]
36. Al-Maliki, A.Y.; Iqbal, K. FLC-Based PID Controller Tuning for Sensorless Speed Control of DC Motor. In Proceedings of the 19th IEEE International Conference on Industrial Technologies (ICIT), Lyon, France, 19–22 February 2018. [CrossRef]
37. Li, H.; Yu, Q. The Wire Beltline Diameter ACO-KF-PID Control Research. In Proceedings of the Prognostics and System Health Management Conference (PHM-Chengdu), Chengdu, China, 19–21 October 2016. [CrossRef]
38. Wu, H.; Chen, S.-X.; Yang, B.-F.; Chen, K. Robust cubature Kalman filter target tracking algorithm based on genernalized M-estiamtion. *Acta Phys. Sin.* **2015**, *64*, 218401. [CrossRef]
39. Golestan, S.; Guerrero, J.M.; Vasquez, J.C.; Abusorrah, A.M.; Al-Turki, Y.A. Single-Phase FLLs Based on Linear Kalman Filter, Limit-Cycle Oscillator, and Complex Bandpass Filter: Analysis and Comparison with a Standard FLL in Grid Applications. *IEEE Trans. Power Electron.* **2019**, *34*, 11774–11790. [CrossRef]
40. Chen, W.-D.; Liu, Y.-L.; Zhu, Q.-G.; Chen, Y. Fuzzy adaptive extended Kalman filter SLAM algorithm based on the improved wild geese PSO algorithm. *Acta Phys. Sin.* **2013**, *62*, 170506. [CrossRef]

Article

Active Vibration-Based Condition Monitoring of a Transmission Line

Liuhuo Wang [1], Chengfeng Liu [2], Xiaowei Zhu [2], Zhixian Xu [2], Wenwei Zhu [3] and Long Zhao [4,*]

1   Guangdong Power Grid Company Limited, China Southern Power Grid, Guangzhou 510050, China; wangliuhuo@gd.csg.cn
2   Yangjiang Power Supply Bureau, Guangdong Power Grid Co. Ltd., Yangjiang 529599, China; liuchengfeng@yj.gd.csg.cn (C.L.); zhuxiaowei@yj.gd.csg.cn (X.Z.); xuzhixian@yj.gd.csg.cn (Z.X.)
3   Planning Research Center, Guangdong Power Grid Co. Ltd., Guangzhou 510699, China; zhuwenwei@gd.csg.cn
4   School of Electronics and Information, Xi'an Polytechnic University, Xi'an 710048, China
*   Correspondence: zhaolong@xpu.edu.cn; Tel.: +86-15686005303

Abstract: In the power system, the transmission tower is located in a variety of terrains. Sometimes there will be displacement, inclination, settlement and other phenomena, which eventually lead to the collapse of the tower. In this paper, a method for monitoring the settlement of a transmission tower based on active vibration response is proposed, which is based on the principle of modal identification. Firstly, a device was designed, which includes three parts: a monitoring host, wireless sensor and excitation device. It can tap the transmission tower independently and regularly, and collect the vibration response of the transmission tower. Then, vibration analysis experiments were used to validate the horizontal vibration responses of transmission towers which can be obtained by striking the transmission towers from either the X direction or Y direction. It can be seen from the frequency response function that the natural frequencies obtained from these two directions are identical. Finally, the transmission tower settlement experiment was carried out. The experimental results show that the third to fifth natural frequencies decreased most obviously, even up to 2.83 Hz. Further, it was found that under different conditions, as long as the tower legs adjacent to the excitation position settle, the natural frequency will decrease more significantly, which is very helpful for engineering application.

Keywords: transmission tower; settlement; monitoring device; natural frequencies

Citation: Wang, L.; Liu, C.; Zhu, X.; Xu, Z.; Zhu, W.; Zhao, L. Active Vibration-Based Condition Monitoring of a Transmission Line. *Actuators* **2021**, *10*, 309. https://doi.org/10.3390/act10120309

Academic Editor: Haim Abramovich

Received: 21 October 2021
Accepted: 17 November 2021
Published: 25 November 2021

Publisher's Note: MDPI stays neutral with regard to jurisdictional claims in published maps and institutional affiliations.

1. Introduction

Transmission towers, like other large-scale structures, may settle due to their weight, soil void compaction and water content, especially in mountainous areas, river beaches, coal mines and other unattended areas. If such settlement cannot be repaired in time, it may cause severe accidents, such as transmission tower collapse and transmission line conductor disconnection [1]. Due to the settlement-induced changes in the stresses in transmission towers [2], many scholars have attempted to analyze the failure behavior of transmission towers through finite element analysis [3,4]. The purpose of such studies is to optimize the tower structure; however, at present, there is no way to avoid the risk of damage [5].

To comprehensively track the structural health of transmission towers in real time, many online monitoring technologies have been proposed and applied. The online monitoring of a transmission tower tilt is currently the most widely used technology for this purpose, especially in China. This approach measures the inclination of a transmission tower through two inclination sensors and provides an early warning when certain conditions are met [6]. This approach is a simple structural health monitoring (SHM) method for transmission towers, which replaces previous techniques using theodolite measurements and enables the remote monitoring of transmission towers. However, in this approach, when the system detects that the tower is inclined, the tower has already been severely deformed, and correcting the tower orientation is not only costly but also creates a large potential safety hazard after maintenance. Automatic recognition from photographs is an intuitive SHM method for engineering systems, which replaces conventional sensors, such as strain gauges, that can only produce results at a discrete number of points [7]. Many engineers and scholars have attempted to use this method to identify changes in transmission tower structures [8], for which robots and unmanned aerial vehicles (UAVs) are widely used [9,10]. There are also methods based on close-range photogrammetry technology and synthetic-aperture radar, which are also used to detect the deformation of transmission towers, avoiding the shortcomings of strain gauges [11,12]. However, these methods have many limitations, such as certain weather conditions (e.g., heavy fog and snow), the camera location, and the photograph background. There are also some transmission towers in some areas that use stress monitoring to monitor the state of the reaction structure, including the use of resistance-type strain gauges and fiber Bragg grating strain gauges [13,14]. This method requires that the number and location of the sensors be adjusted after installation. In addition, this method requires a strong adhesive, which increases the difficulty of field work.

A vibration signal usually contains the structural parameters of an object because its structural change will cause variations in the natural frequency. In engineering, modal analysis is often used to identify structural damage in structures such as bridges [15], wind turbine [16], and transmission wires [17]. Other studies have performed vibration analyses of transmission towers [18], such as the improved neural network used to predict the damage of offshore wind turbine towers [19]. Some scholars have carried out nonlinear buckling analyses of tower line systems, through which they determined the critical wind load and analyzed the dynamic characteristics of tower line systems under different wind loads. The results in the literature show that the lower natural frequency is a useful index for predicting the occurrence of structural instability [20]. Moreover, scholars have established transmission towers in mountainous areas to observe the corresponding wind forces and have obtained the wind characteristics and wind response of transmission towers through experiments [21]. Wireless acceleration sensors are widely used in SHM to monitor the dynamic response of systems [22]. Especially in recent years, there has been growing interest in using micro electro-mechanical system (MEMS) accelerometers—which have a small volume and a high precision—for SHM [23]. Some scholars have studied the vibration response of transmission towers under environmental loads by monitoring the wind speed and vibration acceleration [24,25]. A fiber Bragg grating sensor was designed to measure tower vibrations under hurricane and strong wind conditions [26] and identify structural damage. Most of the above methods detect structural damage based on an analysis of wind-induced vibrations. However, sometimes the vibration of a tower is weak under the action of wind. Therefore, if vibration monitoring is used for tower settlement monitoring, it needs a longer measurement time span to obtain better results, which will lead to excessive power consumption. Moreover, as the data increase, the analysis time increases.

To increase the feasibility of using the vibration method to measure the settlement of a transmission tower, an active vibration-based SHM system was designed for a transmission tower in this paper. This SHM system uses a force-generating device to automatically strike the tower—causing vibrations—and then analyzes the tower settlement from the vibration signal. This method does not rely on the weak signals of wind-induced vibrations, allowing more stable vibration information to be obtained. Finally, a field test was conducted on a 110 kV transmission tower, and how the natural frequency of the tower changed under different settlement conditions was analyzed.

2. Operating Principle and Monitoring System

2.1. Operating Principle

When a transmission tower vibrates, the vibration has multiple degrees of freedom. The Equation of motion of a forced, damped linear system of the transmission tower can be expressed as follows:

$$M\ddot{x}(t) + C\dot{x}(t) + Kx(t) = f(t) \tag{1}$$

where M, C and K are the mass, damping and stiffness matrices of the tower structure, respectively; $x(t)$ is the displacement matrix of the tower vibration signal; $\dot{x}(t)$ is the velocity matrix of the tower vibration signal; $\ddot{x}(t)$ is the acceleration matrix of the tower vibration signal; and $f(t)$ is the force that causes the tower vibration.

Equation (1) shows that when the force is constant, the mass, damping and stiffness matrices can be calculated from the vibration signal [27,28]. The stiffness matrix K and the damping matrix C of the transmission tower will change when tower settlement occurs, especially when the settlement of the tower foundation is uneven. Moreover, previous studies have theoretically shown that different frequency bands from different forces can excite different order modes; nevertheless, this phenomenon will not substantially affect the modal parameters of the system. In actual operation, the main factor causing the vibration of the transmission tower is the natural wind load. However, the effect of the natural wind load is not continuous, and the magnitude of this load varies. In some areas, the wind load is not sufficient to induce obvious transmission tower vibrations. The analysis process is obviously affected by time, so the accuracy of the identification effect is limited. For this reason, an excitation device was designed for the modal identification of transmission towers, which can use the natural frequency of the tower to monitor whether tower settlement has occurred.

The designed monitoring system consists of a monitoring device, a force-generating device, a wireless acceleration sensor and a monitoring center. The monitoring device controls the force-generating device, which excites the tower either at a certain time or in a regular pattern. The vibration data from the acceleration sensor are sent to the monitoring center through a 4G network for analysis.

2.2. Monitoring Device

The monitoring device controls the force-generating device to strike the transmission tower, collects the impact force signal, and collects the transmission tower vibration data from the wireless acceleration sensor. The monitoring device is composed of a power module, a main control module, a 4G wireless module, a Bluetooth module, an analog-to-digital (A/D) sampling module and a driver module, as shown in Figure 1a. The power module, which is composed of a solar panel, a battery and a controller, provides power for the whole device. The main control module is composed of a MSP430F2370 microcontroller, an HC-08 Bluetooth module and a 4G wireless module, which serves as the communication module. The 4G wireless module is used for communication between the monitoring device and the remote monitoring center. The HC-08 Bluetooth module is used to communicate with the wireless acceleration sensor. The force-generating device is driven by current. To achieve this, a driving module is added to the monitoring device, as shown in Figure 1b. The on/off capability of the triode in the figure is controlled by the input/output (IO) port of the microcontroller. For the A/D sampling module, an AD7705 chip is used to record the impact force signal from the excitation device. Figure 1c shows a photograph of the main control board and power control board.

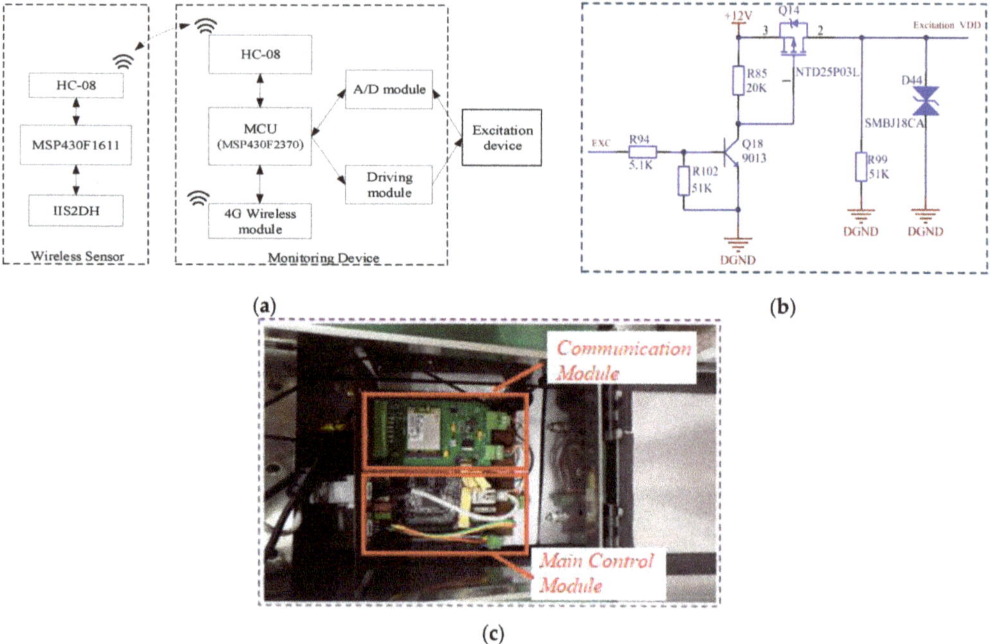

Figure 1. Monitoring device: (**a**) Operating principle block diagram; (**b**) Drive circuit; (**c**) Photograph of the monitoring device.

2.3. Force-Generating Device

The force-generating device strikes the tower under the control of the monitoring device. As shown in Figure 2, the force-generating device designed in this paper is mainly composed of an electromagnet, a high-strength magnet, a helical spring, an impact force sensor, a sleeve, a hammer head, a casing and an expansion wire. The working process of the force-generating device is described hereafter. When a force pulse is needed, the main control module generates an approximately 5-ms pulse through the I/O port. At this time, the electromagnet is magnetized under the action of the current and the coil spring is compressed under the attraction of the high-strength magnet, which causes the hammer fixed on the electromagnet to hit the tower, generating a force pulse of approximately 40 N. Then, under the restoring force of the coil spring, the hammer head returns to the initial position and waits for the next trigger.

The generation and dispersion of the force is very important. Although this SHM method only requires a force of approximately 40 N, the magnitude of the force pulse is affected by the current, the electromagnet and the permanent magnet. For this system, a traction electromagnet and neodymium boron permanent magnet were used. When driving by a power supply at 12 V and 1 A, the system can produce a striking force of approximately 45 N. The spring is the key to the spring back of the hammer head. The spring used in this system is composed of soft stainless steel and has a spring constant of 13.3 N/mm.

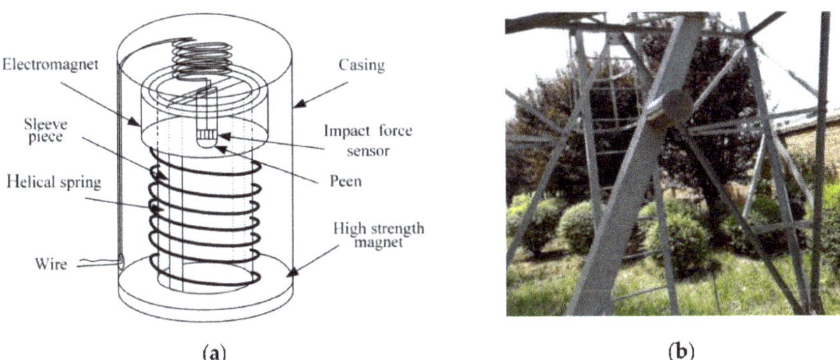

Figure 2. The force generating device: (**a**) Structural of the force generating device; (**b**) A photo of the force generating device.

The hammer head is one of the most important parts of the force-generating device because the system must excite the low-order modes of the tower (which are useful for predicting the occurrence of structural instability) to enable easy identification of tower settlement. If the hammer head was too hard, the frequency band of the excitation signal would be too wide, and the excitation energy per unit frequency would not be sufficiently high, which would affect the signal-to-noise ratio of the system. Moreover, because the acceleration sensor is particularly sensitive to high-frequency signal components, an excessively wide frequency band would make it inconvenient to analyze the low-frequency signal. In previous studies, different hammers were attempted, such as steel hammers, aluminum alloy hammers and nylon hammers, and it was found that the frequency bands created by steel hammers and aluminum alloy hammers were too wide; it was easier to identify the low-order frequency response of the system with nylon hammers. Therefore, the hammer head is made of a nylon material and is attached with a threaded connection, which makes it easy to fix the impact force sensor on the hammer head.

Furthermore, the casing is a hollow cylinder made of 304 stainless steel. The bottom of the casing is a ring-shaped, high-strength, neodymium boron magnet, which allows the whole device to firmly contact the tower. The sleeve is made of a light aluminum alloy, and the soft 304 stainless-steel helical spring is wound around the sleeve at equal intervals. A circular electromagnet is fixed at the top of the spring coil. The impact sensor in this system is a piezoelectric force sensor, which is fixed on the electromagnet through a bar structure and can be directly sampled by the A/D module. Figure 2a shows a schematic of the excitation device.

2.4. Wireless Acceleration Sensor

A wireless acceleration sensor is used to measure the vibration data of the transmission tower, which it wirelessly transmits to the monitoring device. The sensor consists of three parts (see Figure 3). An IIS2DH triaxial accelerometer is used to sense the acceleration in three directions. This accelerometer has the characteristics of ultralow power consumption (2 µA) and high precision (0.98 mg/digit). The HC-08 Bluetooth module is used to receive the data acquisition command sent by the monitoring device and send the acquired acceleration to the monitoring device. The sleep current of this module is only 0.4 µA. A MSP430F1611 microcontroller is connected to the accelerometer through the serial peripheral interface (SPI).

Figure 3. Wireless acceleration sensor.

2.5. Workflow of the Monitoring System

Figure 4 shows a chart of the system workflow. The monitoring device and sensors will enter low power consumption mode as soon as they are powered on for initialization. If the monitoring device receives the data acquisition command from the monitoring center, it will issue a command to the sensor to exit sleep mode. Then, the monitoring device will control the force-generating device to strike the tower and simultaneously collect the vibration signal of the tower and the signal of the excitation device.

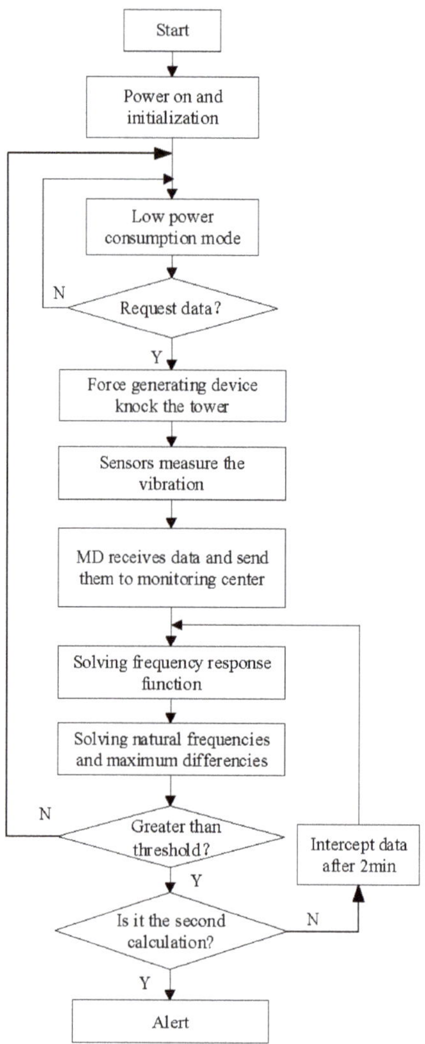

Figure 4. System workflow chart.

Afterwards, the measurement data will be sent to the monitoring center, and the natural frequencies of each order of the tower will be determined. This information is compared with the natural frequency when the structure is operating in normal conditions, the variations in the natural frequency are obtained, and the maximum variation is determined. If the maximum change in natural frequency exceeds the set threshold value, the data after 2 min will be intercepted for analysis. If the result of the second measurement still exceeds the set threshold value, the tower structure will be considered to have changed.

3. Experiment and Analysis

3.1. Experimental Platform

In this paper, the proposed settlement monitoring device, which operates based on the main vibration response of the transmission tower, is tested on site. The experimental site is the ZM-110 kV transmission tower at Xi'an University of Engineering. An LGJ-95/15 conductor is set on the tower. The line runs north to south. The total height of the transmission tower is 19 m, and the root opening is 3.103 m. Figure 5a shows the transmission tower used in this experiment. Figure 5b shows the schematic of the test platform.

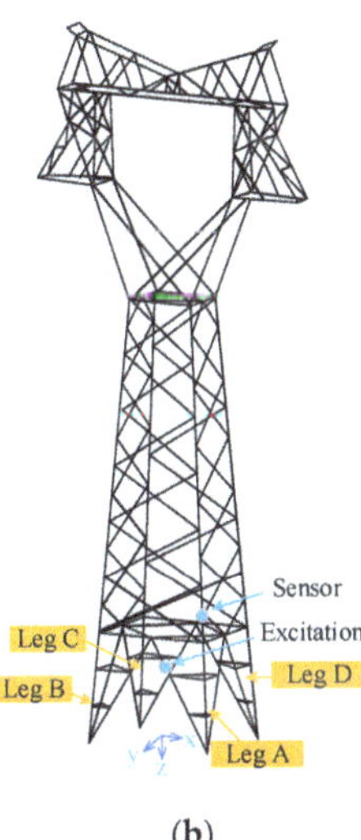

(a) (b)

Figure 5. Test platform: (**a**) A photo of test platform; (**b**) A schematic of the test platform.

3.2. Selection of the Sampling Frequency

Because the low-order natural frequencies of the tower system are less than 50 Hz, the sampling frequency of the device should be greater than 100 Hz according to the Shannon theorem. In addition, the force pulse duration of the self-designed excitation device is 50 ms. Therefore, to ensure the sampling accuracy of the force pulse, the sampling frequency should satisfy the following inequality [29]:

$$f_s = \frac{4}{T_c} \tag{2}$$

where T_c is the pulse width. Hence, the sampling frequency must be greater than 80 Hz. Moreover, the former natural frequencies of the tower are in the dense mode region within 50 Hz, so the frequency resolution must be considered. The frequency resolution can be expressed as follows:

$$\Delta f = \frac{f_s}{N} \tag{3}$$

Combined with the abovementioned factors, the sampling frequency is set to 200 Hz.

3.3. Implementation

On 11 March 2019, field measurements and subsequent analyses were conducted. First, the excitation device was used to apply a sequence of force pulses to the tower in the X, Y and Z directions, during which the acceleration responses of the tower in the X, Y and Z directions were simultaneously extracted by the acceleration sensor. Then, the frequency response function of the tower system was established, and the first five natural frequencies were determined. When the tower foundation subsides, the tower structure and stress will change, which will subsequently change the natural frequencies. A finite element simulation in reference [1] showed that the stresses developed from the settlement and uplift of the tower foundation exhibit a similar trend as the stresses measured in the tower members. Accordingly, an experimental platform for tower base lifting was built, the settlement process of the transmission tower base was manually simulated, and the low-order natural frequencies of the transmission tower system before and after settlement were extracted and analyzed. Figure 6 shows the system vibration acceleration when the excitation device is installed in the X direction. Figure 6a shows the transient variation in the force pulse generated by the excitation device. Here, you can see that the force pulse is about 45 N and its duration is very short. Figure 6b–d represent the acceleration responses of the tower in the X, Y and Z directions, respectively.

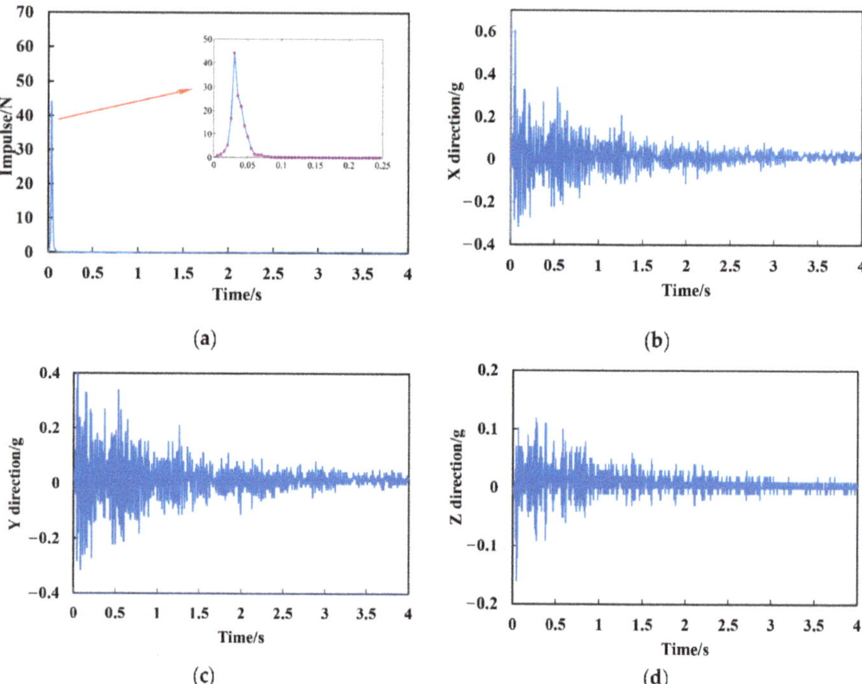

Figure 6. Excitation and response curve: (**a**) The curve of the force; (**b**) Acceleration in the X direction; (**c**) Acceleration in the Y direction; (**d**) Acceleration in the Z direction.

3.4. Identification of the Natural Frequencies

The frequency response function of the vibration can be expressed as follows:

$$[H_0(\omega)] = [X_0(\omega)][\delta_0(\omega)]^{-1} \tag{4}$$

where $\delta_0(\omega)$ is the matrix afterFourier transformation of the measured force, $X_0(\omega)$ is the matrix after Fourier transformation of the measured acceleration, and $H_0(\omega)$ is the frequency response function.

However, the frequency response will be different when the direction of the excitation force changes. Therefore, the objective of the monitoring technology is to determine the strike direction. In this experiment, the directions of the impact were along the X axis, Y axis and Z axis, and the acceleration was simultaneously measured in all three directions.

Figure 7 shows the frequency response function of the tower system where "F-x & A-y" was used to represent the frequency response function in the Y axis when a force pulse is applied in the X direction. Figure 7a–c show the frequency response function curves of the system when the force pulse is applied in the X direction. Figure 7d–f shows the frequency response function curves of the system when the force pulse is applied in the Y direction. Figure 7g–i shows the frequency response function curves of the system when the force pulse is applied in the Z direction.

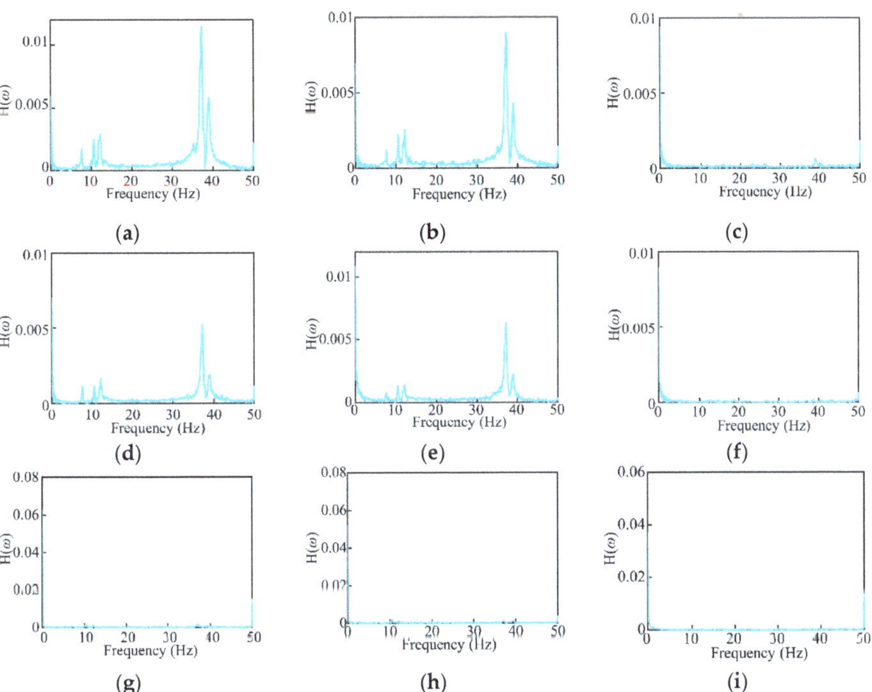

Figure 7. Frequency response functions of the tower system: (**a**) F-x & A-x; (**b**) F-x & A-y; (**c**) F-x & A-z; (**d**) F-y & A-x; (**e**) F-y & A-y; (**f**) F-y & A-z; (**g**) F-z & A-x; (**h**) F-z & A-y; and (**i**) F-z & A-z.

It can be seen from Figure 7 that when the force pulse acts on the X and Y directions of the transmission tower, the modal frequency can be extracted more effectively from the crest abscissa of the frequency response curve of the X axis ((a),(d)) and Y axis ((b),(e)), but the Z axis ((c),(f)), the frequency response curve of almost no crest, unable to extract modal frequency. Moreover, similarly, when the force impulse acts on the Z direction, the frequency response functions of the X axis (g), Y axis (h) and Z axis (i) do not show obvious crest characteristics, which means that it is difficult to identify the natural frequency when the force impulse acts on the transmission tower from the Z direction.

In addition, the vibration of the tower mainly occurs in the X direction and Y direction. This is because, from a macro point of view, the tower is like a cantilever beam fixed at one end, and the vibration mainly occurs in the other two directions. Specifically, on the one hand, the axial stiffness of the tower is much greater than the bending stiffness, which makes the deformation in the z-axis direction very small; on the other hand, the Z direction is constrained by the ground, while the X and Y directions are not constrained. Therefore, it can be seen in Figure 7 that the response mainly occurs in the X direction and Y direction.

The peak values of the frequency response function were extracted as the natural frequencies at different orders. It is worth noting that in a large number of experiments, the natural frequency of 1~2 Hz cannot be obtained. The smallest natural frequency excited by this method is 7.617 Hz. When the force pulses of the normal structure were applied in the X and Y directions, the first five natural frequencies extracted from the peak values of the frequency response function. They are 7.617, 10.55, 12.3, 37.3 and 38.96 Hz. This shows that the first five natural frequencies of the tower can be effectively extracted from the acceleration response in either the X or Y direction with good consistency, regardless of whether the force pulse is applied in the X direction or Y direction.

When the natural frequency is used to judge structural changes, the low-order frequency is usually analyzed because the higher-order frequency is always changeable. To see if these natural frequencies are highly variable in the case where no changes occur in the tower structure, 10 repeated measurements were carried out. The consistency of extracting the natural frequency of the tower system by using the frequency response function was very good, and the maximum error was only 0.09 Hz.

3.5. Tower Leg Displacement Experiment

During the construction of the ZM-110 kV transmission tower at Xi'an University of Engineering, grooves were left in the foundation, which made it very convenient for us to lift the tower legs to simulate their displacement. Figure 8 shows a photograph of the experimental site. Before the test, the tower base was carefully cleaned to avoid introducing unnecessary errors. Because foundation settlement is not easy to achieve, lifting the tower legs is instead. In the experiment, the bolts fixing the tower base were first released, and then an ultrathin hydraulic jack was placed below the tower base to raise the tower legs. When one leg was raised, it corresponded to the settlement of the other legs. For example, if the foot of tower leg A was lifted, this corresponded to the settlement of tower legs B–D. At this time, the whole structure of the tower was deformed, especially the strain in some members of the tower legs [1].

Figure 7 shows that the frequency response function can be obtained by striking the tower in both the X and Y directions. Furthermore, the modal parameters of a mechanical system can theoretically be extracted from the measurement results of an acceleration sensor. Therefore, in this experiment, the excitation device and acceleration sensor were installed in the X direction of tower leg A.

Figure 9 shows the frequency response function curve before and after the settlement of BCD. After the settlement of BCD, the structure of the tower changed, and the reduced stiffness in the bolted connection caused different boundary conditions in leg A, which may result in a downward shift of the frequency. Figure 9 shows that for these two cases, the frequency response function curves were very similar, but after settlement, the natural frequencies of each stage were reduced, especially the fourth and fifth natural frequencies, which sustained the most obvious changes.

Figure 8. Experimental site.

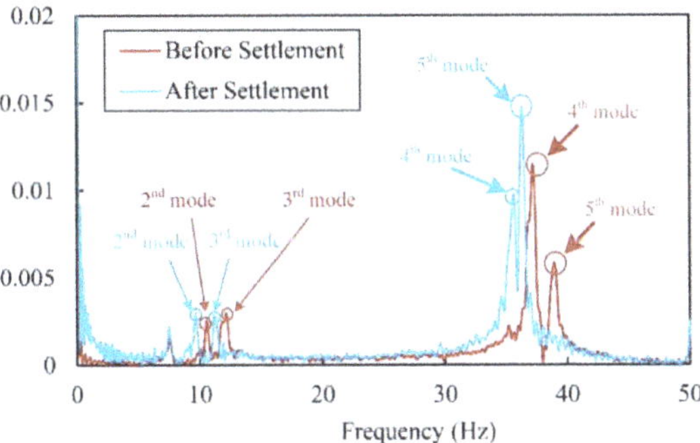

Figure 9. Frequency response function before and after settlement.

In order to further observe the change of natural frequency after settlement, the above experiment process was repeated with sampling frequencies of 100, 200 and 500 Hz, different tower leg settlement experiments were carried out, the natural frequency of the transmission tower of third-, fourth- and fifth-order was extracted, and the box diagram shown in Figure 10 was obtained. In the figure, Normal means that the transmission tower does not settle, while letters A–D and different combinations of letters represent different settling conditions of the transmission tower.

In Figure 10, a–c, d–f and g–i respectively represent the settlement of a single tower leg, two tower legs and three tower legs and the comparison of the third, fourth and fifth natural frequencies when the transmission tower is in normal conditions (the tower leg does not settle).

It can be seen from the figure that the natural frequencies of the settling legs of a single tower vary with respect to normal conditions, and the natural frequencies of the settling legs of different towers are obviously distinguished. For example, the attenuation of the third-, fourth- and fifth-order natural frequencies of A–C and the natural frequencies under normal conditions are 1.1, 0.4, and 0.8 Hz, respectively. For the settlement of the legs of the two towers, the attenuation of the third-, fourth- and fifth-order natural frequencies of the transmission tower is the most obvious when AB, AC and AD are settling, and the attenuation range of the fifth-order natural frequency reaches 3 Hz when AB is settling. For the settlement of the legs of the three towers, the natural frequencies of the third-, fourth- and fifth-order of the transmission tower are obviously attenuated during the

ACD settlement, among which the third- and fifth-order frequencies attenuated by 1.1 and 1.2 Hz.

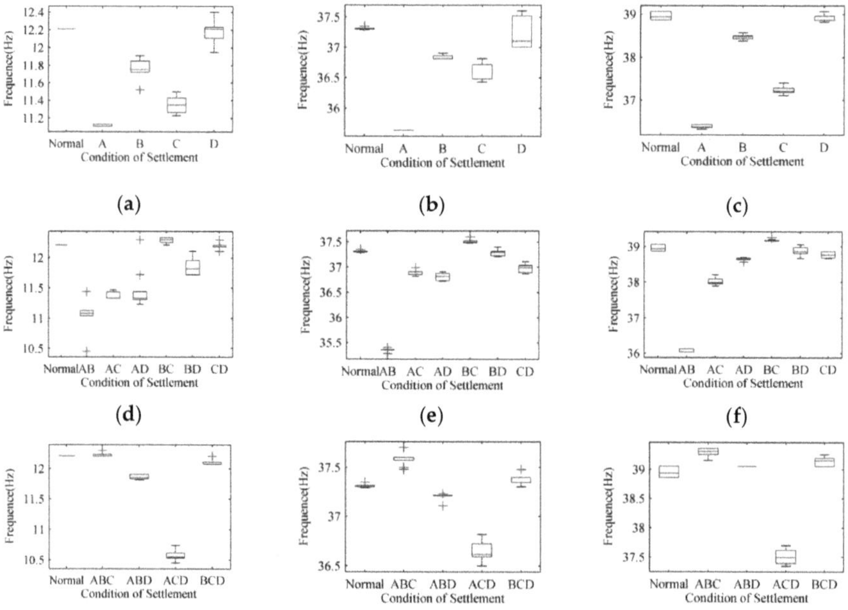

Figure 10. Box diagram: (**a**) 3rd mode at single leg settlement; (**b**) 4th mode at single leg settlement; (**c**) 5th mode at single leg settlement; (**d**) 3rd mode at two legs settlement; (**e**) 4th mode at two legs settlement; (**f**) 5th mode at two legs settlement; (**g**) 3rd mode at three legs settlement; (**h**) 4th mode at three legs settlement; and (**i**) 5th mode at three legs settlement.

For the settlement of these conditions, monitoring frequencies can identify whether there is settling for transmission towers. But some settlement is not linear. For example, the tower leg BC, BD settlement, and tower leg BCD settlement, their average frequencies are very close to the natural frequency of normal towers

To further characterize the settlement behavior, the data were normalized in Figure 10 to obtain the radar chart shown in Figure 11a. Figure 11b shows the same radar chart with the coordinate axis reversed.

As previously mentioned, the excitation was applied to tower leg A, and the sensor was also installed on leg A. Accordingly, among the cases of single-leg settlement, the natural frequencies most obviously decreased in response to the settlement of legs B or D, which are adjacent to leg A. For the cases of two-leg settlement, the natural frequencies exhibited the most obvious decline in response to the settlement of legs BD, whereas marked decreases were also observed with the simultaneous settlement of legs AD, legs BC and legs CD. For the cases of three-leg settlement, the natural frequencies decreased most obviously in response to the settlement of legs B–D. In conclusion, the natural frequencies will decrease obviously when the settlement occurs in the tower legs adjacent to those where the sensors and excitation de-vices are installed. In contrast, the decrease in the natural frequencies will not be as obvious when settlement occurs in the instrumented tower leg or in the tower leg diagonal to the instrumented leg.

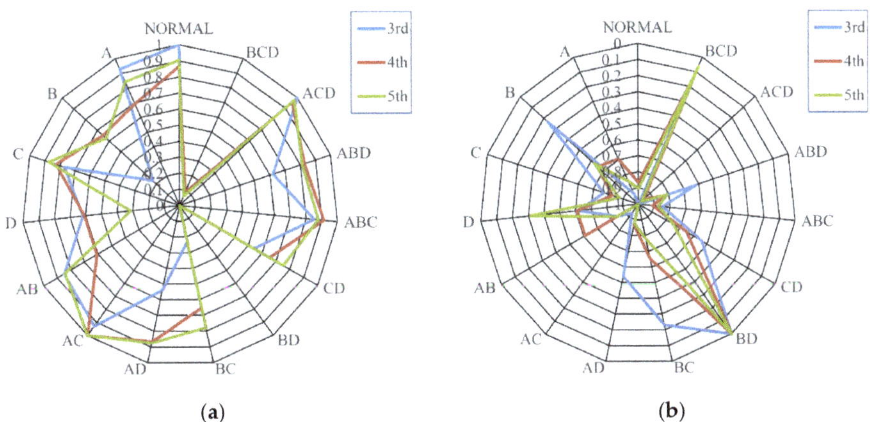

Figure 11. Normalized radar chart: (**a**) Positive sequence; (**b**) Inverted sequence.

In practical application, the natural frequency may fluctuate a little due to external influence, which may cause misjudgment. However, the settlement value of a tower usually changes slowly. Therefore, long-term monitoring is an effective method. Generally, the device can be set to measure once every 12 h. When the natural frequency changes significantly, the measurement frequency increases. If the natural frequency change trend is consistent for two consecutive days, the staff can go to the site for further investigation.

4. Conclusions

In this paper, a settlement monitoring method based on the main vibration response of the transmission tower is proposed and applied to the cat-head tower in the ZM-110kV transmission line of Xi'an Polytechnic University. The effectiveness of the method is verified.

By measuring the vibration response of transmission tower in real time, it is found that the acceleration response in the X and Y directions can be used to construct the frequency response function of a transmission tower structural system when applying force pulse in the X and Y directions. The first five natural frequencies of the tower system can be extracted by this method with a maximum error of 0.09 Hz. However, the Z direction frequency response function curve has no obvious characteristics, so the natural frequency of a transmission tower cannot be extracted from it. It was also found that when the force pulse acts in the Z direction of the transmission tower, the frequency response functions of the X–Z axis also do not show obvious characteristics, which indicates that the X or Y direction force pulse is applied to the transmission tower, and the acceleration response in the X or Y direction can be used to extract the natural frequency of the transmission tower.

Based on the above conclusions, the experiment of simulating the settlement of transmission tower by lifting tower shows that the natural vibration frequency of transmission tower decreases obviously after settlement. For single-pile settlement, the maximum attenuation of the third, fourth and fifth natural frequencies are 0.97, 0.58 and 1.85 Hz, respectively. For leg settlement, the maximum attenuation of the third, fourth and fifth natural frequencies are 1.26, 1.95 and 2.83 Hz, respectively. For the three-leg settlement, the maximum attenuation of third-order, fourth-order and fifth-order natural frequencies is 1.17, 1.75 and 2.63 Hz, respectively. The results show that the variation of natural frequency after settlement can be measured and can be used as an effective basis to judge the settlement of a transmission tower.

This paper also designs a transmission tower settlement monitoring system which is composed of a monitoring center, wireless sensor and excitation device, which can impact the tower and record its vibration response. The design of the excitation device applies a stable force pulse to the transmission tower, so as to overcome the weak vibration caused by natural load and the difficulty of signal extraction.

The above work shows that it is feasible to apply force pulse on the transmission tower to collect and analyze vibration information to monitor the device. The online monitoring device designed according to this principle can realize the on-line monitoring of transmission tower settlement and provide an effective solution for the protection of transmission line safety.

The common faults of towers mainly include settlement, inclination, loose bolts and rod deformation. This paper mainly discusses the variation law of natural frequency after settlement. In the follow-up work, we will further study and simulate more working conditions through more experiments, and explore the evaluation methods of these working conditions.

Author Contributions: Conceptualization, L.Z. and L.W.; methodology, L.Z.; validation, X.Z., C.L. and Z.X.; investigation, W.Z.; resources, W.Z.; data curation, L.Z.; writing—original draft preparation, L.W.; project administration, W.Z.; funding acquisition, L.Z. All authors have read and agreed to the published version of the manuscript.

Funding: This work was supported in part by the Key Research and Development Projects of Shaanxi Province under Grant 2021GY-068.

Institutional Review Board Statement: Not applicable.

Informed Consent Statement: Not applicable.

Data Availability Statement: The raw/processed data required to reproduce these findings cannot be shared at this time as the data also forms part of an ongoing study.

Conflicts of Interest: The authors declare no conflict of interest.

References

1. Huang, X.; Zhao, L.; Chen, Z.; Liu, C. An online monitoring technology of tower foundation deformation of transmission lines. *Struct. Health Monit.* **2019**, *18*, 949–962. [CrossRef]
2. Huang, X.; Chen, Z.; Zhao, L.; Zhu, Y.; Xu, G.; Si, W. Stress simulation and experiment for tower foundation settlement of 110 kV transmission line. *Electr. Power Autom. Equip.* **2017**, *37*, 361–370.
3. Rao, N.P.; Knight, G.S.; Mohan, S.; Lakshmanan, N. Studies on failure of transmission line towers in testing. *Eng. Struct.* **2012**, *35*, 55–70. [CrossRef]
4. Albermani, F.; Kitipornchai, S.; Chan, R. Failure analysis of transmission towers. *Eng. Fail. Anal.* **2009**, *16*, 1922–1928. [CrossRef]
5. An, L.; Jiang, W.; Liu, Y.; Shi, Q.; Wang, Y.; Liu, S. Experimental study of mechanical behavior of angles in transmission towers under freezing temperature. *Adv. Steel Construct.* **2018**, *14*, 461–478.
6. Malhara, S.; Vittal, V. Mechanical State Estimation of Overhead Transmission Lines Using Tilt Sensors. *IEEE Trans. Power Syst.* **2010**, *25*, 1282–1290. [CrossRef]
7. Reagan, D.; Sabato, A.; Niezrecki, C. Feasibility of using digital image correlation for unmanned aerial vehicle structural health monitoring of bridges. *Struct. Health Monit.* **2018**, *17*, 1056–1072. [CrossRef]
8. Huang, X.; Yang, L.; Zhang, Y.; Zhu, Y.; Zhang, G. A Measurement Technology of Space Distance Among Transmission Bundle Conductors Based on Image Sensors. *IEEE Trans. Instrum. Meas.* **2018**, *68*, 4003–4014. [CrossRef]
9. Zhong, M.; Cao, Q.; Guo, J.; Zhou, D. Simultaneous Lever-Arm Compensation and Disturbance Attenuation of POS for a UAV Surveying System. *IEEE Trans. Instrum. Meas.* **2016**, *65*, 2828–2839. [CrossRef]
10. Bian, J.; Hui, X.; Zhao, X.; Tan, M. A monocular vision–based perception approach for unmanned aerial vehicle close proximity transmission tower inspection. *Int. J. Adv. Robot. Syst.* **2019**, *16*, 1729881418820227. [CrossRef]
11. Xiao, Z.; Liang, J.; Yu, D.; Asundi, A. Large field-of-view deformation measurement for transmission tower based on close-range photogrammetry. *Measurement* **2011**, *44*, 1705–1712. [CrossRef]
12. Zeng, T.; Gao, Q.; Ding, Z.; Tian, W.; Yang, Y.; Zhang, Z. Power Transmission Tower Detection Based on Polar Coordinate Semivariogram in High-Resolution SAR Image. *IEEE Geosci. Remote. Sens. Lett.* **2017**, *14*, 2200–2204. [CrossRef]
13. Xia, Y.; Zhang, P.; Ni, Y.-Q.; Zhu, H.-P. Deformation monitoring of a super-tall structure using real-time strain data. *Eng. Struct.* **2014**, *67*, 29–38. [CrossRef]
14. Bang, H.-J.; Kim, H.-I.; Lee, K.-S. Measurement of strain and bending deflection of a wind turbine tower using arrayed FBG sensors. *Int. J. Precis. Eng. Manuf.* **2012**, *13*, 2121–2126. [CrossRef]
15. Rainieri, C.; Magalhães, F.; Gargaro, D.; Fabbrocino, G.; Cunha, Á. Predicting the variability of natural frequencies and its causes by Second-Order Blind Identification. *Struct. Health Monit.* **2019**, *18*, 486–507. [CrossRef]

16. Di Lorenzo, E.; Petrone, G.; Manzato, S.; Peeters, B.; Desmet, W.; Marulo, F. Damage detection in wind turbine blades by using operational modal analysis. *Struct. Health Monit.* **2016**, *15*, 289–301. [CrossRef]
17. Zhao, L.; Huang, X.; Jia, J.; Zhu, Y.; Cao, W. Detection of Broken Strands of Transmission Line Conductors Using Fiber Bragg Grating Sensors. *Sensors* **2018**, *18*, 2397. [CrossRef] [PubMed]
18. Zhang, M.; Zhao, G.F.; Wang, L.L.; Li, J. Wind-Induced Coupling Vibration Effects of High-Voltage Transmission Tow-er-Line Systems. *Shock Vib.* **2017**, *2017*, 1205976.
19. Qiu, B.; Lu, Y.; Sun, L.; Qu, X.; Xue, Y.; Tong, F. Research on the damage prediction method of offshore wind turbine tower structure based on improved neural network. *Measurement* **2020**, *151*, 107141. [CrossRef]
20. Fei, Q.; Zhou, H.; Han, X.; Wang, J. Structural health monitoring oriented stability and dynamic analysis of a long-span transmission tower-line system. *Eng. Fail. Anal.* **2012**, *20*, 80–87. [CrossRef]
21. Okamura, T.; Ohkuma, T.; Hongo, E.; Okada, H. Wind response analysis of a transmission tower in a mountainous area. *J. Wind. Eng. Ind. Aerodyn.* **2003**, *91*, 53–63. [CrossRef]
22. Lynch, J.P.; Loh, K.J. A summary review of wireless sensors and sensor networks for structural health monitoring. *Shock Vib. Dig.* **2006**, *38*, 91–130. [CrossRef]
23. Sabato, A.; Niezrecki, C.; Fortino, G. Wireless MEMS-Based Accelerometer Sensor Boards for Structural Vibration Moni-toring: A Review. *IEEE Sens. J.* **2017**, *17*, 226–235. [CrossRef]
24. Huang, X.; Zhao, Y.; Zhao, L.; Yang, L. A Method for Settlement Detection of the Transmission Line Tower under Wind Force. *Sensors* **2018**, *18*, 4355. [CrossRef] [PubMed]
25. Wang, J.; Wang, Z.; Feng, Y. Measurement and Analysis of Structure Vibration of In-service High Voltage Transmission Tower. *J. North China Electr. Power Univ.* **2016**, *43*, 62–65.
26. Nan, Y.; Xie, W.; Min, L.; Cai, S.; Ni, J.; Yi, J.; Luo, X.; Wang, K.; Nie, M.; Wang, C.; et al. Real-Time Monitoring of Wind-Induced Vibration of High-Voltage Transmission Tower Using an Optical Fiber Sensing System. *IEEE Trans. Instrum. Meas.* **2020**, *69*, 268–274. [CrossRef]
27. Anindya, G.; Mannur, J.S.; Mark, J.S.; Frank, P.P. Structural health monitoring techniques for wind turbine blades. *J. Wind Eng. Ind. Aerodyn.* **2000**, *85*, 309–324.
28. Zhang, H.; Schulz, M.; Naser, A.; Ferguson, F.; Pai, P. Structural Health Monitoring Using Transmittance Functions. *Mech. Syst. Signal Process.* **1999**, *13*, 765–787. [CrossRef]
29. Xiao-feng, H. On the setting of impulse-respouse method. *J. Changde Teach. Univ.* **2003**, *2*, 15.

Article

Design of Two-Axial Actuator for Controlled Vibration Damper for Large Rams

Lukáš Novotný [1,*], Jaroslav Červenka [1], Matěj Sulitka [1], Jiří Švéda [1], Miroslav Janota [1] and Petr Kupka [2]

1   Research Center of Manufacturing Technology, Czech Technical University in Prague,
    128 00 Prague, Czech Republic; j.cervenka@rcmt.cvut.cz (J.Č.); m.sulitka@rcmt.cvut.cz (M.S.);
    j.sveda@rcmt.cvut.cz (J.Š.); m.janota@rcmt.cvut.cz (M.J.)
2   TOS KUŘIM–OS, a.s., 664 34 Kuřim, Czech Republic; petr.kupka@tos-kurim.cz
*   Correspondence: l.novotny@rcmt.cvut.cz

Abstract: Machine tool rams are important constructional elements found on vertical lathes as well as on many other machines. In most cases, a machine tool ram constitutes an assembly with significant dynamic compliance that affects the machine's ability to achieve stable cutting conditions. There are various solutions for increasing a machine tool ram's stiffness and damping. This paper describes an innovative concept of a two-axial electromagnetic actuator for controlled vibration dampers with high dynamic force values. The described solution is purposefully based on the use of standard electric drives. As a result, the size of the actuator is easier to scale to the required application. The solution is designed as a spacer between the end of the ram and the head. The paper presents the actuator concept, construction design, current control loop solution and experimental verification of the controlled vibration damper's function on the test ram in detail. The presented position measurement concept will enable the use of non-contact position sensors for motor commutation as well as for possible use in vibration suppression control. Applications can be expected mainly in the field of vibration suppression of vertical rams of large machine tools.

Keywords: active vibration damper; controlled vibration damper; machine tool rams; vibration damping; actuator for vibration damping

Citation: Novotný, L.; Červenka, J.; Sulitka, M.; Švéda, J.; Janota, M.; Kupka, P. Design of Two-Axial Actuator for Controlled Vibration Damper for Large Rams. *Actuators* **2021**, *10*, 199. https://doi.org/10.3390/act10080199

Academic Editors: Zhao-Dong Xu, Siu-Siu Guo and Jinkoo Kim

Received: 12 July 2021
Accepted: 16 August 2021
Published: 19 August 2021

Publisher's Note: MDPI stays neutral with regard to jurisdictional claims in published maps and institutional affiliations.

1. Introduction

Vibration suppression actuators are usually additional devices that generally introduce an extra force into the system. This force acts against the movement (vibration), thus reducing the amount of energy accumulated in the mechanical system. The effect of actuators on vibration suppression depends on many factors. The development of active vibration suppression methods has a relatively long history, and such methods are well described in the literature and in a number of articles. A very comprehensive overview of the state of the art can be found in publications [1–3]. There are also various energy harvesting methods with great potential for vibration damping. A basic overview of these methods can be found in books [4,5]. Most of the journal publications deal with vibration suppression control methods. Two interesting examples are papers [6,7] Article [6] describes a method for automatic tuning of an active vibration control system using inertial actuators. Article [7] presents adaptive active vibration control for machine tools with highly position-dependent dynamics. The adaptive controller was implemented on an industrial PLC and represents one of the real possibilities for controlling the two-axial actuator presented in this article.

However, the authors of this paper seek to focus mainly on active structural vibration suppression and related actuators, which can be used in particular for vertical rams of large machine tools. The key element in controlled vibration suppression is the solution of the actuator itself. Although it is an essential component of the vibration suppression system, there are relatively few publications that address it (in comparison to those dealing with control). A very comprehensive overview of chatter suppression techniques in metal cutting is provided in papers [8,9]. Nonetheless, these papers do not deal directly with the actuator solutions themselves. However, it provides information on the application of Dynamic Active Stabilizer (DAS®) on a SORALUCE machine (Figure 1). This SORALUCE machine structure, or a very similar structure, is also present in papers [10–12]. There are not many similar applications described on other real (commercial) machine tool structures. These machines from SORALUCE are an exception.

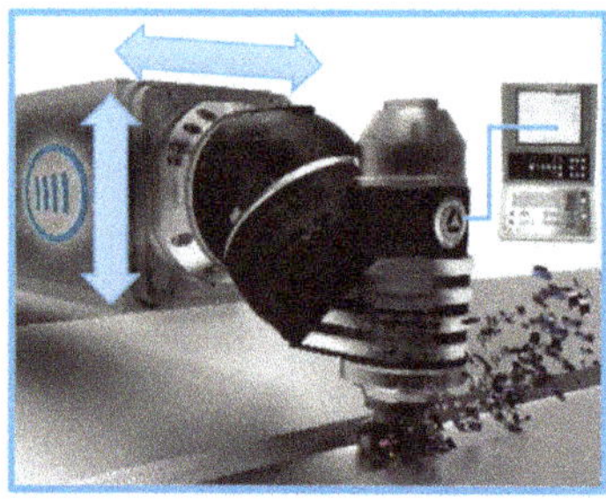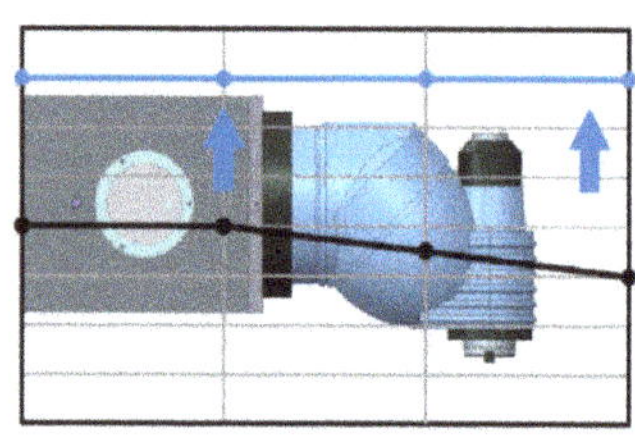

Figure 1. Dynamic Active Stabilizer (DAS®) by SOLARUCE [8]. Reprinted with permission from ref. [8]. Copyright 2016 Elsevier.

The overview of actuators begins with those that work on piezoelectric and magnetostrictive principles. Piezoelectric or other types of actuators are less common for structural vibration suppression. Their use is more common in the domains of vibration suppression of machine spindles, tool holders, workpieces and vibration-assisted machining. A comprehensive overview of devices that are used for vibration-assisted machining is provided in article [13]. Article [14] provides a complex overview of magnetostrictive actuators, including modelling and control issues. Examples of applications of these actuators can be found, for example, in publications [15–18]. The use of actuators that work on the piezoelectric principle for vibration suppression in two axes is described, for example, in articles [19–21]. Although these actuator applications are interesting in this overall overview, none of them directly corresponds to the application of vibration suppression of long, slender and massive structures such as machine tool rams. Generally speaking, actuators suitable for these applications typically work according to the electromagnetic principle.

Typical types of electromagnetic actuators include proof-mass actuators (inertial actuators), which are described, for example, in publication [1]. Solutions that combine several pieces of single-axis actuators can be used for cases where a damping force needs to be applied in several directions. An example of the use of single-axis actuators can be found, for example, in paper [22], which presents a technique to suppress chatter in centerless grinding. Two "ADD-2D-1 kN" actuators by Micromega® (Fernelmont, Belgium) were used for this purpose. They combine a linear motor and a voice coil actuator. The same actuators were also used in [23]. An actuator that functions in a similar way was also used in article [10] to suppress the vibrations of the aforementioned SORALUCE machine. In this case, however, the actuator was designed as a biaxial device. Article [10] focuses on control of the actuator rather than on the actuator concept. Another application on a SORALUCE machine is described in article [11], where a MICA® actuator by Cedrat Technologies (Meylan Cedex, France) is used for active damping. In this case study, a resonant frequency of 36 Hz was suppressed, and the surface quality of the machined part was improved (Figure 2).

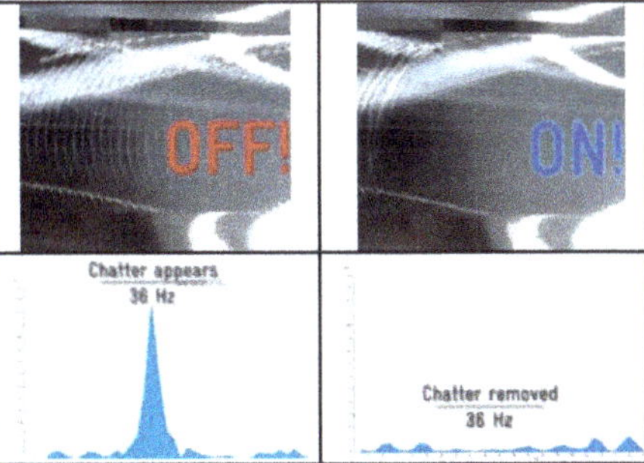

Figure 2. MICA® actuator on a SOLARUCE machine [11].

An interesting concept that employs a two-axial actuator for vibration suppression of boring bars is presented in paper [24] and subsequently used in [25]. These applications describe mounting on a turret of a CNC lathe. Paper [26] describes a planar actuator concept for vibration suppression, which is integrated into the end of the spindle for milling processes. A highly dynamic spindle integrated actuator is also presented in [27]. However, none of these electromagnetic actuators were designed specifically to dampen the vibrations of machine tool rams.

An interesting approach that includes application to the machine is the use of machine drives for vibration suppression (Figure 3). This application is described in article [12] and was also tested on the aforementioned SORALUCE machine. Figure 3 contains a description of a damping vibration method that uses an X-axis feed drive on this horizontal milling machine. A rack and pinion drive is used here, and the compensation signal is entered as an additional velocity command. The control-equipped drive is used to suppress machine oscillations at low frequencies during machining.

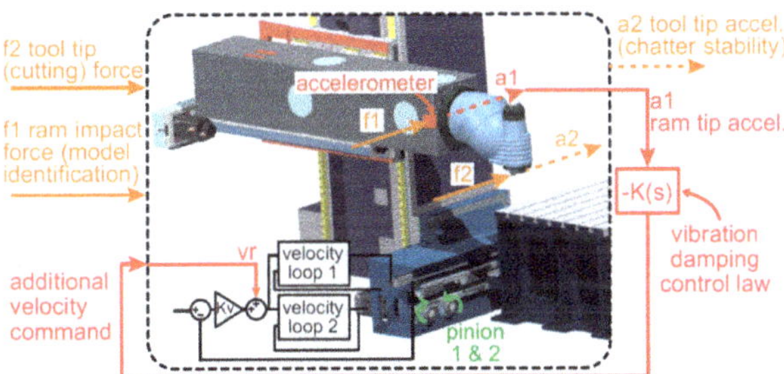

Figure 3. Active vibration suppression using machine drives [12]. Reprinted with permission from ref. [12]. Copyright 2015 Elsevier.

This paper aims to supplement the relatively sparse literature on the construction of actuators for vibration suppression of long, slender and massive machine tool structures, especially vertical rams. It presents a conceptual design of a two-axial electromagnetic actuator as well as a description of its construction and basic parts. The main novelty and originality of this solution is that it purposefully uses standard electric drives, which are available in a wide range of parameters. This makes it easier to adapt the actuator parameters to the required application. In terms of influencing the resulting machine accuracy, it is important to monitor the heat losses of the actuator (thermal deformation of the ram). The machine's thermal influence from the vibration suppression system is not emphasized in any of the cited sources. The concept proposed in this article uses synchronous electric drives (linear motors), which are highly efficient, among other things. The drives used in the case study (hereinafter) allow water cooling to be connected if required. As a result, it is possible to efficiently dissipate heat from a place where it negatively affects machine accuracy. Additionally, it is possible to use higher power levels in motors if necessary and thus achieve higher dynamic forces with actuators. This article also describes the basic concept for controlling the actuator in a current control loop (force control) and demonstration of its function on the ram. This article does not aim to present new control algorithms for vibration suppression. To verify the function of the actuator itself, a simple Direct Velocity Feedback (DVF) control approach is intentionally used here for clarity. The actuator itself works in the current control mode (control of the motor force). It allows the use of virtually any vibration control method. The appropriate control can be chosen with regard to the specific application on the machine.

2. Two-Axial Actuator Conceptual Design

A two-axial actuator is needed to meet the vibration damping requirements of long, slender machine tool parts, such as, in particular, vertical rams on large turning and multifunction machines. Typically, there is a need to increase damping in the proximity of the cutting point. It is often a matter of damping bending oscillations in the two most flexible directions of the ram. The aim is to make the proposed solution scalable in size and power and use drives with standard components. This should make it possible to integrate the solution into machines with different ram solutions.

The design of the two-axial actuator builds on previous experience with one-axis actuator design and operation from the authors' workplace. Conceptually, the design solution is based on the use of an electromagnetic proof mass actuator (voice coil actuator). The aim of designing the two-axial actuator drive is to use standard linear motors and servo inverters. Their specific size and arrangement can be customized for specific applications. In order to eliminate passive resistance and increase the service life of the actuator, the active mass of the actuator is usually mounted with flexures. These must be sufficiently flexible in the direction of the required movement yet sufficiently rigid in other directions. The placement of the active mass of the actuator on the flexures is evident is Figure 4. Actuator drives exert a force on the central part (gray) both in the X direction and in the Y direction (forces Fx and Fy). The movement along the X axis is practically performed only by the central part (gray). The movement along the Y axis is ensured by the central part along with its supporting part (green). Vibration sensors can be located on the outer frame of the actuator (blue) or can be built-in. The disadvantage of using flexures is that there is a small parasitic movement in a direction perpendicular to the desired direction of movement, as indicated in Figure 4a. Although it does not have a significant effect on the resulting actuator parameters, it complicates the selection of the measuring system of the positions of the primary and secondary parts of the motor in relation to each other (Figure 4b).

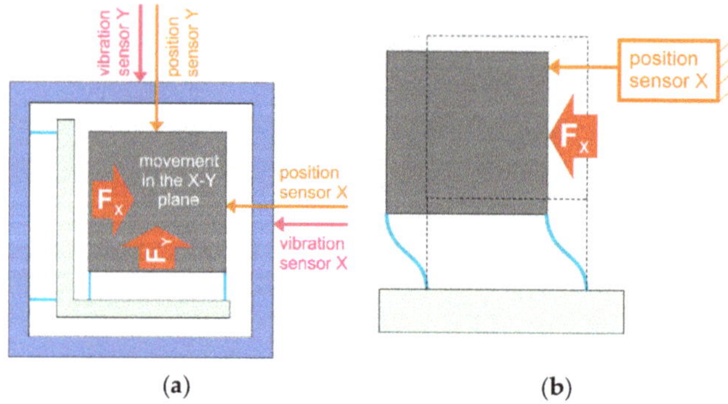

Figure 4. (**a**) A schematic of the proposed 2-axial actuator. (**b**) The mass guided by the flexures performs a small parasitic movement.

3. Case Study of the Actuator Design

The constructional design solution of the actuator's central part with a description of its basic components is provided by assembling in Figures 5 and 6a. The basis is a moving mass in the middle section of the actuator (Figure 5a), to which the secondary parts of the linear motors of the *X*-axis and the *Y*-axis are mounted perpendicular to each other (a type VUES L3S100S-816 in the case study). In Figure 5a, only drive X is visible. Drive Y is mounted on the other (bottom) side. The mobility of the central part's mass is ensured by mounting on the flexures against the intermediate piece–the spacer. Together, they form a movable mass along the *Y*-axis (Figure 5b). This spacer is further mounted on the flexures against the actuator's base frame (Figure 6a). Mobility along both axes is limited by rubber end stops. The central part of the actuator can also be designed as a version with a through center (Figure 6b).

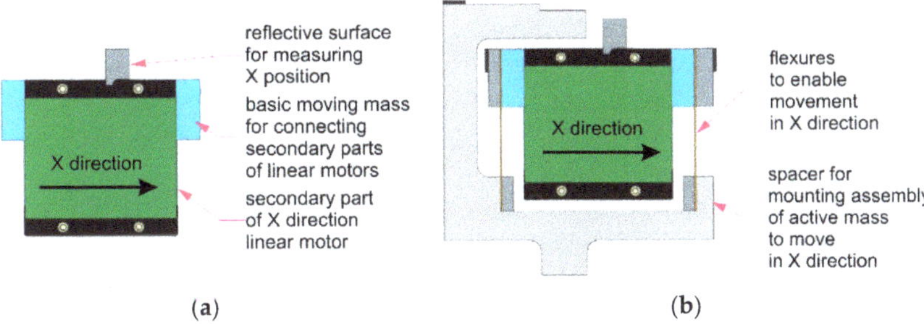

(a) (b)

Figure 5. (**a**) The active mass in the X direction; the secondary part of motor Y is located on the opposite part of the body, perpendicular to direction X; (**b**) The active mass in the Y direction.

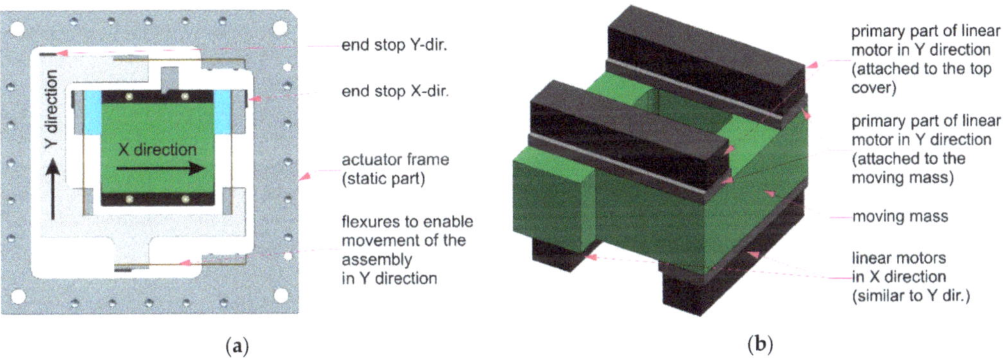

(a) (b)

Figure 6. (**a**) The suspension of active mass with secondary parts of the motor in the Y direction (mounted on the other side); (**b**) An example of the design solution of the central body with a through hole.

During the design process, significant attention was paid to the suspension of active mass on the flexures. The design (case study) was performed for specific linear motors with a continuous traction force of 440 N (see description below). The suspension was optimized using the parametric FEM model. Traction forces from the linear motors in the X and Y directions enter the model as loads. The gravitational force and attractive forces from permanent magnets between the primary and secondary parts of the linear motors then act perpendicular to them (Z direction). The aim of the optimization was to achieve low natural frequencies of the active body suspension in the flexible direction while requiring high rigidity in other directions to prevent the flexures from collapsing at the limit loads (loss of stability). This would cause contact between the primary and secondary parts of one of the motors, which could lead to damage to the actuator in extreme cases. In this case study, a value of 0.2 mm was chosen as a safe design value of maximum deformation in the Z direction. The aim was to choose the flexure parameters so that the lowest natural frequencies are approx. around 15 Hz. An example of the calculated first eigenshapes with flexures that meet the selected conditions is shown in Figure 7a,b. Another condition was that the stress in any part of the flexures must not exceed the safe level for the given material (spring steel EN 1CS67) at maximum deformation. The calculations also included topological optimization of the flexure shape, which aimed to minimize the natural frequency of the suspension while maintaining the maximum rigidity in the Z direction and minimum stress in the flexures. However, the benefits of such optimization were not unambiguous, and thus, finally, simple rectangular flexures were chosen. The flexure (plate) thickness parameter was chosen from available metallurgical products.

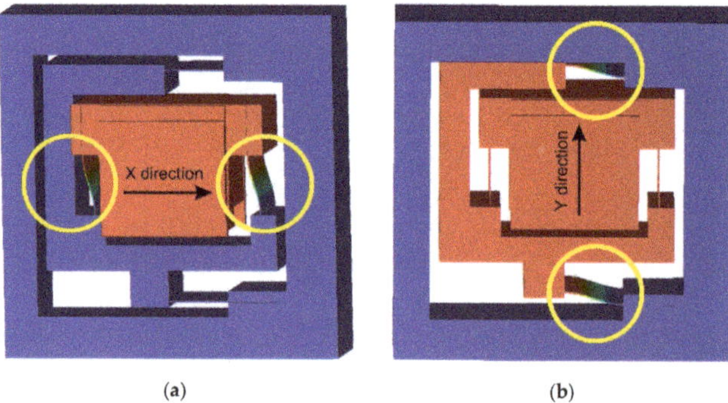

Figure 7. (**a**) The lowest eigenshape in the X direction is at a frequency of 16 Hz (total size 340 mm × 340 mm); (**b**) The lowest eigenshape in the Y direction is at a frequency of 15 Hz (total size 340 mm × 340 mm). Flexure deformation is highlighted.

The central part of the actuator contains only the secondary parts of the linear motors with permanent magnets. The primary parts of the motor (water-cooled type VUES L3SK075P-1215) are attached to the actuator covers along with an absolute non-contact position sensor (Micro-Epsilon ILD1320-50). The actuator cover assembly is shown in Figure 8a. In Figure 8b, there is a 3D view of the actuator center assembly along with the location of the primary parts of the linear motors (for clarity).

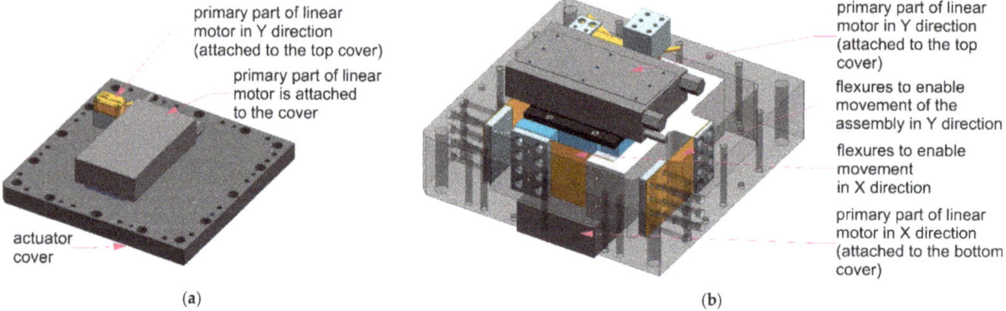

Figure 8. (**a**) The design of actuator cover assembly with primary part of the linear motor and laser absolute position sensor. (**b**) The assembly of the central part of the actuator with the indicated location of the primary parts of the linear motor, mounted on the upper and lower covers (not visible in the image).

Figure 9a provides a view of the mounted assembly central part of the actuator. Figure 9b shows the mounted assembly of one of the covers. Due to the large attractive forces between the primary and secondary parts of the linear motors, the installation of the covers on the central part must occur simultaneously from both sides while both covers are gradually closed using the push-off screws (Figure 10a). The assembled actuator is supplemented with industrial connectors to enable better cabling connection options (Figure 10b).

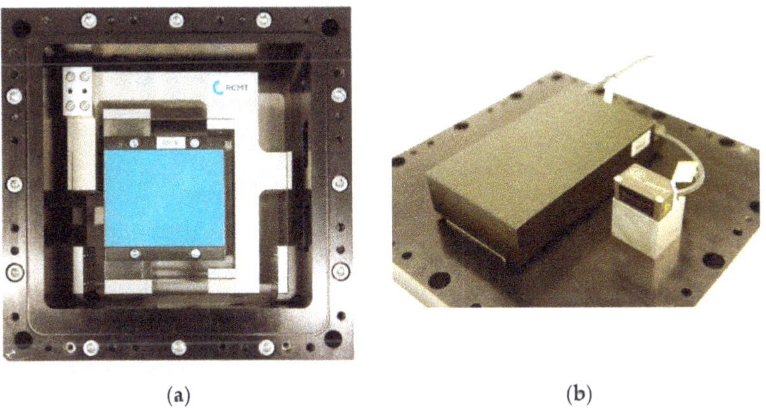

(a)   (b)

Figure 9. (a) The central part of the actuator assembly. (b) One of the two actuator covers with the installed primary part of the linear motor and the laser absolute position sensor.

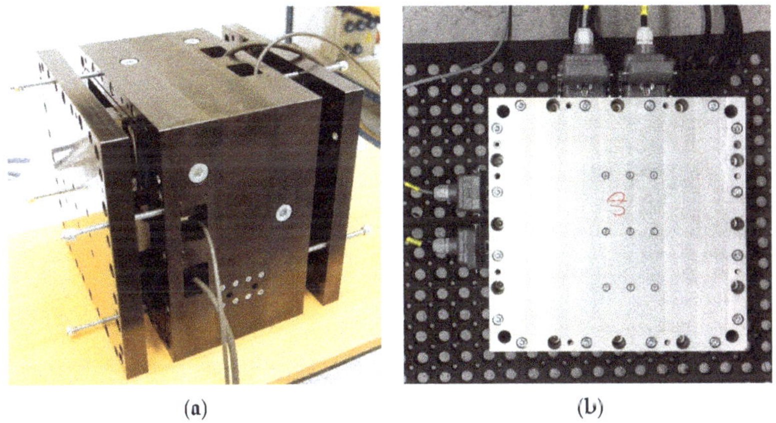

(a)   (b)

Figure 10. (a) Symmetrical mounting of the actuator cover; (b) the complete actuator, equipped with connectors.

The specific actuator (case study) was designed to enable testing on a test ram of real dimensions. The external dimensions of the actuator are $340 \times 340 \times 270$ mm^3 (X, Y, Z), the total weight of one complete actuator is 80 kg, and the intended operating frequency range of the actuator is from 20 to 300 Hz. The basic parameters for the X and Y axes of the designed actuator are summarized in Table 1. For the purposes of testing on a model ram, two structurally identical actuators were manufactured (see below). The actuator drives are designed to allow the connection of water cooling. This corresponds (in theory) to a higher continuous achievable traction force. However, water cooling is considered mainly due to the reduction of the actuator's thermal effect on the machine ram.

Table 1. Basic actuator parameters for *X* and *Y* axes.

Parameter	*X*-Axis	*Y*-Axis
Nominal continuous traction force of the lin. electric motor	440 N [1]	
Continuous traction force of the lin. motor with optional water cooling	750 N	
Peak force of the linear electric motor	1000 N	
Force constant of the linear motors	120 N/A	
Current control loop bandwidth (approx.)	1 kHz	
Maximum stroke of the active mass	$\pm$10 mm	
Actual stroke of the active mass (intentionally reduced by soft stops)	$\pm$7 mm	
e je Weight of the moving mass	15.6 kg	19.7 kg
Weight of the actuator frame	64.4 kg	60.3 kg
Stiffness of flexures	1.58×10^5 N/m	1.75×10^5 N/m
Damping coefficient (approx.)	471 N·s/m	557 N·s/m
First eigenfrequency	16 Hz	15 Hz

[1] For harmonic force, a value of approx. 620 N may be considered.

The mathematical model of the actuator can be relatively simply described analytically (e.g., in [1]). However, for the sake of clarity, this article shows a model created in the Matlab Simscape environment. This model is presented in Figure 11 and can be understood as a subsystem that will have the same structure for both the *X*-axis and the *Y*-axis. The values in the actuator model can be substituted according to Table 1. The actuator subsystem created in this way can be connected both to a complex model of the entire machine structure (e.g., in state-space form-reduced FEM model) and to a vibration control model. To close the control loop, this model would be supplemented by a virtual acceleration sensor, which would be placed on the "actuator frame mass" element or on a suitable element of the machine tool structure model (in close proximity to the actuator). This paper does not deal directly with control algorithms for vibration dampers. A simple controller, a Direct Velocity Feedback (DVF) as described in [1], is intended to be used to verify the actuator's function as a damper. The force entering the actuator model in Figure 11 is obtained from the controller by multiplying the value of the motor current and the motor force constant.

In practice, the motor's force constant may differ slightly from the catalogue value. The biggest differences may be related to motor overload and warm-up. However, machine tool ram warming is a significant potential source of machine inaccuracy. The actuator for the vibration damper is intentionally designed with motors that allow water cooling. The primary purpose is not to achieve a higher force but a lower thermal effect. The operation of the actuator is realistically assumed to be at maximum continuous catalogue values, which are permissible for the motor without cooling. Under these conditions, the motor's force constant can be considered approximately as a constant, which provides a conversion between the motor's current and action force with sufficient accuracy (for a given purpose).

The actuator model in Figure 11 has a transfer function, which is plotted in Figure 12 after the substitution of the *X*-axis parameters from Table 1 (for illustration). For low frequencies, the transmission of force to the machine structure is small. It grows around the resonance of the actuator (16 Hz), and subsequently, the transmission amplitude stabilizes at value 1 (0 dB). The value of the phase shift is practically negligible for higher frequencies. The actuator then acts as a good source of force. For the case study application in this article, the main focus is on vibration suppression around 70 Hz (see text below). In this area, the actuator already works with minimal distortion. For this reason, no special attention needs to be paid to force compensation according to the transfer function in Figure 12 during the experiments described below.

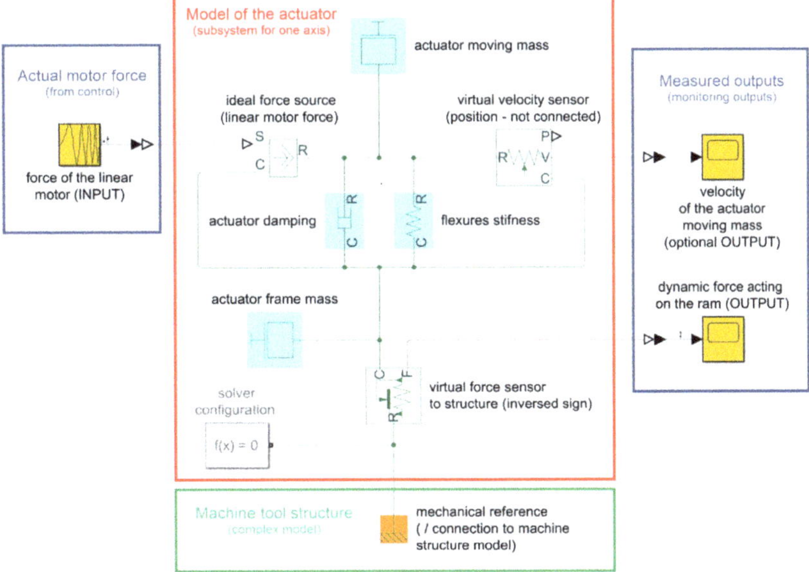

Figure 11. An actuator single-axis model that can be used as a subsystem in a complex machine tool model (including vibration control by linear motor force). Model parameters can be substituted according to Table 1.

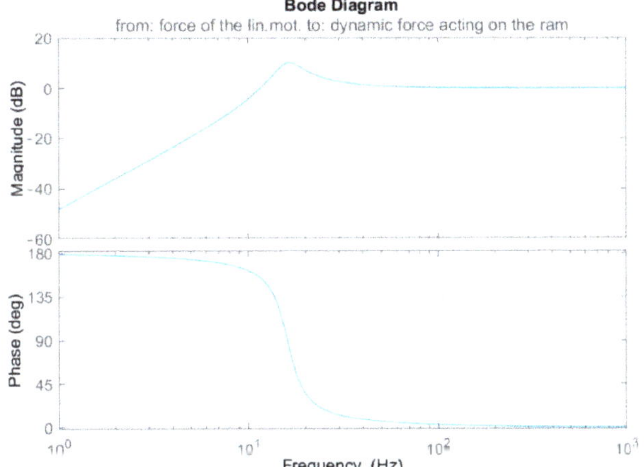

Figure 12. Bode diagram for the actuator's *X*-axis: transfer function between the force of the linear motor and the force transmitted to the machine structure. The actuator no longer shows significant distortion for the monitored area around 70 Hz (amplitude 1.05; phase 4 degrees).

The great advantage of the proposed actuator design is the ability to measure the absolute position of the actuator moving mass (laser sensor, Figure 8a). The model's stiffness parameters can be verified (in this case) by measuring static deformation at a specific linear motor force. In the normal vibration suppression mode, the flexure deformations are usually relatively small and can be considered linear. The damping coefficient can be approximately estimated using the position response to the force step demand.

4. Solving the Problem of Actuator Control in the Current Control Loop

In order for linear motors to function properly, it is necessary to ensure the measurement of the positions of the motor's primary and secondary parts in relation to each other (commutation of linear motors). Due to the parasitic movement of the mass on the flexures mentioned above, this is relatively problematic to ensure. Standard position sensors for linear motors (optical sensors and other types) are able to tolerate this movement only to a very small extent, which is not enough in practical terms. Hall sensors are an exception, but they cannot be used for common components in this case due to the installation dimensions. For these reasons, an absolute laser noncontact position sensor with a range of 50 mm (specifically a Micro-Epsilon ILD1320-50) was used to measure the position of the active mass. At the time of writing this paper, the sensors available on the market do not have an output that can be fed as a feedback signal to the servo inverter. The sensor selected for this case study has a current output of 4–20 mA, which must be converted to another type of signal that is compatible with commonly available drives. The most universal signal is likely the TTL signal from incremental position sensors, which is supported by virtually all servo inverter manufacturers. However, no signal converter with the required parameters is available on the market. For this reason, a custom converter was designed on the FPGA part of the NI cRIO-9064 computer. The block diagram of the converter for one channel is shown in Figure 13. The output is a standard incremental encoder signal in the RS422 standard (TTL) with an adjustable length period (position increment) and a reference mark. This solution allows the actuator drives to be commutated after start-up and further operated in standard mode in the current control loop (which corresponds practically to force control).

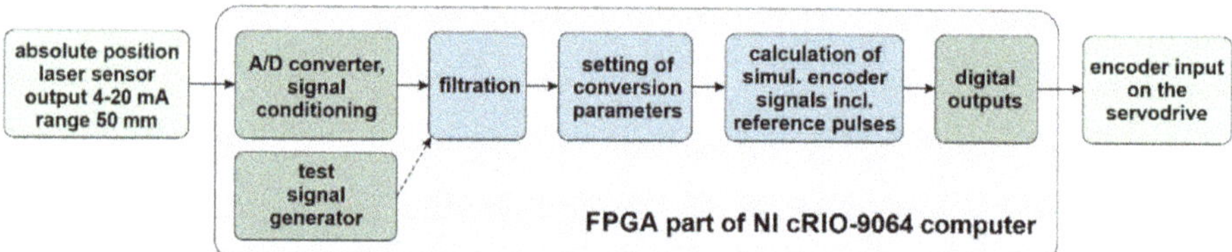

Figure 13. A block diagram of one channel signal converter for position measurements. The input is an analog current signal, and the output is an incremental position signal with a reference mark in the RS422 standard (TTL).

The authors of this paper also have experience with other applications where the RS422 encoder was used in the vibrating system. Due to the higher vibration frequencies and higher resolution of the encoder, there was an occasional loss of pulse reading. This problem was prevented here by the fact that the sensor (laser) works as an absolute. A possible loss of pulses when reading the simulated encoder signal would be quickly corrected by using a reference pulse, which is generated in the middle stroke position. During testing, this signal conversion system proved to be sufficiently robust. No commutation problem or encoder error was detected on the servo inverter.

5. Experimental Verification of Actuator Function

A test ram measuring $270 \times 270 \times 2000$ mm^3 was used to verify the actuator function. The ram was attached to a cast-iron cube in an inverse vertical position and then fixed to the floor. Both manufactured actuators were attached to the upper clamping plate of the ram. The arrangement of the measurements is visible in Figure 14a. A view of the electrical cabinet for the actuator drives is shown in Figure 14b.

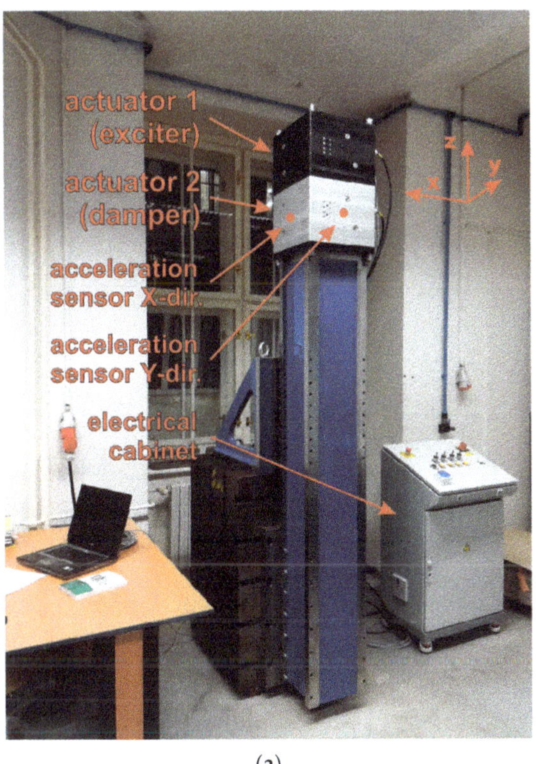

(**a**)

(**b**)

Figure 14. (**a**) The set-up of experiment to verify actuator function; (**b**) the electrical cabinet with servo inverters for two actuators.

The actuator control block diagram in Figure 15 consists of two parts. The first part involves measuring, i.e., the solution of the aforementioned simulated signal of the incremental encoder. The second part concerns the input of the current setpoint signal for each of the four controlled linear motors. Conventional servo amplifiers (Kollmorgen AKD-P00607-NBCC, Radford, VA, USA) were used for control. The setpoint can be entered into the selected drives either in digital form via the EtherCAT bus or using the analog voltage input of the inverter. Although analog input can be problematic because it is more susceptible to electrical noise, it nonetheless seems to be more suitable for vibration suppression applications because it has less transport delay. This is also the case selected in Figure 15.

This paper does not deal directly with control algorithms for vibration dampers. A simple controller, a Direct Velocity Feedback (DVF) as described in [1], was used to verify the function of the actuators as dampers. The integration of signals from accelerometers (Kistler PiezoBeam 8640A5T, Winterthur, Switzerland) was used to close the feedback. The control loop and test signal generation were implemented on a NI cRIO-9063 computer (Figure 16). The algorithm was the same for both directions, and the control cycle frequency was 20 kHz. The control of actuator 1 (exciter) and control of actuator 2 (damper) run on the same computer but completely independently of each other. The gain of the vibration damper controllers (DVF) was experimentally adjusted so that the amplitude decreased at the monitored frequencies in the region around 70 Hz (Figure 17). At the same time, it was important to maintain the stability of the regulation and not cause a significant spillover effect on the surrounding frequencies.

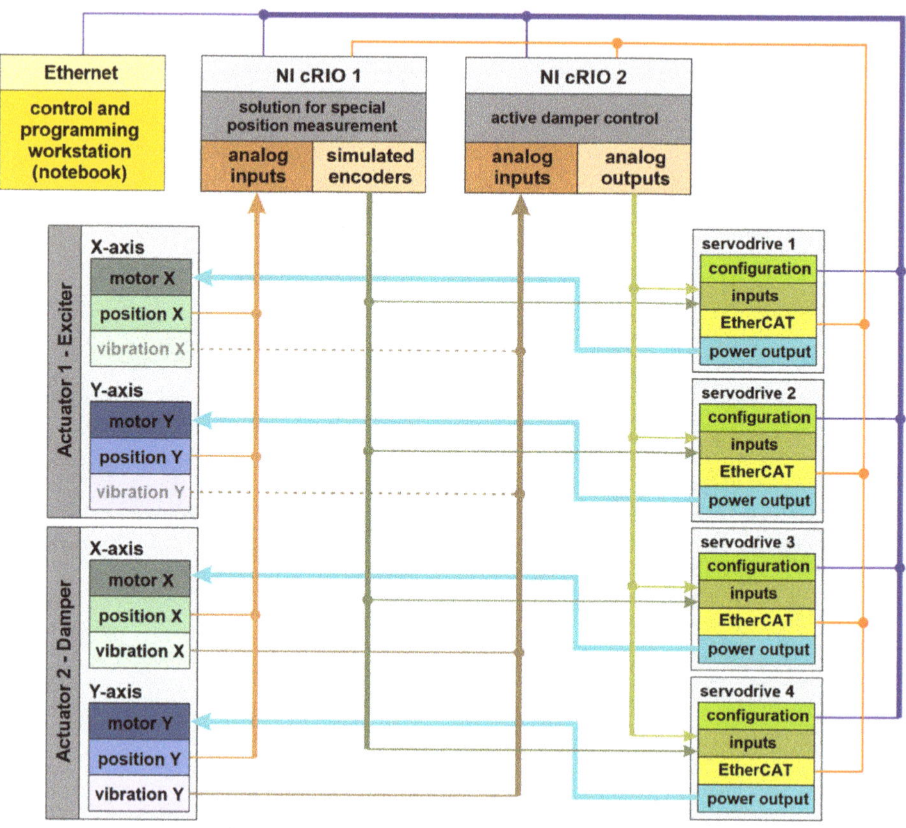

Figure 15. Block diagram of control of two independent two-axial actuators.

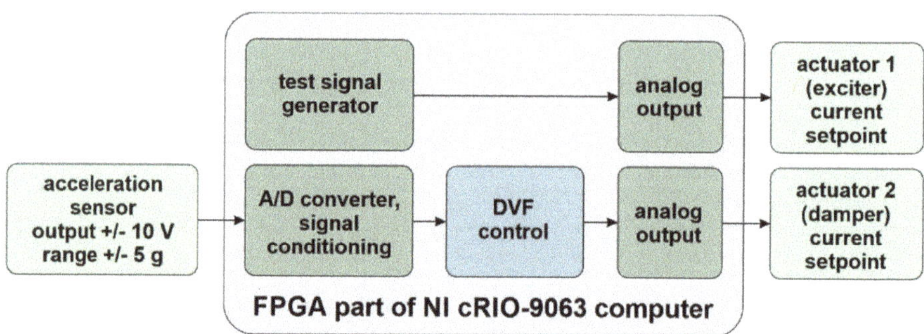

Figure 16. The control loop scheme for damper and exciter control (same for both directions).

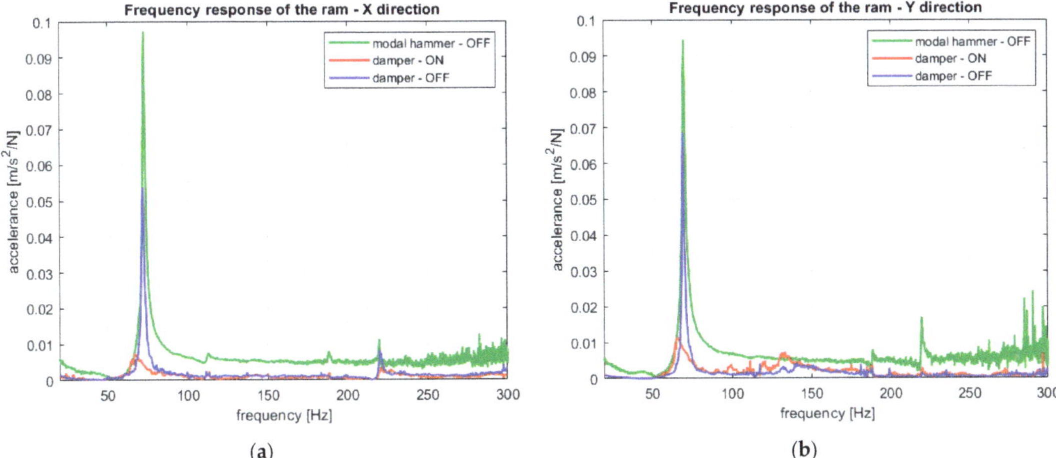

Figure 17. (**a**) The amplitude frequency response of the ram in the X direction; (**b**) the amplitude frequency response of the ram in the Y direction.

As a reference measurement, a classical measurement of the amplitude frequency transfer function of the ram was taken (Figure 17—green lines). The excitation was carried out using a modal hammer (PCB 086D05, PCB Piezotronics, Depew, NY, USA) at the location of the accelerometers on actuator 2. A separate acceleration sensor (Kistler K-Shear 8702B50M1, Winterthur, Switzerland) was used to measure the response. This sensor was placed in close proximity to the accelerometers for active control feedback. The actuators were switched off while this measurement was taken. Further measurements were taken with the amplitude frequency response excitation progressively point by point using actuator 1 as an exciter in the given frequency range (20–300 Hz). The actuator at the end of the ram was chosen for excitation, as this corresponds to the real cutting force position. A frequency step of 0.2 Hz was selected for the measurement. The conversion of excitation to force was performed using the force constant of the linear motors (120 N/A for the given motors). The amplitude of the excitation force during the measurement was, after recalculation, approximately 70 N in the entire frequency range, which corresponds to approximately 15% of the actuator's nominal value. This value was chosen as the safe limit for overcoming the resonance range around 70 Hz (excitation by the actuator nominal value was only possible in the safe area outside the resonance).

One measurement was performed with actuator 2 switched off (Figure 17—blue lines). During the second measurement, the vibration damper-actuator 2 with the Direct Velocity Feedback (DVF) control was switched on (Figure 17—red lines). In the *X*-axis of the ram, there is only one dominant bending natural frequency in the given frequency range at the value of 72 Hz (eigenshape in Figure 18a). In the *Y* axis it is 70 Hz (the eigenshape in Figure 18b). Furthermore, frequencies corresponding to torsional modes predominate. These cannot be significantly affected by the damper. With the use of a damper, it was possible to reduce the amplitude of the most important bending mode to approximately 15% of the original value in both directions.

The frequency transfer function graphs show the differences between the measurement excited by a modal hammer and point by point by actuator 1. These differences are attributed to the error that is introduced into the measurement by converting the excitation force from the actuator current through the catalogue value of the force constant. The conversion of the force transmission to the ram (according to the Bode diagram in Figure 12) in this presented frequency range has no significant effect. A completely different level of excitation force also plays an important role. When excited by the actuator, the measurement occurs at large excitation amplitudes, and vibrations stabilize at individual frequencies.

For these reasons, the differences in the measured transmissions are understandable, and the agreement can be considered acceptable. This is especially true for the 70 Hz region, where the most significant bending modes are located.

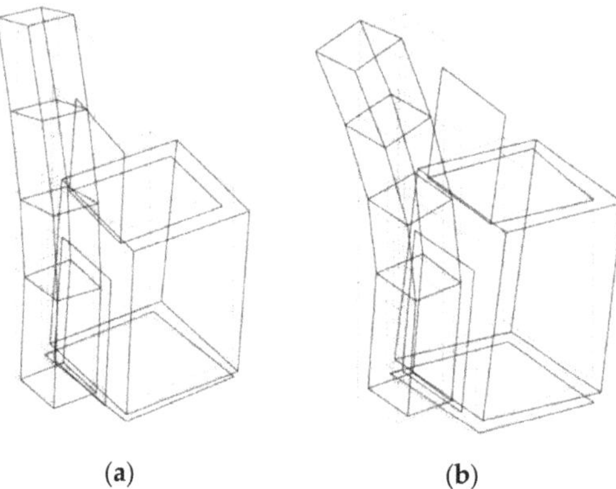

Figure 18. (**a**) The eigenshape of the ram in the X-axis (72 Hz); (**b**) the eigenshape of the ram in the Y-axis (70 Hz).

6. Conclusions

This paper describes a two-axial actuator concept for vibration dampers, which is designed specifically for damping bending modes of vertical rams of large turning and multifunctional machine tools. In the introductory section of the paper, an overview of relevant actuators for vibration suppression was presented. The field of research was expanded to include relevant actuators for vibration-assisted machining.

The actuator presented in this article was designed as an intermediate piece (spacer) between the end of the ram and the tool head. The main novelty and originality of the presented solution are that it purposefully uses standard electric drives, which are available in a wide range of parameters. This makes it easier to adapt the actuator parameters to the required application. In terms of influencing the resulting machine accuracy, it is also very important to monitor the heat losses of the actuator (thermal deformation of the ram). The concept proposed in this article uses synchronous electric drives (linear motors), which excel, among other things, in high efficiency. The drives used in the case study allow the connection of water cooling if required. It is thus possible to efficiently dissipate heat from a place where it negatively affects the accuracy of the machine. At the same time, it is theoretically possible to use higher power of motors if necessary and thus achieve higher dynamic forces with the same actuator. If a pair of linear motors are used for each direction, it is possible to keep the actuator center free (e.g., for spindle drive shafts). This option is also not common for other actuators found in the cited references.

The article presents a case study of the specific actuator for a nominal continuous force of 440 N and a range of operating frequencies from 20 to 300 Hz. It also summarizes the basic parameters of the assembled device and a model that will allow the simulation of the actuator with the given parameters within a complex model of the machine. From the presented model, it is possible to approximately determine the actuator's basic dynamic characteristic between the force exerted by the linear motor and the force transmitted through the elastic suspension (flexures) to the structure of the machine tool.

The paper also addresses the concept of noncontact measurement of the moving mass of the actuator, guided on the flexures. Information on the position of the moving mass can be used for the commutation of linear motors and, if necessary, for the control of the vibration damper itself.

The actuator was installed on a real-size test ram, and its function as a vibration damper was verified. This article does not aim to present any new control algorithms for vibration suppression. To verify the function of the actuator, a simple Direct Velocity Feedback (DVF) control approach is intentionally used here for clarity. The actuator itself works in the current control mode (controlling the motor force). It allows the use of virtually any method of vibration control. The appropriate control can be chosen with regard to the specific application on the machine.

In the presented case study, two NI cRIO computers were used to control the complete application. One of the computers solved the functions of special position measurement, while the other ran the control of the actuators. In a practical application on a machine tool, only one two-axial actuator would be used, and only one NI cRIO control computer would suffice. The number of inputs and outputs of one computer, as well as the computing power, would be sufficient to control a complete application. Communication between the NI cRIO and the machine tool control system may not be in real-time (at high rates). It may thus proceed via standard communication protocols. The presented solution is applicable in practice and can be integrated into existing industrial control systems. Applications can be mainly expected in the field of vibration suppression of vertical rams of large machine tools.

Author Contributions: L.N.: conceptualization, data curation, formal analysis, methodology, supervision, validation, visualization and writing—original draft; J.Č.: conceptualization, data curation, visualization and writing—original draft; M.S.: formal analysis, funding acquisition, project administration and supervision; J.Š.: project administration and validation; M.J.: data curation and visualization; P.K.: project administration and resources. All authors have read and agreed to the published version of the manuscript.

Funding: Funding support from the Czech Ministry of Education, Youth and Sports under the project CZ.02.1.01/0.0/0.0/16_026/0008404 "Machine Tools and Precision Engineering" financed by the OP RDE (ERDF). The project is also co-financed by the European Union.

Institutional Review Board Statement: Not applicable.

Informed Consent Statement: Not applicable.

Data Availability Statement: Not applicable.

Conflicts of Interest: The authors declare no conflict of interest.

References

1. Preumont, A. *Vibration Control of Active Structures: An Introduction*; Springer: Dordrech, The Netherlands, 2011; ISBN 9789400720336.
2. Fuller, C.; Elliott, C.S.; Nelson, P.A. *Active Control of Vibration*; Academic Press: Cambridge, MA, USA, 1996; ISBN 9780080525914.
3. Hansen, C.H. *Active Control of Noise and Vibration*; CRC Press: Boca Raton, FL, USA, 2013; ISBN 9781466563360.
4. Elvin, N.; Erturk, A. *Advances in Energy Harvesting Methods*; Springer: New York, NY, USA, 2013; ISBN 9781461457053.
5. Kaźmierski, T.J.; Beeby, S. *Energy Harvesting Systems: Principles, Modeling and Applications*; Springer: New York, NY, USA, 2010; ISBN 9781441975669.
6. Zaeh, M.F.; Kleinwort, R.; Fagerer, P.; Altintas, Y. Automatic tuning of active vibration control systems using inertial actuators. *CIRP Ann.* **2017**, *66*, 365–368. [CrossRef]
7. Kleinwort, R.; Plata, J.; Zaeh, M.F. Adaptive active vibration control for machine tools with highly position-dependent dynamics. *Int. J. Autom. Technol.* **2018**, *12*, 631–641. [CrossRef]
8. Munoa, J.; Beudaert, X.; Dombovari, Z.; Altintas, Y.; Budak, E.; Brecher, C.; Stepan, G. Chatter suppression techniques in metal cutting. *CIRP Ann.* **2016**, *65*, 785–808. [CrossRef]
9. Yue, C.; Gao, H.; Liu, X.; Liang, S.Y.; Wang, L. A review of chatter vibration research in milling. *Chin. J. Aeronaut.* **2019**, *32*, 215–242. [CrossRef]
10. Munoa, J.; Mancisidor, I.; Loix, N.; Uriarte, L.G.; Barcena, R.; Zatarain, M. Chatter suppression in ram type travelling column milling machines using a biaxial inertial actuator. *CIRP Ann.* **2013**, *62*, 407–410. [CrossRef]
11. Meneroud, P.; Porchez, T.; Munoa, J.; Rowe, S.; Pages, A.; Benoit, C.; Belly, C. Moving Iron Controllable Actuator: Performances in Closed Loop. *Proc. Actuator* **2014**, *14*, 530–533.
12. Munoa, J.; Beudaert, X.; Erkorkmaz, K.; Iglesias, A.; Barrios, A.; Zatarain, M. Active suppression of structural chatter vibrations using machine drives and accelerometers. *CIRP Ann.* **2015**, *64*, 385–388. [CrossRef]
13. Zheng, L.; Chen, W.; Huo, D. Review of vibration devices for vibration-assisted machining. *Int. J. Adv. Manuf. Technol.* **2020**, *108*, 1631–1651. [CrossRef]

14. Apicella, V.; Clemente, C.S.; Davino, D.; Leone, D.; Visone, C. Review of Modeling and Control of Magnetostrictive Actuators. *Actuators* **2019**, *8*, 45. [CrossRef]
15. Zheng, L.; Chen, W.; Huo, D.; Lyu, X. Design, Analysis, and Control of a Two-Dimensional Vibration Device for Vibration-Assisted Micromilling. *IEEE/ASME Trans. Mechatron.* **2020**, *25*, 1510–1518. [CrossRef]
16. Lu, M.; Zhao, D.; Lin, J.; Zhou, X.; Zhou, J.; Chen, B.; Wang, H. Design and analysis of a novel piezoelectrically actuated vibration assisted rotation cutting system. *Smart Mater. Struct.* **2018**, *27*, 095020. [CrossRef]
17. Titsch, C.; Li, Q.; Kimme, S.; Drossel, W.-G. Proof of Principle of a Rotating Actuator Based on Magnetostrictive Material with Simultaneous Vibration Amplitude. *Actuators* **2020**, *9*, 81. [CrossRef]
18. Wang, G.; Zhou, X.; Ma, P.; Wang, R.; Meng, G.; Yang, X. A novel vibration assisted polishing device based on the flexural mechanism driven by the piezoelectric actuators. *AIP Adv.* **2018**, *8*, 015012. [CrossRef]
19. Tůma, J.; Šimek, J.; Mahdal, M.; Pawlenka, M.; Wagnerova, R. Piezoelectric actuators in the active vibration control system of journal bearings. *J. Phys. Conf. Ser.* **2017**, *870*, 012017. [CrossRef]
20. Pugi, L.; Abati, A. Design optimization of a planar piezo-electric actuation stage for vibration control of rotating machinery. *Meccanica* **2020**, *55*, 581–596. [CrossRef]
21. Sallese, L.; Grossi, N.; Scippa, A.; Campatelli, G. Numerical investigation of chatter suppression in milling using active fixtures in open-loop control. *J. Vib. Control.* **2018**, *24*, 1757–1773. [CrossRef]
22. Barrenetxea, D.; Mancisidor, I.; Beudaert, X.; Munoa, J. Increased productivity in centerless grinding using inertial active dampers. *CIRP Ann.* **2018**, *67*, 337–340. [CrossRef]
23. Peukert, C.; Pöhlmann, P.; Merx, M.; Müller, J.; Ihlenfeldt, S. Modal-Space Control of a Linear Motor-Driven Gantry System. *MM Sci. J.* **2019**, *12*, 3285–3292. [CrossRef]
24. Lu, X.; Chen, F.; Altintas, Y. Magnetic actuator for active damping of boring bars. *CIRP Ann.* **2014**, *63*, 369–372. [CrossRef]
25. Chen, F.; Liu, G. Active damping of machine tool vibrations and cutting force measurement with a magnetic actuator. *Int. J. Adv. Manuf. Technol.* **2017**, *89*, 691–700. [CrossRef]
26. Wan, S.; Li, X.; Su, W.; Yuan, J.; Hong, J.; Jin, X. Active damping of milling chatter vibration via a novel spindle system with an integrated electromagnetic actuator. *Precis. Eng.* **2019**, *57*, 203–210. [CrossRef]
27. Königsberg, J.; Reiners, J.; Ponick, B.; Denkena, B.; Bergmann, B. Highly dynamic spindle integrated magnet actuators for chatter reduction. *Int. J. Autom. Technol.* **2018**, *12*, 669–677. [CrossRef]

Article

Vibrational Amplitude Frequency Characteristics Analysis of a Controlled Nonlinear Meso-Scale Beam

Zu-Guang Ying [1,*] and Yi-Qing Ni [2]

[1] Department of Mechanics, School of Aeronautics and Astronautics, Zhejiang University, Hangzhou 310027, China

[2] Hong Kong Branch of National Rail Transit Electrification and Automation Engineering Technology Research Center, Department of Civil and Environmental Engineering, The Hong Kong Polytechnic University, Hung Hom, Kowloon, Hong Kong; ceyqni@polyu.edu.hk

* Correspondence: yingzg@zju.edu.cn

Abstract: Vibration response and amplitude frequency characteristics of a controlled nonlinear meso-scale beam under periodic loading are studied. A method including a general analytical expression for harmonic balance solution to periodic vibration and an updated cycle iteration algorithm for amplitude frequency relation of periodic response is developed. A vibration equation with the general expression of nonlinear terms for periodic response is derived and a general analytical expression for harmonic balance solution is obtained. An updated cycle iteration procedure is proposed to obtain amplitude frequency relation. Periodic vibration response with various frequencies can be calculated uniformly using the method. The method can take into account the effect of higher harmonic components on vibration response, and it is applicable to various periodic vibration analyses including principal resonance, super-harmonic resonance, and multiple stationary responses. Numerical results demonstrate that the developed method has good convergence and accuracy. The response amplitude should be determined by the periodic solution with multiple harmonic terms instead of only the first harmonic term. The damping effect on response illustrates that vibration responses of the nonlinear meso beam can be reduced by feedback control with certain damping gain. The amplitude frequency characteristics including anti-resonance and resonant response variation have potential application to the vibration control design of nonlinear meso-scale structure systems.

Keywords: amplitude frequency characteristics; damping effect; vibration control; meso-scale beam; nonlinear vibration; harmonic balance solution; updated cycle iteration algorithm

Citation: Ying, Z.-G.; Ni, Y.-Q. Vibrational Amplitude Frequency Characteristics Analysis of a Controlled Nonlinear Meso-Scale Beam. *Actuators* **2021**, *10*, 180. https://doi.org/10.3390/act10080180

Academic Editors: Zhao-Dong Xu, Siu-Siu Guo and Jinkoo Kim

Received: 30 June 2021
Accepted: 31 July 2021
Published: 3 August 2021

Publisher's Note: MDPI stays neutral with regard to jurisdictional claims in published maps and institutional affiliations.

1. Introduction

The meso-scale beam is an important component of many precise instruments such as micro-sensors or micro-actuators [1]. The large amplitude motion of the beam subjected to strong loading, e.g., a cantilever beam under strong support motion loading, leads to complicated nonlinear vibration [2], which will degrade the mechanical performance of the instruments. Thus, vibration control of the beam is highly desirable, and in this connection, the vibration response and amplitude frequency characteristics of the controlled nonlinear beam need to be studied. Some investigations on the nonlinear dynamics of meso beams have been reported, but there is a paucity of research on its vibration control.

For a meso-scale beam, a conventional mechanical model such as that based on Bernoulli–Euler theory can be applied. However, the foundational frequency of the beam is high due to its small length and in general, it is larger than the loading frequency. Thus, first several vibration modes will be involved in the vibration response, albeit the first mode vibration is dominant. Increasing structural damping is a common strategy, in particular, for resonant response mitigation [3,4]. However, damping produced by passive control is limited, and active feedback control is necessary to execute large artificial damping. The controlled beam can be converted into a multi-degree-of-freedom (MDOF) nonlinear system. Feedback control of MDOF systems has been studied [5–14]. However, the control effectiveness of a large damping gain on vibration mitigation of the nonlinear meso beam needs to be studied further.

Most of the investigations on nonlinear dynamics of the meso beam are based on experiments, while theoretical studies on the nonlinear vibration and control are limited. Periodic loading is a class of important real excitations, and periodic vibration and amplitude frequency relation can represent dynamic characteristics of the beam system with various frequencies. The harmonic balance method is suitable for periodic vibration analysis of the nonlinear beam system to obtain amplitude frequency relation by uniform procedure in a wide frequency band [2,15–20]. It converts a nonlinear system equation into a set of algebraic equations for harmonic coefficients of periodic response, and the response based on basic harmonic components can be improved by using high harmonic components. However, general forms of harmonic balance equations, general analytical expressions of periodic response solutions, and an effective iteration procedure for amplitude frequency relation need to be developed further for the controlled nonlinear meso beam.

In the present paper, the vibration response and amplitude frequency characteristics of a controlled nonlinear meso-scale beam under periodic loading are studied. A method including a general analytical expression for harmonic balance solution to periodic vibration and an updated cycle iteration algorithm for amplitude frequency relation of periodic response is developed for the nonlinear beam system. In Section 2, a vibration equation for the controlled beam is established, and a general expression of nonlinear terms for periodic response with multiple harmonic terms is derived. A general analytical expression for a harmonic balance solution to periodic vibration is obtained. In Section 3, an updated cycle iteration procedure for amplitude frequency relation and periodic response solution is proposed. Then, a periodic vibration response with various frequencies is calculated uniformly using the method by only changing the frequency. An accurate response can be obtained by using adequate high harmonic components. In Section 4, numerical results on the periodic vibration response of the controlled nonlinear beam system are given to demonstrate the good convergence and accuracy of the developed method. The amplitude frequency characteristics including principal, super, and fractional harmonic resonances are perceived. The effect of damping feedback gain on vibration response reduction of the controlled nonlinear beam system is explored.

2. General Analytical Solution to Controlled Nonlinear Beam

The meso-scale beam is an important component of many precise instruments such as micro-sensors or micro-actuators. A conventional mechanical model such as that based on Bernoulli–Euler theory is suitable to the meso beam. The large amplitude motion of the beam subjected to such as strong support loading leads to complicated nonlinear vibration, which will degrade the performance of the instruments. The beam vibration needs to be controlled, and vibration characteristics of the controlled beam need to be identified. A meso beam with control under transverse excitation, as shown in Figure 1, is addressed first in this study. Its nonlinear vibration equation in dimensionless form can be expressed as [21].

$$\alpha_1 \frac{\partial^2 w}{\partial t^2} + \alpha_2 \frac{\partial w}{\partial t} + \alpha_3 \frac{\partial^4 w}{\partial y^4} + \alpha_4 \frac{\partial^2 w}{\partial y^2} \left(\frac{\partial w}{\partial y}\right)^2 = f_e(y,t) + u_c\left(w, \frac{\partial w}{\partial t}\right) \tag{1}$$

where w is the non-dimensional transverse displacement, α_1 is the mass coefficient, α_2 is the damping coefficient, α_3 is the linear stiffness coefficient, α_4 is the nonlinear stiffness coefficient, f_e is the non-dimensional transverse loading, u_c is the feedback control, $y \in [0,1]$ is the non-dimensional longitudinal coordinate, and t is the time variable. Equation (1) is a partial differential equation and can be converted into a set of ordinary differential equations by using the Galerkin method. Vibration modes of the beam are determined based on boundary constraint conditions, and the displacement w is expanded into series using the modes. The displacement expansion is

$$w(y,t) = \sum_{i=1}^{N_w} \phi_i(y)q_i(t) = \mathbf{\Phi Q} \tag{2}$$

where N_w is the number of modes, φ_i is the ith mode, and q_i is the ith modal displacement. The mode vector $\mathbf{\Phi}$ and modal displacement vector $\mathbf{Q}$ are, respectively

$$\mathbf{\Phi} = [\phi_1\phi_2\dots\phi_{N_w}], \ \mathbf{Q} = [q_1q_2\dots q_{N_w}]^{\mathrm{T}} \tag{3}$$

For instance, a cantilever beam has its vibration modes

$$\phi_i(y) = (\sin\beta_i - \sinh\beta_i)(\sin\beta_iy - \sinh\beta_iy) + (\cos\beta_i + \cosh\beta_i)(\cos\beta_iy - \cosh\beta_iy) \tag{4}$$

where the parameter β_i is determined by the algebraic equation

$$\cos\beta_i \cosh\beta_i = -1 \tag{5}$$

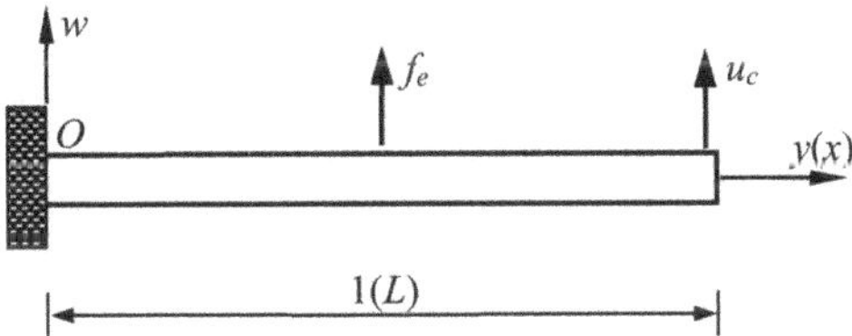

Figure 1. Meso-scale beam with control u_c under loading f_e.

The first three solutions to Equation (5) are 1.875, 4.694, and 7.855. The ratio of the second or third natural frequency to the first natural frequency of the cantilever beam is larger than that of the corresponding simply supported beam.

Substituting expression (2) into Equation (1), pre-multiplying the equation by $\mathbf{\Phi}^{\mathrm{T}}$ and integrating it with respect to y yields ordinary differential equations in the matrix form

$$\mathbf{M}\frac{\mathrm{d}^2\mathbf{Q}}{\mathrm{d}t^2} + \mathbf{C}_0\frac{\mathrm{d}\mathbf{Q}}{\mathrm{d}t} + \mathbf{K}_0\mathbf{Q} + \mathbf{G}_0(\mathbf{Q}) = \mathbf{F}(t) + \mathbf{U}(\mathbf{Q}, \frac{\mathrm{d}\mathbf{Q}}{\mathrm{d}t}) \tag{6}$$

where $\mathbf{M}$ is the modal mass matrix, $\mathbf{C}_0$ is the modal damping matrix, $\mathbf{K}_0$ is the modal linear stiffness matrix, $\mathbf{G}_0$ is the modal nonlinear restoring force vector, $\mathbf{F}$ is the modal loading vector, and $\mathbf{U}$ is the modal control vector. Their expressions are

$$\mathbf{M} = \int_0^1 \mathbf{\Phi}^{\mathrm{T}}(y)\alpha_1(y)\mathbf{\Phi}(y)\mathrm{d}y, \ \mathbf{C}_0 = \int_0^1 \mathbf{\Phi}^{\mathrm{T}}(y)\alpha_2(y)\mathbf{\Phi}(y)\mathrm{d}y$$
$$\mathbf{K}_0 = \int_0^1 \mathbf{\Phi}^{\mathrm{T}}(y)\alpha_3(y)\frac{\mathrm{d}^4\mathbf{\Phi}(y)}{\mathrm{d}y^4}\mathrm{d}y, \ \mathbf{F} = \int_0^1 \mathbf{\Phi}^{\mathrm{T}}(y)f_e(y,t)\mathrm{d}y$$
$$\mathbf{U} = \int_0^1 \mathbf{\Phi}^{\mathrm{T}}(y)u_c(\mathbf{\Phi Q}, \mathbf{\Phi}\frac{\mathrm{d}\mathbf{Q}}{\mathrm{d}t})\mathrm{d}y \tag{7}$$
$$\mathbf{G}_0(\mathbf{Q}) = \int_0^1 \mathbf{\Phi}^{\mathrm{T}}(y)\alpha_4(y)[\frac{\mathrm{d}^2\mathbf{\Phi}(y)}{\mathrm{d}y^2}\mathbf{Q}][\frac{\mathrm{d}\mathbf{\Phi}(y)}{\mathrm{d}y}\mathbf{Q}]^2\mathrm{d}y$$

In general, $\mathbf{M}$, $\mathbf{C}_0$, and $\mathbf{K}_0$ are diagonal matrices. $\mathbf{G}_0$ is a nonlinear function vector of modal displacement $\mathbf{Q}$. The feedback control u_c is a function of displacement w and velocity, and thus, the modal control $\mathbf{U}$ is a function of the modal displacement $\mathbf{Q}$ and modal velocity. Nonlinear control for the nonlinear system can be expressed as

$$\mathbf{U} = -\mathbf{C}_c\frac{\partial\mathbf{Q}}{\partial t} - \mathbf{K}_c\mathbf{Q} - \mathbf{G}_c(\mathbf{Q}) \tag{8}$$

where $\mathbf{C}_c$ is the linear damping feedback gain matrix, $\mathbf{K}_c$ is the linear stiffness feedback gain matrix, and $\mathbf{G}_c$ is the nonlinear feedback control vector. By incorporating the control (8) into the uncontrolled system, Equation (6) is rewritten as

$$\mathbf{M}\frac{d^2\mathbf{Q}}{dt^2} + \mathbf{C}\frac{d\mathbf{Q}}{dt} + \mathbf{KQ} + \mathbf{G}(\mathbf{Q}) = \mathbf{F}(t) \tag{9}$$

where $\mathbf{C} = \mathbf{C}_0 + \mathbf{C}_c$, $\mathbf{K} = \mathbf{K}_0 + \mathbf{K}_c$, and $\mathbf{G} = \mathbf{G}_0 + \mathbf{G}_c$. The damping and stiffness of system (9) can be adjusted by feedback control, and thus, the system vibration or beam vibration can be controlled.

Equation (9) is a nonlinear ordinary differential matrix equation for the controlled beam in multiple degrees-of-freedom coupling vibration. The lth element of the nonlinear term can be expanded into the power series

$$[\mathbf{G}(\mathbf{Q})]_l = \sum_{j=2}^{N_j} \left(\sum_{i_1,i_2,\dots,i_j=1}^{N_w} k^N_{l,i_1 i_2,\dots i_j} q_{i_1} q_{i_2} \cdots q_{i_j} \right) \tag{10}$$

where N_j is an integer and $k^N_{l,i1i2\dots ij}$ is the nonlinear stiffness coefficient. For the beam with cubic nonlinearity (1) and corresponding control, the lth element of the nonlinear term becomes

$$[\mathbf{G}(\mathbf{Q})]_l = \sum_{i,j,k=1}^{N_w} k^N_{l,ijk} q_i q_j q_k \tag{11}$$

where $k^N_{l,ijk}$ is the cubic nonlinear stiffness coefficient. The nonlinear restoring force vector is expressed as

$$\mathbf{G}(\mathbf{Q}) = \mathbf{K}^N \otimes \mathbf{Q}_i \otimes \mathbf{Q}_j \otimes \mathbf{Q}_k \tag{12}$$

where $\mathbf{K}^N$ is a four-dimensional tensor of nonlinear stiffness coefficients and $\otimes$ denotes a generalized product of tensor and vector.

Periodic vibration and amplitude frequency relation represent the frequency response characteristics of the controlled beam system with various frequencies, which are important for dynamic optimization design. The harmonic balance method is suitable for nonlinear periodic vibration analysis to obtain amplitude frequency relation by a uniform procedure in a wide frequency band [2]. A general analytical expression and effective iteration solution procedure for system (9) will be developed for the first time in the following. The periodic loading is expressed as

$$\mathbf{F}(t) = \sum_{k=1}^{N_f} (\mathbf{F}_{sk} \sin k\omega t + \mathbf{F}_{ck} \cos k\omega t) = \mathbf{F}_s \mathbf{\Phi}_s + \mathbf{F}_c \mathbf{\Phi}_c \tag{13}$$

where ω is the loading frequency, $\mathbf{F}_{sk}$ and $\mathbf{F}_{ck}$ are the harmonic coefficient vectors of the loading, and N_f is an integer. The harmonic vectors $\mathbf{\Phi}_s$ and $\mathbf{\Phi}_c$ with coefficient matrices $\mathbf{F}_s$ and $\mathbf{F}_c$ are, respectively

$$\begin{aligned}
\mathbf{\Phi}_s &= [\sin \omega t \, \sin 2\omega t \dots \sin N_q \omega t]^{\mathrm{T}}, \quad \mathbf{\Phi}_c = [\cos \omega t \, \cos 2\omega t \dots \cos N_q \omega t]^{\mathrm{T}} \\
\mathbf{F}_s &= [\mathbf{F}_{s1} \mathbf{F}_{s2} \dots \mathbf{F}_{sN_q}], \quad \mathbf{F}_c = [\mathbf{F}_{c1} \mathbf{F}_{c2} \dots \mathbf{F}_{cN_q}]
\end{aligned} \tag{14}$$

The integer N_q is larger than N_f, and thus, later columns of the matrices $\mathbf{F}_s$ and $\mathbf{F}_c$ are zeros. The periodic response of the vibration system to the periodic loading is expanded as

$$\mathbf{Q}(t) = \sum_{k=1}^{N_q} (\mathbf{A}_k \sin k\omega t + \mathbf{B}_k \cos k\omega t) = \mathbf{A}_Q \mathbf{\Phi}_s + \mathbf{B}_Q \mathbf{\Phi}_c \tag{15}$$

where $\mathbf{A}_k$ and $\mathbf{B}_k$ are harmonic coefficient vectors of the response. The coefficient matrices $\mathbf{A}_Q$ and $\mathbf{B}_Q$ are, respectively

$$\mathbf{A}_Q = [\mathbf{A}_1 \mathbf{A}_2 \ldots \mathbf{A}_{N_q}], \ \mathbf{B}_Q = [\mathbf{B}_1 \mathbf{B}_2 \ldots \mathbf{B}_{N_q}] \tag{16}$$

As a result of the high-order harmonic effect, the term number N_q in the expansion for response is larger than N_f for loading. The lth element of the response vector $\mathbf{Q}$ is

$$q_l(t) = \sum_{k=1}^{N_q} (a_{lk} \sin k\omega t + b_{lk} \cos k\omega t) \tag{17}$$

where a_{lk} and b_{lk} are harmonic coefficients of the response. The response coefficient vectors $\mathbf{A}_k$ and $\mathbf{B}_k$ have the expressions

$$\mathbf{A}_k = [a_{1k} a_{2k} \ldots a_{N_w k}]^{\mathrm{T}}, \ \mathbf{B}_k = [b_{1k} b_{2k} \ldots b_{N_w k}]^{\mathrm{T}}. \tag{18}$$

By substituting expression (15) into nonlinear term (12), using harmonic function properties and neglecting the higher harmonic part, the nonlinear term can be obtained as

$$\mathbf{G}(\mathbf{Q}) = \mathbf{K}_{3s}^N(\mathbf{A}_Q, \mathbf{B}_Q)\mathbf{\Phi}_s + \mathbf{K}_{3c}^N(\mathbf{A}_Q, \mathbf{B}_Q)\mathbf{\Phi}_c \tag{19}$$

where $\mathbf{K}_{3s}^N$ and $\mathbf{K}_{3c}^N$ are the function matrices of the response coefficients $\mathbf{A}_Q$ and $\mathbf{B}_Q$. The lth element of the vector $\mathbf{G}$ is

$$[\mathbf{G}(\mathbf{Q})]_l = \sum_{m=1}^{N_q} (k_{3s,lm}^N \sin m\omega t + k_{3c,lm}^N \cos m\omega t) \tag{20}$$

where $k_{3s,lm}^N$ and $k_{3c,lm}^N$ are the harmonic coefficients of the nonlinear term. The coefficients can be further expressed as

$$k_{3s,lm}^N = \sum_{i,j,k=1}^{N_w} k_{l,ijk}^N d_m(i,j,k), \ k_{3c,lm}^N = \sum_{i,j,k=1}^{N_w} k_{l,ijk}^N e_m(i,j,k) \tag{21}$$

where d_m and e_m are functions of the response coefficients a_{lk} and b_{lk}. They are

$$\begin{aligned}
d_m &= \frac{\omega}{\pi} \int_0^{2\pi/\omega} q_i q_j q_k \sin \omega t \, dt \\
&= \frac{1}{4} \sum_{l_1,l_2,l_3=1}^{N_q} \sum_{\lambda_1,\lambda_2,\lambda_3=1}^{2} \alpha_{il_1}(\lambda_1)\alpha_{jl_2}(\lambda_2)\alpha_{kl_3}(\lambda_3)g_s(\mu_1 l_1, \mu_2 l_2, \mu_3 l_3)
\end{aligned} \tag{22}$$

$$\begin{aligned}
e_m &= \frac{\omega}{\pi} \int_0^{2\pi/\omega} q_i q_j q_k \cos \omega t \, dt \\
&= \frac{1}{4} \sum_{l_1,l_2,l_3=1}^{N_q} \sum_{\lambda_1,\lambda_2,\lambda_3=1}^{2} \alpha_{il_1}(\lambda_1)\alpha_{jl_2}(\lambda_2)\alpha_{kl_3}(\lambda_3)g_c(\mu_1 l_1, \mu_2 l_2, \mu_3 l_3)
\end{aligned} \tag{23}$$

with

$$\begin{aligned}
\alpha_{il_1}(\lambda_1) &= \left\{ \begin{array}{l} a_{il_1}, \ \lambda_1 = 1 \\ b_{il_1}, \ \lambda_1 = 2 \end{array} \right., \ \alpha_{jl_2}(\lambda_2) = \left\{ \begin{array}{l} a_{jl_2}, \ \lambda_2 = 1 \\ b_{jl_2}, \ \lambda_2 = 2 \end{array} \right., \ \alpha_{kl_3}(\lambda_3) = \left\{ \begin{array}{l} a_{kl_3}, \ \lambda_3 = 1 \\ b_{kl_3}, \ \lambda_3 = 2 \end{array} \right. \\
g_s &= \tfrac{1}{2}[1 - (-1)^{\lambda_1+\lambda_2+\lambda_3}]\sum_{S_\mu} \delta_{m+S_\mu,0} s_g(S_\mu), \ g_c = \tfrac{1}{2}[1 + (-1)^{\lambda_1+\lambda_2+\lambda_3}]\sum_{C_\mu} \delta_{m+C_\mu,0} s_g(C_\mu) \\
s_g(S_\mu) &= (-1)^{\Sigma \lambda_{j2}/4}\textstyle\prod \mu_{j_1}, \ s_g(C_\mu) = (-1)^{\Sigma \lambda_{j2}/2}\textstyle\prod \mu_{j_1}
\end{aligned} \tag{24}$$

where μ_1, μ_2, and μ_3 are $+1$ or -1, S_μ and C_μ denote all combinations of $(\mu_1 l_1, \mu_2 l_2, \mu_3 l_3)$, j_1 is for $\lambda_{j1} = 1$ and j_2 is for $\lambda_{j2} = 2$, and δ is the Kronecker delta function. Expressions (22) and (23) indicate that d_m and e_m are cubic power functions of the response coefficients a_{lk} and b_{lk}. As such, general expression of the nonlinear term has been obtained in Equations (19)–(24) for the periodic response with multiple harmonic terms, which is given first for the meso beam.

According to the harmonic balance method, substituting response (15) with expression (19) and loading (13) into Equation (9) and balancing each harmonic term yields algebraic equations for response coefficients $\mathbf{A}_Q$ and $\mathbf{B}_Q$ as

$$\mathbf{KA}_Q - \mathbf{MA}_Q\Omega^2 - \mathbf{CB}_Q\Omega = \mathbf{F}_s - \mathbf{K}^N_{3s} \tag{25}$$

$$\mathbf{KB}_Q - \mathbf{MB}_Q\Omega^2 + \mathbf{CA}_Q\Omega = \mathbf{F}_c - \mathbf{K}^N_{3c} \tag{26}$$

where the vibration frequency matrix is

$$\Omega = \mathrm{diag}[\omega 2\omega \ldots N_q\omega] \tag{27}$$

Equations (25) and (26) are algebraic matrix equations, and they can be rearranged as equations for the response coefficient vector of a_{lk} and b_{lk} to solve. However, equations for the mth column $\mathbf{A}_m$ of the matrix $\mathbf{A}_Q$ and the mth column $\mathbf{B}_m$ of the matrix $\mathbf{B}_Q$ are obtained from Equations (25) and (26) as

$$\mathbf{KA}_m - m^2\omega^2\mathbf{MA}_m - m\omega\mathbf{CB}_m = \mathbf{F}_{sm} - \mathbf{K}^N_{3s,\cdot m} \tag{28}$$

$$m\omega\mathbf{CA}_m + \mathbf{KB}_m - m^2\omega^2\mathbf{MB}_m = \mathbf{F}_{cm} - \mathbf{K}^N_{3c,\cdot m} \tag{29}$$

where $\mathbf{K}^N_{3s,\cdot m}$ is the mth column vector of the matrix $\mathbf{K}^N_{3s}$ while $\mathbf{K}^N_{3c,\cdot m}$ is the mth column vector of the matrix $\mathbf{K}^N_{3c}$. By solving Equations (28) and (29), response coefficient vectors $\mathbf{A}_m$ and $\mathbf{B}_m$ with elements a_{lm} and b_{lm} are obtained as

$$\left\{ \begin{array}{c} \mathbf{A}_m \\ \mathbf{B}_m \end{array} \right\} = \Delta_m^{-1} \left\{ \begin{array}{c} \mathbf{F}_{sm} - \mathbf{K}^N_{3s,\cdot m} \\ \mathbf{F}_{cm} - \mathbf{K}^N_{3c,\cdot m} \end{array} \right\} \tag{30}$$

where

$$\Delta_m(\omega) = \left[\begin{array}{cc} \mathbf{K} - m^2\omega^2\mathbf{M} & -m\omega\mathbf{C} \\ m\omega\mathbf{C} & \mathbf{K} - m^2\omega^2\mathbf{M} \end{array} \right]. \tag{31}$$

Equation (30) provides a set of general harmonic balance solution formulae for periodic vibration of the nonlinear system (9) representative of the controlled meso beam. By the equation, the harmonic response coefficients $\mathbf{A}_m$ and $\mathbf{B}_m$ and then $\mathbf{A}_Q$ and $\mathbf{B}_Q$ can be obtained, and thus, the periodic vibration response of the system is determined by Equation (15) and the beam displacement is calculated by Equation (2). The vibrational amplitude frequency characteristics of the controlled beam are obtained by means of the responses under various frequencies. The accuracy of the responses can be improved by increasing numbers of expansion terms in Equations (2) and (15). Convergence of the response expansion is determined by relative differences of the harmonic response coefficients, i.e., $|(a_{lm} - a'_{lm})/a_{lm}|$ and $|(b_{lm} - b'_{lm})/b_{lm}|$. However, because $\mathbf{K}^N_{3s,\cdot m}$ and $\mathbf{K}^N_{3c,\cdot m}$ depend on response coefficients $\mathbf{A}_m$ and $\mathbf{B}_m$, Equation (30) is nonlinear and will be solved by iteration.

3. Iteration Procedure for Vibration Response

Equations (28) and (29) or (30) include $2 \times N_w \times N_q$ algebraic equations for harmonic response coefficients a_{lm} and b_{lm} ($l = 1, 2, \ldots, N_w$; $m = 1, 2, \ldots, N_q$). The equations can be rewritten generally as

$$f_{a,lm}(\mathbf{A}_Q, \mathbf{B}_Q) = 0 \tag{32}$$

$$f_{b,lm}(\mathbf{A}_Q, \mathbf{B}_Q) = 0 \tag{33}$$

where $f_{a,lm}$ and $f_{b,lm}$ are functions of $\mathbf{A}_Q$ and $\mathbf{B}_Q$. $\mathbf{A}_Q$ and $\mathbf{B}_Q$ are matrices by assembling a_{lm} and b_{lm}, respectively, as expressions (16) and (18). Based on Equations (32) and (33), iterative equations are given according to the generalized Newton iteration algorithm as

$$a_{lm}^{k+1} = a_{lm}^{k} - \theta^{k}\frac{\partial f^{k}}{\partial a_{lm}^{k}}, \quad b_{lm}^{k+1} = b_{lm}^{k} - \theta^{k}\frac{\partial f^{k}}{\partial b_{lm}^{k}} \tag{34}$$

where superscript "k" denotes the kth step of iteration, functions f and θ are

$$f^{k} = \sum_{l=1}^{N_w}\sum_{m=1}^{N_q}[f_{a,lm}(\mathbf{A}_Q^{k},\mathbf{B}_Q^{k})^{2} + f_{b,lm}(\mathbf{A}_Q^{k},\mathbf{B}_Q^{k})^{2}]$$
$$\theta^{k} = \frac{f^{k}}{\sum_{l=1}^{N_w}\sum_{m=1}^{N_q}[(\partial f^{k}/\partial a_{lm}^{k})^{2}+(\partial f^{k}/\partial b_{lm}^{k})^{2}]}. \tag{35}$$

Derivatives of the function f are given by

$$\frac{\partial f^{k}}{\partial a_{lm}^{k}} = \frac{f^{k}(a_{lm}^{k}+\Delta a_{lm}^{k},b_{lm}^{k})-f^{k}(a_{lm}^{k},b_{lm}^{k})}{\Delta a_{lm}^{k}}$$
$$\frac{\partial f^{k}}{\partial b_{lm}^{k}} = \frac{f^{k}(a_{lm}^{k},b_{lm}^{k}+\Delta b_{lm}^{k})-f^{k}(a_{lm}^{k},b_{lm}^{k})}{\Delta b_{lm}^{k}} \tag{36}$$
$$\Delta a_{lm}^{k} = \beta a_{lm}^{k}, \quad \Delta b_{lm}^{k} = \beta b_{lm}^{k}$$

where the parameter β is set as a small value, e.g., 10^{-5}. Equation (34) presents a general iteration algorithm for the harmonic balance solution to periodic vibration of the nonlinear system (9). However, a computational expense of the iterative Equation (34) is large due to coupling high dimensions.

To improve computational efficiency, based on the partially uncoupled Equation (30), another form of the iterative equation is obtained as

$$\left\{\begin{array}{c} \mathbf{A}_m^{k+1} \\ \mathbf{B}_m^{k+1} \end{array}\right\} = \Delta_{mk}^{-1}\left\{\begin{array}{c} \mathbf{F}_{sm} - \mathbf{K}_{3s,\cdot m}^{N_0}(\mathbf{A}_Q^{k},\mathbf{B}_Q^{k}) \\ \mathbf{F}_{cm} - \mathbf{K}_{3c,\cdot m}^{N_0}(\mathbf{A}_Q^{k},\mathbf{B}_Q^{k}) \end{array}\right\} \tag{37}$$

where $\mathbf{A}_m$ and $\mathbf{B}_m$ are vectors respectively by assembling a_{lm} and b_{lm}, $\mathbf{K}_{3s,\cdot m}^{N_0}$ is a part of $\mathbf{K}^{N}_{3s,\cdot m}$ independent of $\mathbf{A}_m$ and the other part $\mathbf{K}_{3s,\cdot m}^{N_A}\mathbf{A}_m$ depends on $\mathbf{A}_m$, $\mathbf{K}_{3c,\cdot m}^{N_0}$ is a part of $\mathbf{K}^{N}_{3c,\cdot m}$ independent of $\mathbf{B}_m$ and the other part $\mathbf{K}_{3c,\cdot m}^{N_B}\mathbf{B}_m$ depends on $\mathbf{B}_m$, and the matrix

$$\Delta_{mk} = \begin{bmatrix} \mathbf{K} - m^{2}\omega^{2}\mathbf{M} + \mathbf{K}_{3s,\cdot m}^{N_A}(\mathbf{A}_Q^{k},\mathbf{B}_Q^{k}) & -m\omega\mathbf{C} \\ m\omega\mathbf{C} & \mathbf{K} - m^{2}\omega^{2}\mathbf{M} + \mathbf{K}_{3c,\cdot m}^{N_B}(\mathbf{A}_Q^{k},\mathbf{B}_Q^{k}) \end{bmatrix}. \tag{38}$$

The iteration solution procedure for solving Equation (37) or (34) is proposed as follows: first, select initial values of the response coefficient matrices $\mathbf{A}_Q$ and $\mathbf{B}_Q$, e.g., as a solution to the linearized system; second, substituting the initial values into Equation (37) or (34) yields the next step values of the response coefficient vectors $\mathbf{A}_m$ and $\mathbf{B}_m$, and $\mathbf{A}_m$ and $\mathbf{B}_m$ are used to update the last step values and to calculate the response coefficient vectors $\mathbf{A}_{m+1}$ and $\mathbf{B}_{m+1}$ ($m = 1, 2, \ldots, N_q - 1$), that is constantly updated cycle; third, examine the convergence of the response coefficient matrices $\mathbf{A}_Q$ and $\mathbf{B}_Q$ by using relative differences of adjacent step values of the response coefficients, i.e.,

$$\left|a_{lm}^{k+1} - a_{lm}^{k}\right| \leq \gamma_I\left|a_{lm}^{k}\right|, \quad \left|b_{lm}^{k+1} - b_{lm}^{k}\right| \leq \gamma_I\left|b_{lm}^{k}\right| \tag{39}$$

where γ_I is a small value representing iteration accuracy and $k = 1, 2, \ldots$ If the current step values of the response coefficients do not satisfy condition (39), continue to carry out the next step until convergence. This iteration procedure is called updated cycle iteration. Finally, the harmonic response coefficients $\mathbf{A}_Q$ and $\mathbf{B}_Q$ are obtained and used to calculate periodic responses (15) and (2), by which the amplitude frequency characteristics of the vibration response of the controlled beam system can be determined. The vibration response with various frequencies can be calculated using the uniform method by only changing frequency. Accurate response will be obtained by using adequate high harmonic components.

The above method, including a general analytical expression for a harmonic balance solution to periodic vibration and an updated cycle iteration algorithm for amplitude frequency relation of periodic response, is a uniform method for a nonlinear system of the controlled mseo-scale beam under various frequencies, and it is suitable for obtaining amplitude frequency relation in a wide frequency band. It is also applicable to various periodic vibration analyses including principal resonant response, super-harmonic resonant response, and multiple stationary responses. The dynamic stability of the periodic vibration can be solved by using the direct eigenvalue analysis approach for systems with periodically varying parameters [20,21].

When loading frequency ω keeps away from the resonant frequencies of the nonlinear system, a non-resonant response as a function of frequency ω can be obtained by using the above method. When the loading frequency ω is close to a resonant frequency of the nonlinear system, a principal resonant response as function of frequency ω can be obtained by using the uniform method. In the case of multiple response solutions, a resonant response obtained depends on initial iteration values and vibration stability. When the loading frequency ω is close to $1/n$-times resonant frequency of the nonlinear system (n is an integer larger than 1), a super-harmonic resonance near $n\omega$ can occur, and in the meantime, a response component with frequency ω also exists. In this case, the vibration response obtained has the frequency of ω.

The vibration response of the nonlinear system (9) or beam (1) is composed of multiple harmonic components, and its amplitude will be obtained by periodic solution under maximum condition. For example, the displacement response amplitude $\max\{|w(y_0, t)|\}$ on y_0 is determined by periodic solution varying with time in a period under the condition $\partial w(y_0, t)/\partial t = 0$. As a consequence, the amplitude frequency characteristics of periodic vibration can be determined.

4. Numerical Results and Discussion

To illustrate the application of the developed method and amplitude frequency characteristics, consider a meso-scale cantilever beam with control under a periodic loading. Its nonlinear vibration equation in dimensionless form is as Equations (1) and (9), in which α_i ($i = 1, 2, 3, 4$) are constants for the uniform beam. The foundational frequency of the meso beam is high and generally larger than the loading frequency. Thus, first several vibration modes will be involved in vibration response, while the first mode vibration is dominant. Increasing structural damping is a common strategy for resonant response suppression. Damping produced by passive control is limited, whereas active feedback control can produce large artificial damping, which is incorporated in damping coefficient of system (9). The effect of the damping on vibration response characteristics will be explored for control at the midpoint. The periodic support loading is $\mathbf{F} = \mathbf{F}_{s1} \sin\omega t$ in system (9). After normalized with respect to mass, non-dimensional damping coefficient, linear stiffness coefficient, nonlinear stiffness coefficient, and loading amplitude for the first mode are denoted by c, k_1, k_3, and F_{s1}, respectively. The other coefficients can be determined by these coefficients based on Equations (4), (7), and (8). Using the developed method, harmonic response coefficients (18) and periodic vibration response (2) of the beam system are obtained, which are used to determine amplitude frequency characteristics. In calculation, basic parameter values are $k_1 = 1.0$, $c = 0.1$, $k_3 = 0.12$, $F_{s1} = 0.2$, and $y_0 = 1$ unless otherwise specified. Numerical results on the non-dimensional vibration response and amplitude frequency relation are shown in Figures 2–14, which have been verified through direct numerical simulation.

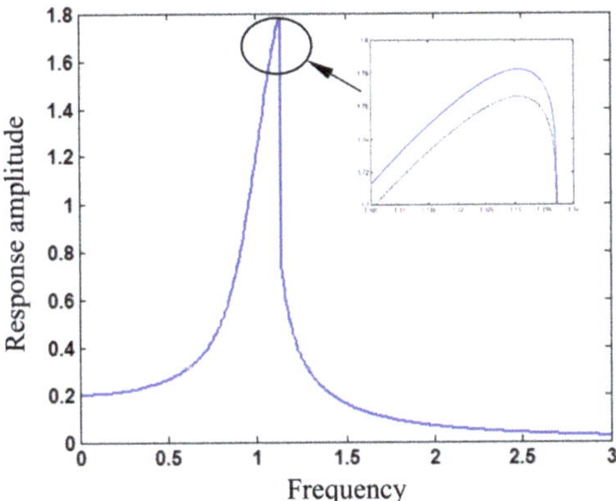

Figure 2. Amplitude (w_{Am}) frequency (ω) relation around principal resonance for loading amplitude $F_{s1} = 0.2$ (blue solid line: response based on periodic solution; black dotted line: response based on first harmonic term).

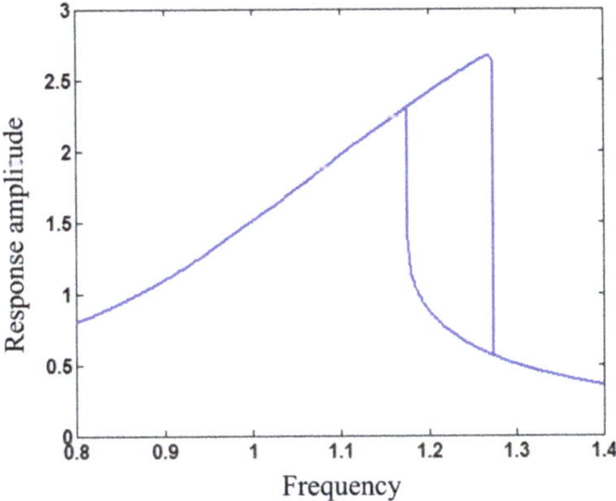

Figure 3. Amplitude (w_{Am}) frequency (ω) relation around principal resonance for loading amplitude $F_{s1} = 0.34$.

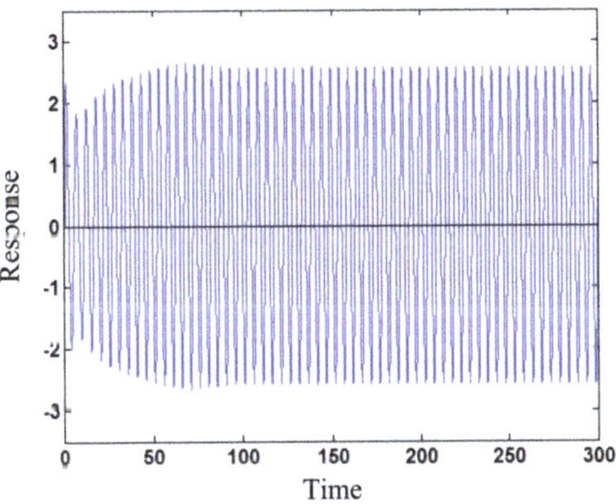

Figure 4. Larger amplitude response (w) varying with time (t) for loading frequency $\omega = 1.24$ by numerical simulation from initial displacement 1.0 and velocity 2.5.

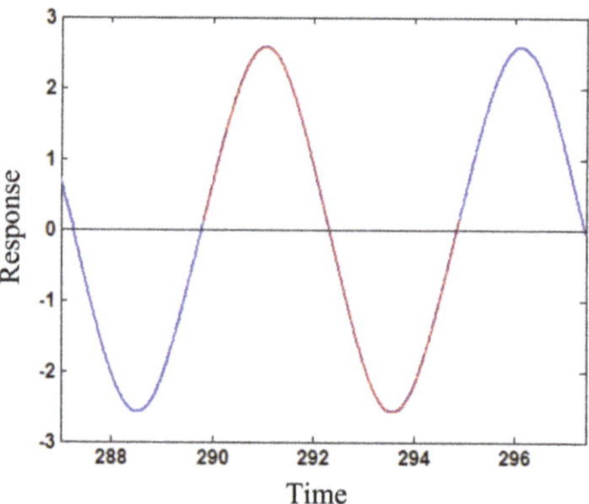

Figure 5. Stationary larger-amplitude response (w) varying with time (t) for loading frequency $\omega = 1.24$ (blue solid line: numerical simulation; red dotted line: the developed method).

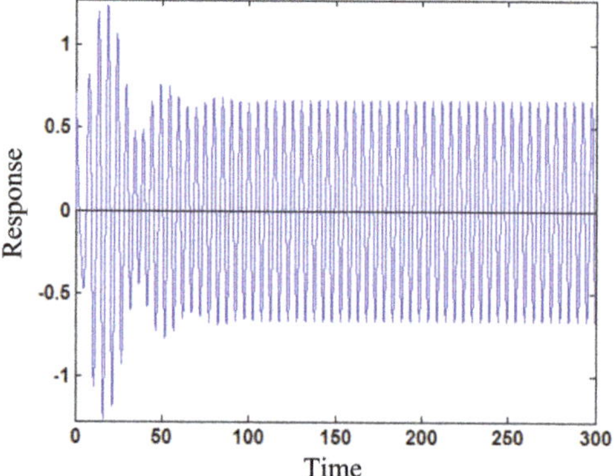

Figure 6. Smaller amplitude response (w) varying with time (t) for loading frequency $\omega = 1.24$ by numerical simulation from initial displacement 1.0 and velocity 0.

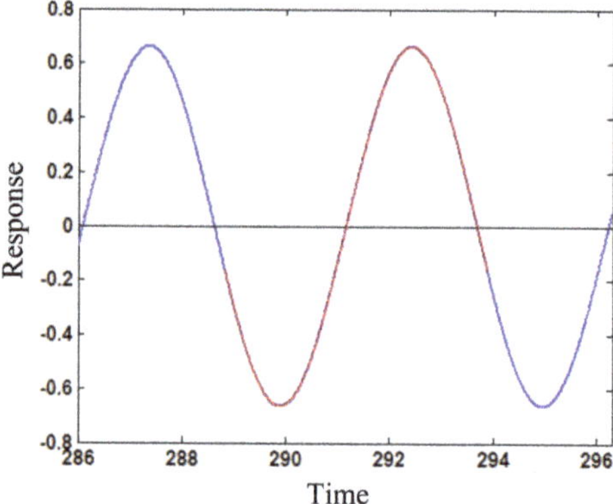

Figure 7. Stationary smaller-amplitude response (w) varying with time (t) for loading frequency $\omega = 1.24$ (blue solid line: numerical simulation; red dotted line: the developed method).

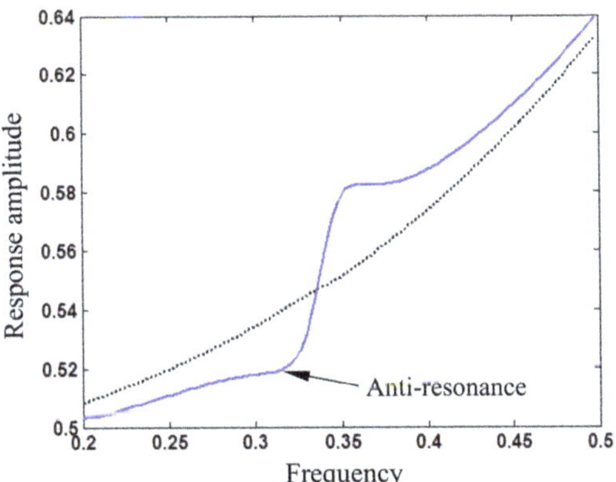

Figure 8. Amplitude (w_{Am}) frequency (ω) relation around super-harmonic resonance for loading amplitude $F_{s1} = 0.5$ (blue solid line: response based on periodic solution; black dotted line: response based on first harmonic term).

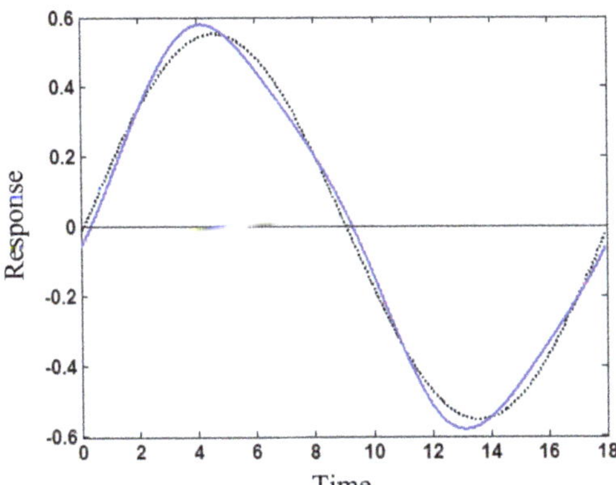

Figure 9. Stationary response (w) varying with time (t) for loading frequency $\omega = 0.35$ (blue solid line: response based on periodic solution; black dotted line: response based on first harmonic term).

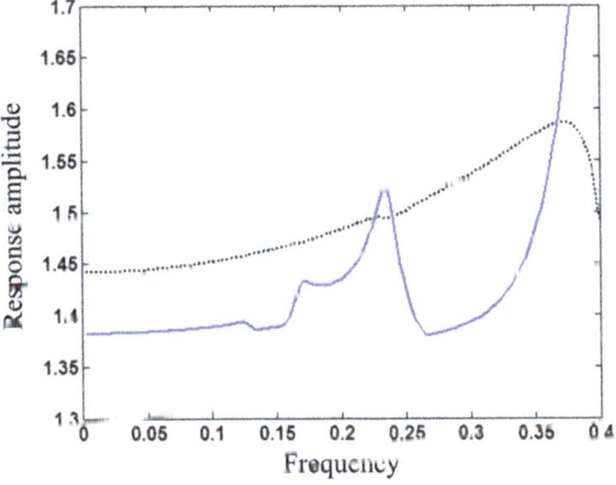

Figure 10. Amplitude (w_{Am}) frequency (ω) relation around fractional harmonic resonance for loading amplitude $F_{s1} = 1.7$ (blue solid line: response based on periodic solution; black dotted line: response based on first harmonic term).

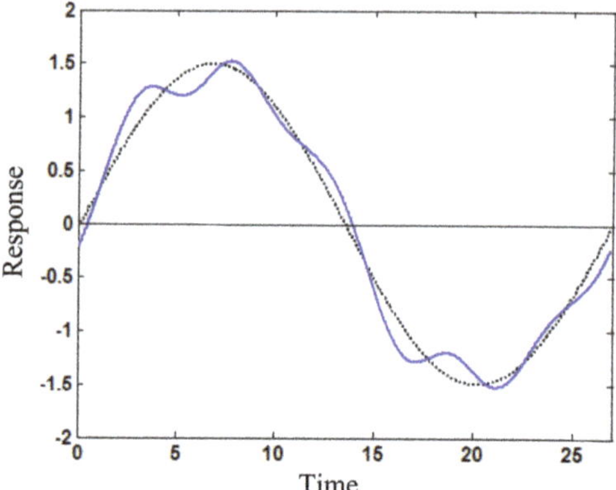

Figure 11. Stationary response (w) varying with time (t) for loading frequency $\omega = 0.23$ (blue solid line: response based on periodic solution; black dotted line: response based on first harmonic term).

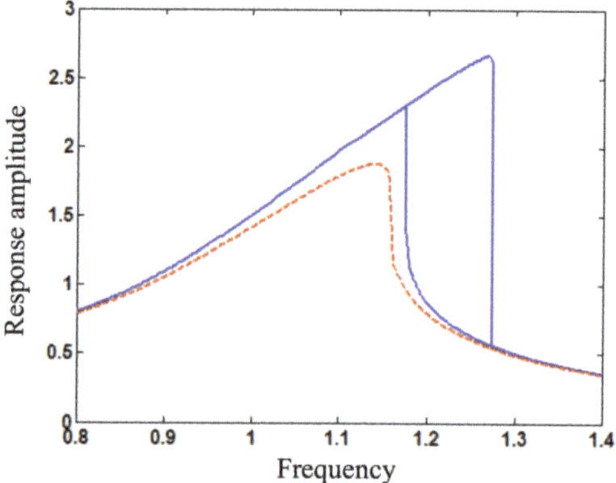

Figure 12. Amplitude (w_{Am}) frequency (ω) relation around principal resonance for loading amplitude $F_{s1} = 0.34$ and different damping coefficients c (blue solid line: $c = 0.1$; red dashed line: $c = 0.16$).

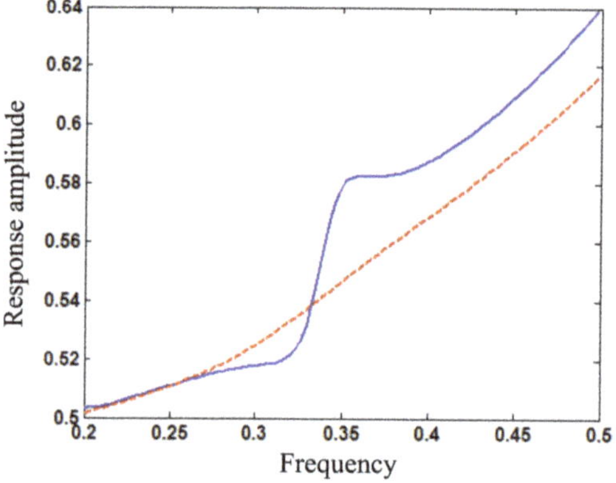

Figure 13. Amplitude (w_{Am}) frequency (ω) relation around super-harmonic resonance for loading amplitude $F_{s1} = 0.5$ and different damping coefficients c (blue solid line: $c = 0.1$; red dashed line: $c = 0.45$).

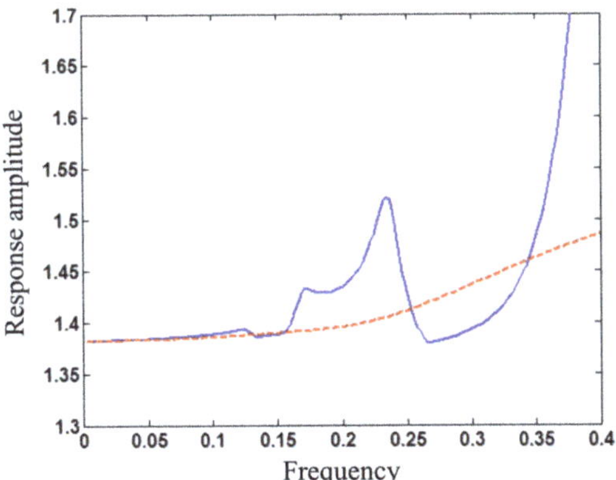

Figure 14. Amplitude (w_{Am}) frequency (ω) relation around fractional harmonic resonance for loading amplitude $F_{s1} = 1.7$ and different damping coefficients c (blue solid line: $c = 0.1$; red dashed line: $c = 1.25$).

Figure 2 shows the amplitude frequency relation around principal resonance for periodic vibration response of the beam under $F_{s1} = 0.2$. The amplitude frequency relation around principal resonance based on the first expansion term of the response is also given for comparison. The non-dimensional resonant frequency is close to 1.13, and the non-dimensional resonant response amplitude is 1.78 determined by the developed periodic solution, which is larger than 1.76 based on only the first expansion term. The difference will be shown further in the subsequent results, which increases with loading amplitude. Thus, the response amplitude should be determined by periodic solution (2) with (15) instead of only the first harmonic term. If the resonant response amplitude is larger than 1, the nonlinear term of the system cannot be regarded as small in resonance.

As the loading amplitude increases to $F_{s1} = 0.34$, Figure 3 shows the amplitude frequency relation around principal resonance for periodic vibration response of the beam. It is seen that two stationary responses exist in a frequency interval [1.17, 1.27], which can be caused also by increasing the nonlinear stiffness coefficient. If the loading amplitude or nonlinear stiffness increases, the multiple responses region will be enlarged in amplitude and frequency and move right. In particular, the initial state region, which tends to smaller amplitude response with time, is larger than that for the larger amplitude response. Thus, the smaller amplitude response has a larger stability probability than the larger amplitude response; that is, the smaller amplitude response is more likely to occur practically. Figure 4 shows the larger amplitude response varying with time for the loading frequency $\omega = 1.24$ by numerical simulation from initial displacement 1.0 and velocity 2.5. The stationary response amplitude is 2.58. Figure 5 illustrates that the stationary response by numerical simulation accords completely with that by using the developed method. Figure 6 shows the smaller amplitude response (stationary amplitude 0.66) varying with time for the loading frequency $\omega = 1.24$ by numerical simulation from initial displacement 1.0 and velocity 0. Figure 7 illustrates again that the stationary response by numerical simulation agrees completely with that by using the developed method.

As the loading amplitude increases to $F_{s1} = 0.5$, Figure 8 shows the amplitude frequency relation around super-harmonic resonance for the periodic vibration response of the beam. It is seen that the amplitude frequency curve of the response based on only the first harmonic term does not display any peak and valley, but the curve of the response determined by multiple harmonic terms of periodic solution (15) has a peak and a valley. In particular, the response valley or anti-resonant response near frequency $\omega = 0.32$ in the nonlinear system is obtained for the first time. For frequency $\omega > 0.34$, the response amplitude determined by the periodic solution is larger than that based on only the first harmonic term, and for frequency $\omega < 0.33$, the response amplitude determined by the periodic solution is smaller than that based on only the first harmonic term. Thus, it is supported further that the response amplitude should be determined by periodic solution (2) with (15) instead of only the first harmonic term. The dynamic characteristics including anti-resonance have potential application to vibration control design of meso-scale nonlinear systems. Figure 9 illustrates that the stationary response varying with time in amplitude and shape determined by the periodic solution is different from that based on only the first harmonic term, where the loading frequency $\omega = 0.35$. The response amplitude is 0.58 for the former and 0.55 for the latter. The relative difference is 5%, but the variation of dynamic characteristics is remarkable, as shown in Figure 8. The response contains a super-harmonic resonance component with frequency 3ω and basic harmonic component with frequency ω, which is relatively large. The response peak increases with the loading amplitude.

As the loading amplitude increases to $F_{s1} = 1.7$, Figure 10 shows the amplitude frequency relation around fractional harmonic resonance (e.g., $9\omega/2$) for the periodic vibration response of the beam. Figure 11 illustrates that the stationary response varying with time particularly in shape determined by the periodic solution is largely different from that based on only the first harmonic term. The loading frequency is $\omega = 0.23$, and the stationary response amplitude is 1.52. The relative difference of amplitudes based on the periodic solution and first harmonic term for frequency $\omega = 0.4$ is 30.4%. The response peak also increases with the loading amplitude.

For the controlled nonlinear beam system under the harmonic loading with amplitude $F_{s1} = 0.34$, when the damping coefficient increases to exceed 0.16, the two stationary responses around principal resonance in Figure 3 disappear and become a new pattern of response, as shown in Figure 12. Thus, the principal resonant responses can be controlled by increasing the damping gain. For the controlled beam system under the harmonic loading with amplitude $F_{s1} = 0.5$, when the damping coefficient increases to exceed 0.45, the peak response around super-harmonic resonance and the anti-resonant response in Figure 8 disappear, as shown in Figure 13. Thus, large damping or feedback gain is required to mitigate the beam vibration with frequency smaller than the principal resonant frequency. However, for the controlled beam system under the harmonic loading with amplitude $F_{s1} = 1.7$, the peak response around fractional harmonic resonance in Figure 10 disappears only when the damping coefficient increases up to 1.25, as shown in Figure 14. In fact, it is unpractical to produce such large damping in passive control, but the large damping or feedback gain can be generated by active control as artificial damping. Thus, the resonant responses of the nonlinear meso beam can be reduced by feedback control with certain damping gain.

For an actual meso beam, its basic geometrical and physical parameters are obtained first to determine the parameters of model (1) such as α_i ($i = 1,2,3,4$), as shown in reference [21] and by Equations (4), (7), and (8). Then, the model parameters are updated by test results compared with analytical results on amplitude frequency relation. The updated model can be used for dynamic analysis and control design of the beam.

5. Conclusions

The method including a general analytical expression for harmonic balance solution to periodic vibration and an updated cycle iteration algorithm for amplitude frequency relation of periodic response has been developed for the controlled nonlinear meso-scale beam. The general expression of nonlinear terms in the nonlinear system for periodic response with multiple harmonic terms is derived. The general analytical expression for harmonic balance solution to periodic vibration of the nonlinear system is obtained. The updated cycle iteration procedure for amplitude frequency relation and periodic response solution is proposed. The developed method has the following main advantages: (1) it is a unified method for nonlinear systems with various frequencies, and periodic vibration responses for different frequencies can be calculated uniformly by only changing frequency;

(2) it is suitable for obtaining the amplitude frequency relation of nonlinear systems in a wide frequency band and for effective numerical computation; (3) it can take into account the effect of higher harmonic components on vibration response; and (4) it is applicable to various periodic vibration analyses including principal resonant response, super-harmonic resonant response, and multiple stationary responses.

The amplitude frequency characteristics of the controlled nonlinear meso beam and the effect of damping feedback gain on vibration response have been studied. Numerical results on vibration response of the beam with various damping and loading amplitudes have demonstrated that: (1) the developed method has good convergence and accuracy with increasing response expansion terms, and the response amplitude should be determined by the periodic solution with multiple harmonic terms instead of only the first harmonic term; (2) the amplitude frequency characteristics around the principal resonance of the controlled beam system show two stationary responses for large loading amplitude and nonlinear stiffness coefficient, the resonant frequency is dependent on the loading and response amplitudes, the region of multiple responses can be enlarged in amplitude and frequency as the loading amplitude or nonlinear stiffness increases, and the smaller amplitude response has a larger stability probability than the larger amplitude response in terms of the initial states region of stable vibration; (3) the amplitude frequency curve of vibration response near super-harmonic resonance has a peak and a valley for a large loading amplitude, the response valley or anti-resonant response in the nonlinear system is obtained for the first time, and the dynamic characteristics have potential application to vibration control design of nonlinear meso structures; (4) the peak response with frequency around super-harmonic resonance or fractional harmonic resonance increases with loading amplitude; and (5) the principal resonant responses of the nonlinear beam system can be controlled by increasing damping gain, the response around super-harmonic resonance can be mitigated by feedback control with large damping gain, and the response around fractional harmonic resonance with frequency smaller than the principal resonant frequency cannot be mitigated effectively by passive control, but it can be controlled by active feedback with large artificial damping. Thus, the vibration responses of the nonlinear meso beam can be reduced by feedback control with certain damping gain. The vibrational amplitude frequency characteristics including anti-resonance and resonant response variation have potential application to vibration control or dynamic optimization design of nonlinear meso-scale structure systems.

However, the developed method is limited to periodic vibration analysis, and a method for non-periodic vibration analysis needs to be developed further. The analysis method is expected to apply to an actual meso-scale structure for dynamic optimization design and experimental verification.

Author Contributions: Conceptualization, Z.-G.Y. and Y.-Q.N.; methodology, Z.-G.Y. and Y.-Q.N.; software, Z.-G.Y. ; validation, Z.-G.Y. ; writing—original draft preparation, Z.-G.Y. ; writing—review and editing, Y.-Q.N. and Z.-G.Y. ; project administration, Z.-G.Y. and Y.-Q.N.; funding acquisition, Z.-G.Y. and Y.-Q.N. Both authors have read and agreed to the published version of the manuscript.

Funding: This research was funded by the National Natural Science Foundation of China (grant number 12072312), the Research Grants Council of the Hong Kong Special Administrative Region (grant number R-5020-18), and the Innovation and Technology Commission of the Hong Kong Special Administrative Region (grant number K-BBY1).

Institutional Review Board Statement: Not applicable.

Informed Consent Statement: Not applicable.

Data Availability Statement: Not applicable.

Conflicts of Interest: The authors declare no conflict of interest

References

1. Wang, X.F.; Wei, X.Y.; Pu, D.; Huan, R.H. Single-electron detection utilizing coupled nonlinear microresonators. *Microsyst. Nanoeng.* **2020**, *6*, 78. [CrossRef]
2. Nayfeh, A.H.; Mook, D.T. *Nonlinear Oscillations*; John Wiley & Sons: New York, NY, USA, 1979.
3. Soong, T.T.; Spencer, B.F. Supplemental energy dissipation: State-of-the-art and state-of-the-practice. *Eng. Struct.* **2002**, *24*, 243–259. [CrossRef]
4. Casciati, F.; Rodellar, J.; Yildirim, U. Active and semi-active control of structures—Theory and applications: A review of recent advances. *J. Intell. Mater. Syst. Struct.* **2012**, *23*, 1181–1195. [CrossRef]
5. Stengel, R.F. *Optimal Control and Estimation*; John Wiley & Sons: New York, NY, USA, 1994.
6. Fleming, W.H.; Soner, H.M. *Controlled Markov Processes and Viscosity Solutions*; Springer: New York, NY, USA, 2006.
7. Fisco, N.R.; Adeli, H. Smart structures: Part I—Active and semi-active control. *Sci. Iran. A* **2011**, *18*, 275–284. [CrossRef]
8. Spencer, B.F.; Nagarajaiah, S. State of the art of structural control. *ASCE J. Struct. Eng.* **2003**, *129*, 845–856. [CrossRef]
9. Ying, Z.G.; Ni, Y.Q.; Duan, Y.F. Parametric optimal bounded feedback control for smart parameter-controllable composite structures. *J. Sound Vib.* **2015**, *339*, 38–55. [CrossRef]
10. Xu, Z.D.; Xu, F.H.; Chen, X. Vibration suppression on a platform by using vibration isolation and mitigation devices. *Nonlinear Dyn.* **2016**, *83*, 1341–1353. [CrossRef]
11. Ying, Z.G.; Zhu, W.Q. Optimal bounded control for nonlinear stochastic smart structure systems based on extended Kalman filter. *Nonlinear Dyn.* **2017**, *90*, 105–114. [CrossRef]
12. Wang, Q.; Li, H.N.; Zhang, P. Vibration control of a high-rise slender structure with a spring pendulum pounding tuned mass damper. *Actuators* **2021**, *10*, 44. [CrossRef]
13. Xu, Z.D.; Dong, Y.R.; Chen, S.; Guo, Y.Q.; Li, Q.Q.; Xu, Y.S. Development of hybrid test system for three-dimensional viscoelastic damping frame structures based on Matlab-OpenSees combined programming. *Soil Dyn. Earthq. Eng.* **2021**, *144*, 106681. [CrossRef]
14. Xu, Z.D.; Yang, Y.; Miao, A.N. Dynamic analysis and parameter optimization of pipelines with multidimensional vibration isolation and mitigation device. *ASCE J. Pipeline Syst. Eng. Pract.* **2021**, *12*, 04020058. [CrossRef]
15. Kim, Y.B. Multiple harmonic balance method for aperiodic vibration of a piecewise-linear System. *ASME J. Vib. Acoust.* **1998**, *120*, 181–187. [CrossRef]
16. Huseyin, K.; Lin, R. An intrinsic multiple-scale harmonic balance method for non-linear vibration and bifurcation problems. *Int. J. Non-Linear Mech.* **1991**, *26*, 727–740. [CrossRef]
17. Chatterjee, A. Harmonic balance based averaging: Approximate realizations of an asymptotic technique. *Nonlinear Dyn.* **2003**, *32*, 323–343. [CrossRef]
18. Lai, S.K.; Lim, C.W.; Wu, B.S.; Wang, C.; Zeng, Q.C.; He, X.F. Newton-harmonic balancing approach for accurate solutions to nonlinear cubic-quintic Duffing oscillators. *Appl. Math. Model.* **2009**, *33*, 852–866. [CrossRef]
19. Zhou, B.; Thouverez, F.; Lenoir, D. A variable-coefficient harmonic balance method for the prediction of quasi-periodic response in nonlinear systems. *Mech. Syst. Signal Process.* **2015**, *64–65*, 233–244. [CrossRef]
20. Ying, Z.G.; Ni, Y.Q.; Fan, L. Parametrically excited stability of periodically supported beams under longitudinal harmonic excitations. *Int. J. Struct. Stab. Dyn.* **2019**, *19*, 1950095. [CrossRef]
21. Ying, Z.G.; Ni, Y.Q. A multimode perturbation method for frequency response analysis of nonlinearly vibrational beams with periodic parameters. *J. Vib. Control.* **2019**, *26*, 1260–1272. [CrossRef]

Article

Factors Affecting the Dependency of Shear Strain of LRB and SHDR: Experimental Study

Chao-Yong Shen *,†, Xiang-Yun Huang †, Yang-Yang Chen and Yu-Hong Ma

Earthquake Engineering Research & Test Center, Guangzhou University, Guangzhou 510405, China;
eertchxy@gzhu.edu.cn (X.-Y.H.); yychen@gzhu.edu.cn (Y.-Y.C.); myhzth@gzhu.edu.cn (Y.-H.M.)
* Correspondence: shency@gzhu.edu.cn
† These authors comtributed equally to this work.

Abstract: In this research we conducted a sensitivity experimental study where we explored the dependency of the shear strain on the seismic properties of bearings, namely lead rubber bearing (LRB) and super high damping rubber bearing (SHDR). The factors studied were vertical pressure, temperature, shear modulus of the inner rubber (G value), loading frequency, and loading sequence. Six specimens were adopted, i.e., three LRBs and three SHDR bearings. A series of test plans were designed. The seismic characteristics of the bearings were captured through a cyclic loading test, which included post-yield stiffness, characteristic strength, area of a single cycle of the hysteretic loop, equivalent stiffness, and equivalent damping ratio. A whole analysis of variances was then conducted. At the same time, to explore certain phenomena caused by the factors, an extended discussion was carried out. Test results showed that the temperature is the most dominant feature, whereas the G value is the least contributing factor, with the effect of the loading frequency and the loading sequence found between these two. The increment of the post-yielded stiffness for LRB from 100% to 25% is a significant reduction from a low temperature to high one. The slope of the characteristic strength versus the shear strain for LRB under high temperature is larger than the one under low temperature.

Keywords: factors; dependency of shear strain; LRB; SHDR; seismic characteristics; cyclic loading test

Citation: Shen, C.-Y.; Huang, X.-Y.; Chen, Y.-Y.; Ma, Y.-H. Factors Affecting the Dependency of Shear Strain of LRB and SHDR: Experimental Study. *Actuators* **2021**, *10*, 98. https://doi.org/10.3390/act10050098

Academic Editor: André Preumont

Received: 23 March 2021
Accepted: 27 April 2021
Published: 7 May 2021

1. Introduction

By lengthening the natural period of an isolation structure and dissipating the earthquake energy, the earthquake action of the isolation structure can be reduced considerably [1]. At present, this innovative technology is used widely in buildings and bridges, and it has successfully withstood many real earthquakes [2,3]. Elastomeric bearings are mostly applied in this technology, which includes linear natural rubber bearing (LNR), lead rubber bearing (LRB), elastic sliding bearing (ESB), and high damping rubber bearing (HDR). In China, LNR and LRB are most extensively used, followed by ESB, and lastly, HDR. There is little damping for LNR, which is a bearing constructed of alternating elastomeric layers bonded to intermediate steel plates; here, the elastomeric rubber is natural. When lead is poured into the central hole of LNR, LNR becomes LRB. There is a certain yielded strength for lead; when lateral shear displacements occur, it is easy for lead to yield, and this provides the hysteresis dissipating function. ESB is made up of two parts, namely one natural bearing and one pair of sliding friction plates; the former provides a restorative force, and the latter provides sliding dissipating damping. In LNR, the natural rubber is added into a special material, such as active carbon; after it is vulcanized under high temperature, dissipative damping can be provided after the shear deformation of bearing takes place, where LNR becomes HDR. Depending on the amount of the damping ratio [4], there can be two kinds of bearings, HDR and super high damping rubber bearing (SHDR); the former is less than 18%, and the latter is more than 18%. Because bearings not only support a great amount of vertical force but also bear a large horizontal deformation during an earthquake, the safety of bearings is most important for the whole isolation structure. Accurately grasping the mechanical properties, especially the horizontal properties under different shear strain states, is especially significant. In the past years, investigations about the horizontal mechanical properties of elastomeric bearings by testing were carried out, and some useful numerical models were also put forward.

Roeder C. et al. in 1990 [5] conducted research on the low-temperature performance of elastomeric bearings. Nakano O. et al. in 1993 [6] explored the temperature dependency of isolation bearings. Kim D.K. et al. in 1996 [7] researched the effects of the temperature and the strain rate on the seismic performance of LNR and LRB. Liu W.G. et al. in 2002 [8] researched the dependence and durability of shear properties of LRB by tests, including vertical pressure dependence, shear strain dependence, frequency dependence, temperature dependence, and the durability of aging and creeping. Yakut A. et al. in 2002 [9] evaluated LNR performance at low temperatures. Yakut A. et al. in 2002 [10] also conducted research about parameters' effect on the performance of LNR at low temperatures. Li H. et al. in 2006 [11] conducted experimental research about the properties of LRB at extremely low temperatures, such as −50 °C. Li L. et al. in 2009 [12] studied the temperature dependence of shear properties of LRB tested from −40 °C to 40 °C. Fuller K. et al. in 2010 [13] explored the effect of low-temperature crystallization on the mechanical properties of LNR and HDR. Shirazi A. in 2010 [14] researched the thermal degradation of the performance of LNR. Cardone D. et al. in 2011 [15] researched the effect of temperature on the horizontal properties of LNR and HDR. Shen C.Y. et al. in 2012 [16] conducted test research about the dependency of HDR. Shen C.Y. et al. in 2014 [17] investigated various dependencies of seismic performance of LNR and LRB, in which the hardness was ultra-low. Basit Q. in 2016 [18] looked into the performance of LRB in a full-size field experience bridge at low temperature. Wang J.Q. et al. in 2016 [19] conducted an experimental study of the vertical pressure dependency on the shear properties of LRB, and in 2016 the authors of [20] also investigated the same properties of HDR. Rohola R. et al. in 2019 [21] performed numerical research on the effect of the number of cores of LNR. Zhang R.J. et al. in 2020 [22] performed test research on temperature dependency and shear strain dependency of high-performance rubber bearings. Rohola R. et al. in 2020 [23] presented research on the analysis of static and dynamic stability of LNR. Javad S. et al. in 2020 [24] conducted a numerical analysis on LNR with a steel ring. Radkia, S. et al. in 2020 [25] investigated the effects of isolators on structures considering soil-structure interaction. Sheikhi, J. et al. in 2021 [26] performed a numerical analysis on LNR with steel and shape memory alloys.

In the above literatures, most of the researchers paid more attention to the temperature on the seismic performance of elastomeric bearings, especially low temperature, such as Roeder C. et al., Yakut A. et al., Li. H. et al., Li L. et al., and Basit Q. At the same time, there is literature on the dependency of elastomeric bearings. The studies by Liu W.G. et al., Shen C.Y. et al., and Kim D.K. et al. focused on temperature and shear dependency. Wang J.Q. gave more attention to vertical pressure and shear strain dependency, while Zhang R.J. focused on temperature and shear strain dependency. In most of the above literatures, the authors only focused on a single factor for one dependency, such as shear strain dependency; the only additional factor for some was the different shear strains. In a real earthquake, the vertical load supported on the bearing would be changed because of the tension and compression action caused by the horizontal seismic load. Because there are tremendous differences in the frequencies of the seismic waves during an actual earthquake, using a single loading frequency in a test might be not proper. The shear modulus of the inner rubber of the bearing might also be different and can be from 0.196 to 2.0 MPa. Moreover, in China there is a great difference in temperature between the north and the south. For the shear strain dependency, using only one single value for the vertical pressure, the loading frequency, shear modulus, and temperature condition cannot give a true reflection of the feature. Although Wang J.Q. and Zhang R.J. respectively considered the vertical pressure and the temperature effect on the shear strain dependency, other factors were neglected. In [27,28], although the integrated expression formula was given for the shear strain dependency of LRB and SHDR separately, the shear strain was only considered as a single factor; the former was based on Olies's company test data in Japan, and the latter was based on test data from Bridgestone's company test data in Japan. As a whole, in the above literatures, for the shear strain dependency of LRB or SHDR, only one or two factors affecting it was considered, which may be inadequate.

In this paper, the seismic performance of the shear strain dependency of LRB and SHDR is investigated comprehensively, in which various factors are considered. Three isolators of LRB and three of SHDR are made and tested. Factors including vertical pressure, loading frequency, shear modulus of inner rubber, temperature, and loading sequence, etc., are explored, especially the effect of high temperature, which we paid more attention to.

2. Experimental Setup

There were a total of six isolators to be made, i.e., three LRBs and three SHDRs. They have the same inner structure, and only the shear modulus of the inner rubber (i.e., G value; see Table 1 and Figure 1) was different. The specimens used in the test were square; the effective length of each side was 300 mm (i.e., the length of the inner steel plate), and the height was about 104.8 mm. According to the literature [29], they belonged to type I. Three different G values were selected, i.e., 0.8, 1.0, and 1.2 MPa. In the literature [30], it was said that when the shear strain was more than 250%, nonlinearity might appear. For this reason, the values of 25%, 50%, 100%, 150%, 175%, and 200% were adopted in the test. The method of the test was referred to in the literature [29], i.e., the method of three cycles was adopted, and the mechanical properties of the bearings were decided by the third cycle. To ensure integrity of the curve of the third cycle, four cycles were completed in each test. The factors of vertical pressure, frequency of loading, temperature, G value, and loading sequence were all investigated. All of the test programs are shown in Table 2. The post-yield stiffness (K_d), the characteristic strength (Q_d), the area of a single cycle of the hysteretic loop (W), the equivalent stiffness (K_h), and the equivalent damping ratio (H_{eq}) were investigated in this paper for both LRB and SHDR. K_d, Q_d, W, and K_h are marked in Figure 2, and H_{eq} was derived by Equation (1) [1]:

$$H_{eq} = \frac{W}{2\pi K_h \gamma T_r} \tag{1}$$

where W is the area of a single cycle of the hysteretic loop (see the dashed area in Figure 2), r is the shear strain, and T_r is the total thickness of the inner rubber of the bearing.

Table 1. Detailed structures of LRB or SHDR.

Effective Side Length of Square (mm)	Shear Modulus of Inner Rubber (MPa)	Diameter of Lead (mm)	Number of Layers of Inner Rubber	Thickness of Single Inner Rubber (mm)	Thickness of Single Inner Steel (mm)	Thickness of Cover Rubber (mm)	Thickness of Cover Steel (mm)
300	0.8/1.0/1.2	4 × 42.5	7	6	3.8	10	20

Table 2. Test program.

Test Case	Specimen		Values Used for Factors
	LRB	SHDR	
Different frequency of loading	300 × 300 (G0.8)	300 × 300 (G0.8)	0.05 Hz, 0.25 Hz
Different sequence of loading	300 × 300 (G0.8)	300 × 300 (G0.8)	Shear strain increasing or decreasing
Different pressure	300 × 300 (G0.8)	300 × 300 (G0.8)	6 MPa, 12 MPa
Different shear modulus of inner rubber	300 × 300 (G0.8/G1.0/G1.2)	300 × 300 (G0.8/G1.0/G1.2)	0.8 MPa, 1.0 Mpa, 1.2 MPa
Different temperature	300 × 300 (G0.8)	300 × 300 (G0.8)	16 °C, 40 °C
Shear strain			25%, 50%, 100%, 150%, 175%, 200%

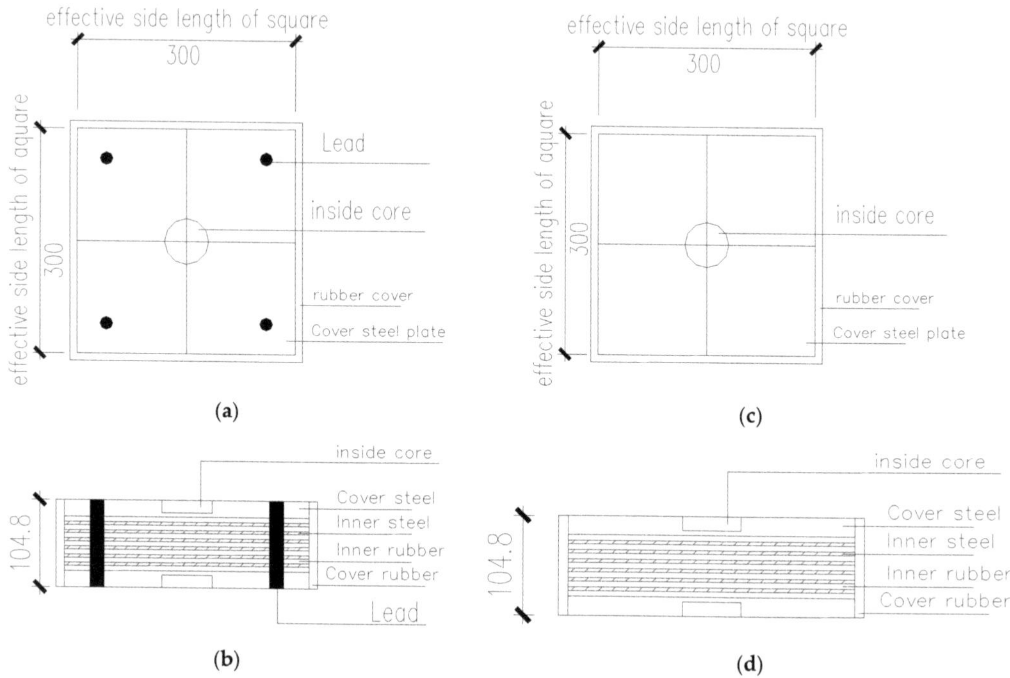

Figure 1. Details of isolator: (**a**) planar view (LRB); (**b**) top-down view (LRB); (**c**) planar view (SHDR); (**d**) top-down view (SHDR).

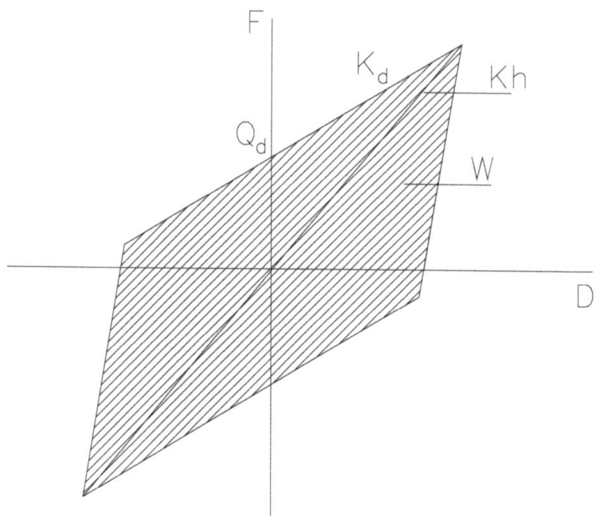

Figure 2. Meanings of K_d, Q_d, K_h and W in the hysteresis loop of bearing.

All of the tests were done in the Earthquake Engineering Research and Test Center at Guangzhou University. Figure 3 shows the test equipment and specimen used for testing.

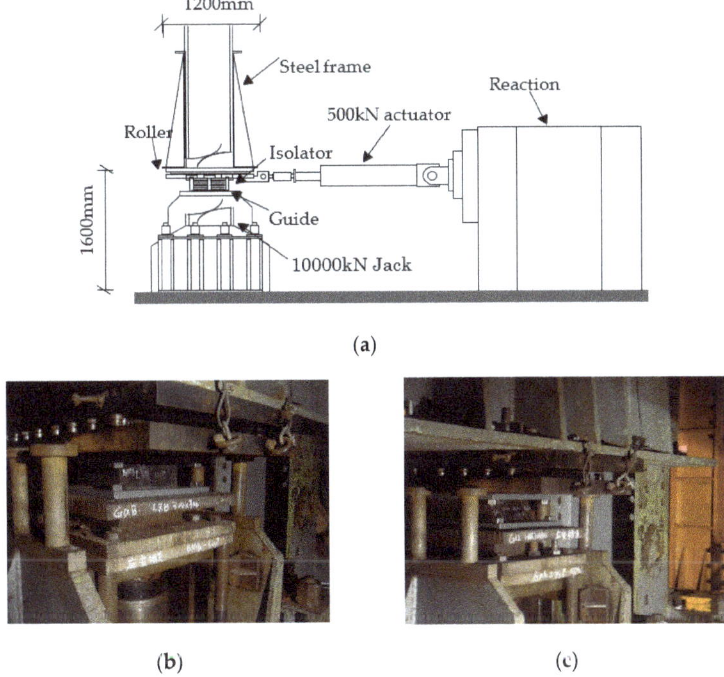

Figure 3. Test equipment and isolator used for testing. (**a**) Top-down view of test equipment; (**b**) LRB in testing; (**c**) SHDR in testing.

3. General Observation of Tests

Of the five parameters, i.e., K_d, Q_d, W, K_h, and H_{eq}, the first four were obtained directly from the third cycle of the hysteretic loop tested, while the last one, H_{eq}, was derived from W and K_h (shown in Equation (1)). The effect of various factors (such as vertical pressure, loading frequency, temperature, etc.) on these parameters is discussed in the following subsections.

3.1. Post-Yield Stiffness

To better observe the variation trend of the dependency of shear strain under different conditions, in each condition, the value of K_d at 100% was adopted as the referenced value, and all other data are divided by this value. The curves of the normalized post-yield stiffness of LRB, which varies with the shear strain under different conditions, are plotted in Figure 4a. It can be clearly observed that when the shear strain is more than 75%, the influence of each condition is very little; when the shear strain is less than 75%, an apparent impact under different conditions can be observed. The temperature may be the biggest impact factor. Especially at 25%, the normalized value at 40 °C is about 1.658, while at 16 °C it is about 2.697; the ratio of the former to the latter is 61%, showing a 39% reduction. The second impact factor may be the loading frequency. Although the trend is still decreasing, the reduction is about half of that of the temperature. The third one of impact is the loading sequence, where the trend seen is further reduction. The fourth one is the pressure. At 25%, there is about a 15% reduction from 6 to 12 Mpa, but at 50% there is a slight magnification. The G value may be the last impact factor, that is, the one with the least influence, although there is some difference seen near the 50% shear strain when the G values were 1.2 and 1.0 MPa.

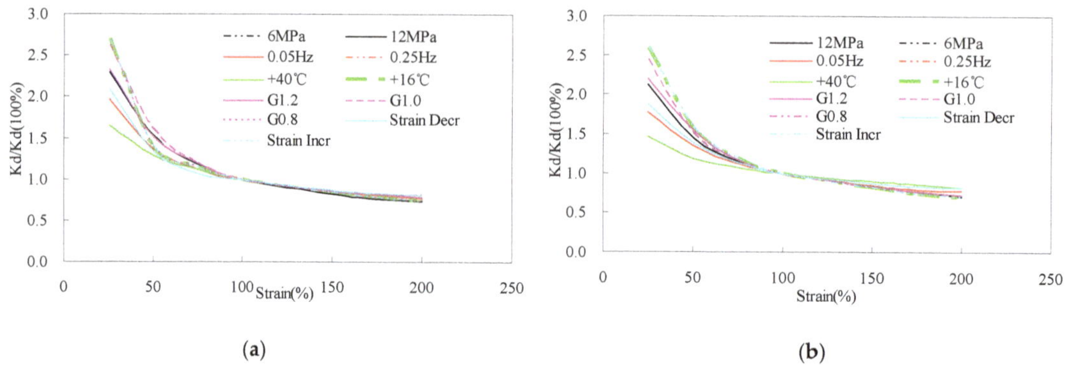

Figure 4. The mechanic properties K_d of the bearings under different conditions: (**a**) LRB; (**b**) SHDR.

The influence of the different factors on the post-yield stiffness of SHDR is also shown in Figure 4b. The same variation trend can be found in a comparison of Figure 4b with Figure 4a. The influence is mainly found for small shear strains, i.e., 25% and 50%; there is little effect on large shear strains. The temperature is also the biggest impact factor here. At a 25% shear strain, the normalized value of K_d decreased from 2.60 at 16 °C to 1.48 at 40 °C, with a reduction of about 43%, which is slightly more than the reduction of that of LRB caused by temperature. Similar to LRB, the second impact factor here is the loading frequency. From 0.25 to 0.05 Hz, the reduction of the normalized value of K_d of SHDR is slightly more than that of LRB under the same shear strain stage. The third and the fourth impact factors are the loading sequence and pressure—same as that for LRB—and it can be seen that their influence on SHDR is slightly more than that on LRB. The last impact factor is still the G value, except for the 50% shear strain; at other shear strain stages, there is a similar effect on both LRB and SHDR.

3.2. Characteristic Strength

For LRB, Figure 5a shows the influence of different factors on the shear strain dependency of the characteristic strength parameter Q_d. Other than K_d, no matter how small or large the shear strain stages are, there is a certain effect on Q_d for each factor. In all of the factors, the temperature is also the one with the largest impact. At the 25% shear strain, the normalized value of Q_d at 16 °C is about 0.695, while at 40 °C it is about 0.448, with a reduction of about 36%, which is slightly less than the reduction of K_d. At the 200% shear strain, the normalized value is 1.4 at 40 °C and 1.06 at 16 °C, with the former being about 1.32 times more than the latter, resulting in a change of about 30%. The second impact factor is still the loading frequency at most of the shear strain stages, except for the 200% shear strain. At the 25% shear strain, the normalized value at 0.05 Hz is 0.578, while at 200% it is 1.246; both changes in relation to the value at 0.25 Hz are about 17%. The third one of impact is the loading sequence. The changes at most of the shear stages are slightly less than that those under the loading frequency. The last one of impact may be the pressure or G value, as there is little difference between them; the biggest change in these two factors happens at the 10% shear strain at a G value of 6 MPa and 0.8 MPa, respectively.

For SHDR, Figure 5b shows the influence of the different factors on the shear strain dependency of Q_d. In comparison to LRB, the difference of the factors' effect on SHDR becomes less. Excluding the 200% strain, the temperature and loading sequence may be the largest impact factors. The influence of the loading frequency is small at the small shear strain stages and is large at the large shear strain stages. There is a similar small effect for the G value and pressure.

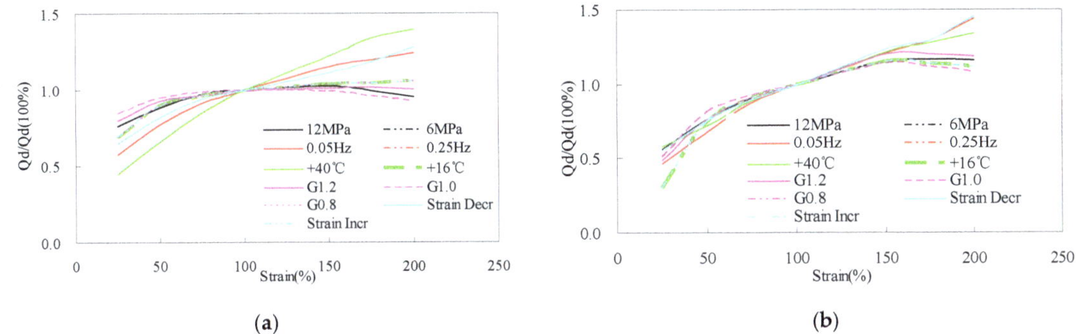

(a)   **(b)**

Figure 5. The mechanic properties Q_d of the bearings under different conditions: (**a**) LRB; (**b**) SHDR.

3.3. Area of Single Cycle of Hysteretic Loop

Figure 6a shows how the different factors affect the shear strain dependency of the area of a single cycle of LRB's hysteretic loop. It can be observed that for all of the factors, there is little effect at the small shear strain stages, while some influence can be found at the large shear strain stages, i.e., 150%, 175%, 200%. For all of the factors, the temperature may still be the biggest one of impact, such as at the 200% strain where there is about a 25% increment from 16 to 40 °C. The loading sequence and loading frequency may be the second ones of impact. In these two factors, there is little difference, as the maximum increment is about 20% at 200%. The third ones of impact may be the pressure and G value; there are few changes from 6 to 12 Mpa or from the G value of 0.8 to 1.0 MPa and 1.2 MPa.

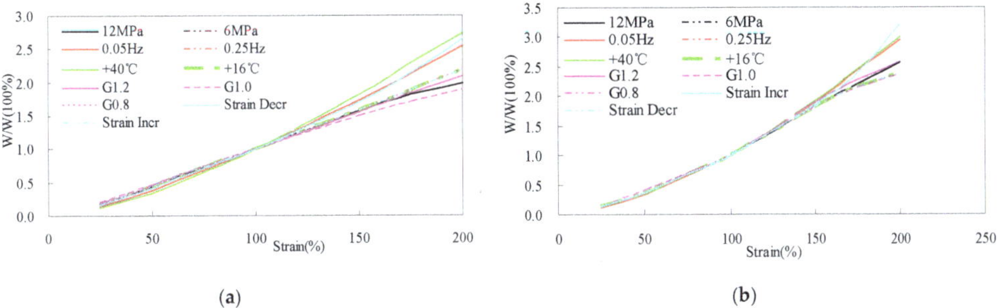

(a)   **(b)**

Figure 6. The mechanic properties W of bearings under different conditions: (**a**) LRB; (**b**) SHDR.

For SHDR, the different factors' effects on W are shown in Figure 6b. In comparison to LRB, the similarity is that the effect of all factors can be found at large strain stages; in terms of difference, the biggest factor of impact is not the temperature but the loading sequence. The second factor of impact is the temperature and the loading frequency, with the maximum change of about 25% at the 200% strain. The G value and pressure may be the third factors of impact; there is little difference for the two factors, with the maximum change being at 8% at the 200% strain.

3.4. Equivalent Stiffness

Figure 7a shows the different factors' effect on the equivalent stiffness of LRB. Some effect at the small shear strain stages can be observed; however, there is little effect at the large shear strain stages. The temperature is still the biggest impact factor, with the greatest induction being 28% at the 25% shear strain. The loading frequency and the loading sequence may be the second impact factors; the influence difference between them is very little, and the largest change caused by the two factors is about 18% at the 25% shear strain. The G value and pressure may be the third impact factors, with the largest change caused by them being about 10% in all of the shear strains.

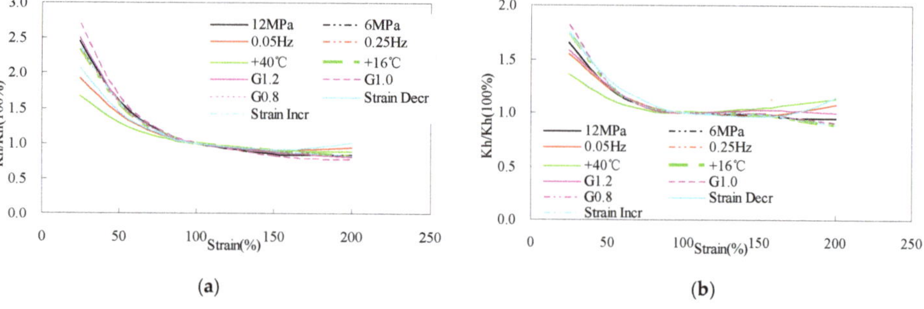

Figure 7. The mechanic properties K_h of bearings under different conditions: (**a**) LRB; (**b**) SHDR.

For SHDR, the different factors' effects on the equivalent stiffness parameter are shown in Figure 7b. Compared to Figure 7a, there is an apparent difference at large shear strain stages. For example, at the 200% strain for SHDR, there is a certain effect on the shear strain dependency, especially under different temperatures, loading frequencies, and loading sequences. However, the G value and pressure still show little influence at large shear strain stages. In the small shear strain stages, influence at the 25% strain may be the largest. In all of the factors, the temperature may be still the largest impact factor. From 16 to 40 °C, the reduction of the normalized value of K_h at the 25% shear strain is about 22%. The loading frequency and the remaining factors do not seem to cause much impact in comparison to the temperature; the biggest change caused by these factors is about half that of the temperature, i.e., about 10%.

3.5. Equivalent Damping Ratio

According to Equation (1), the equivalent damping ratio can be calculated from the area of a single cycle of the hysteretic loop and the equivalent stiffness. Figure 8a shows the influence of the different factors on the equivalent damping ratio of LRB. At the large shear strain stage, an obvious influence can be found for the temperature, such as at the 200% strain stage, where there is about an 18% change for the normalized value of H_{eq} from 16 to 40 °C. At the smaller shear strain stages, there are some changes at the 50% shear strain under the temperature's effect, but there is little effect at the 25% shear strain. However, the loading sequence plays the biggest impact role at the 25% strain, where there is an 8% change, while the change caused by other factors is about 6%. For SHDR, in Figure 8b, different phenomena can be found; only the temperature plays the single role in affecting the shear strain dependency of H_{eq}.

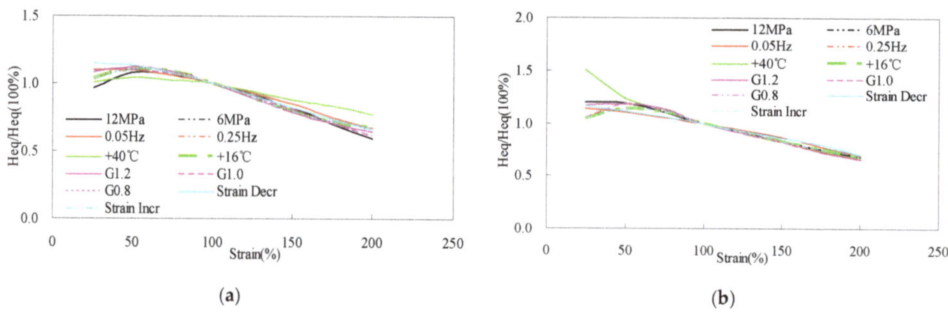

Figure 8. The mechanic properties H_{eq} of bearings under different condition: (**a**) LRB; (**b**) SHDR.

3.6. Fitted Curve

According to the observations in Sections 3.1–3.5, in most of the conditions, the temperature shows a significant influence on the dependence of shear strain, whether it be for LRB or SHDR, while other factors show a relatively lesser effect. When analyzing the isolation structure, two main parameters, namely, the post-yield stiffness and characteristic strength, need special consideration. For the data to be conveniently adopted by engineers, excluding those for the temperature factor, all of the test data for K_d and Q_d were collected. The corresponding fitted curves were plotted in Figure 9, and the fitted equations for LRB were also given in Equations (2) and (3), SHDR in Equations (4) and (5):

$$K_d(T)/K_d(100\%) = \gamma^{-0.5404} \tag{2}$$

$$Q_d(T)/Q_d(100\%) = \begin{cases} 1.054056\gamma^{0.2479} & (\gamma \leq 75\%) \\ \gamma^{0.0649} & (\gamma > 75\%) \end{cases} \tag{3}$$

$$K_d(T)/K_d(100\%) = \gamma^{-0.5124} \tag{4}$$

$$Q_d(T)/Q_d(100\%) = \begin{cases} 1.083229\gamma^{0.6064} & (\gamma \leq 75\%) \\ \gamma^{0.3285} & (\gamma > 75\%) \end{cases} \tag{5}$$

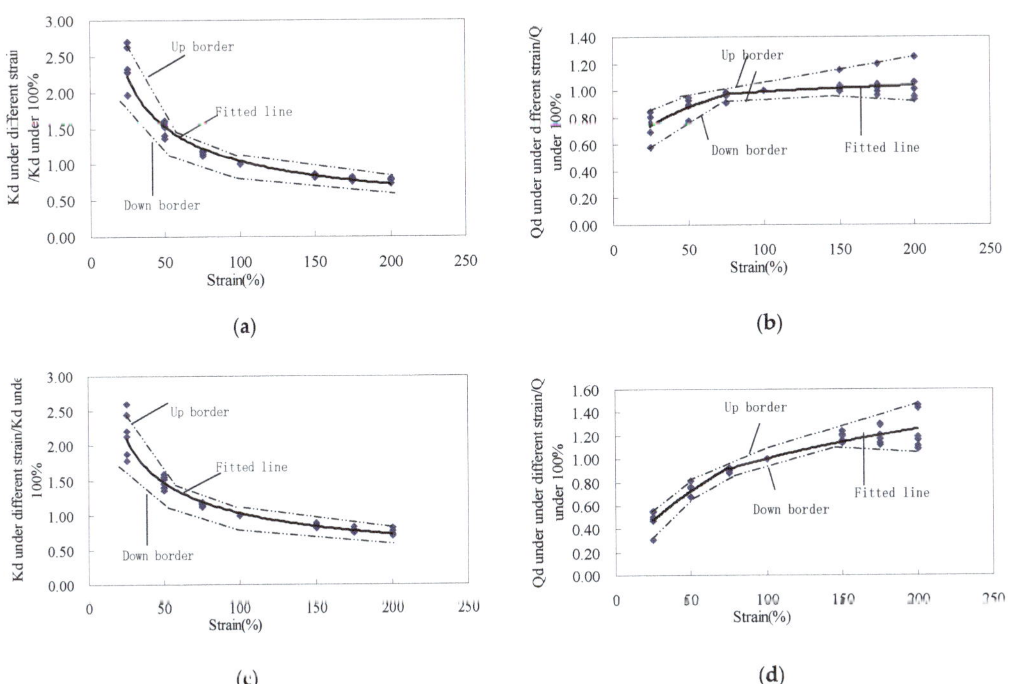

Figure 9. The test data and fitted curve of LRB and SHDR: (**a**) K_d for LRB; (**b**) Q_d for LRB; (**c**) K_d for SHDR; (**d**) Q_d for SHDR.

Figure 9a,c shows the difference between the fitted value of K_d and the test values, which are given separately for LRB and SHDR. The main difference occurs at the small shear strain stages, especially at the 25% strain, but the maximum deviation is about 13%. The difference for Q_d between the fitted value and test values is shown in Figure 9b for LRB and Figure 9d for SHDR. In contrast to K_d, the main deviation appears at the large shear strain stages for both types of bearings. The maximum deviation is about 19% and is slightly more than that of K_d.

The curves of the two shear strain dependency parameters K_d and Q_d for LRB (i.e., Equations (6) and (7)) and SHDR (i.e., Equations (8) and (9)) from [27,28] are plotted in Figure 10a–d. To compare the differences between them and the fitted curves (i.e., Equations (2)–(5)), the expression from Equations (2)–(5) is also plotted in Figure 10.

$$K_d(\gamma) = K_{d100\%} \times C_{kd}(\gamma) = K_{d100\%} \times \begin{cases} 0.779\gamma^{-0.43} & (\gamma < 0.25) \\ \gamma^{-0.25} & (0.25 \leq \gamma < 1.0) \\ \gamma^{-0.12} & (1.0 \leq \gamma) \end{cases} \tag{6}$$

$$Q_d(\gamma) = Q_{d100\%} \times C_d(\gamma) = Q_{d100\%} \times \begin{cases} 2.036\gamma^{0.41} & (\gamma < 0.1) \\ 1.106\gamma^{0.145} & (0.1 \leq \gamma < 0.5) \\ 1 & (0.5 \leq \gamma) \end{cases} \tag{7}$$

$$K_d(r) = \frac{A}{T_r}(1 - u)(1.77 - 2.404r + 1.8r^2 - 0.63r^3 + 0.0846r^4) \tag{8}$$

$$Q_d(r) = u\,Ar(1.77 - 2.404r + 1.8r^2 - 0.63r^3 + 0.0846r^4) \tag{9}$$

where $u = 0.3685 + 0.1106r - 0.08498r^2 + 0.013958r^3$.

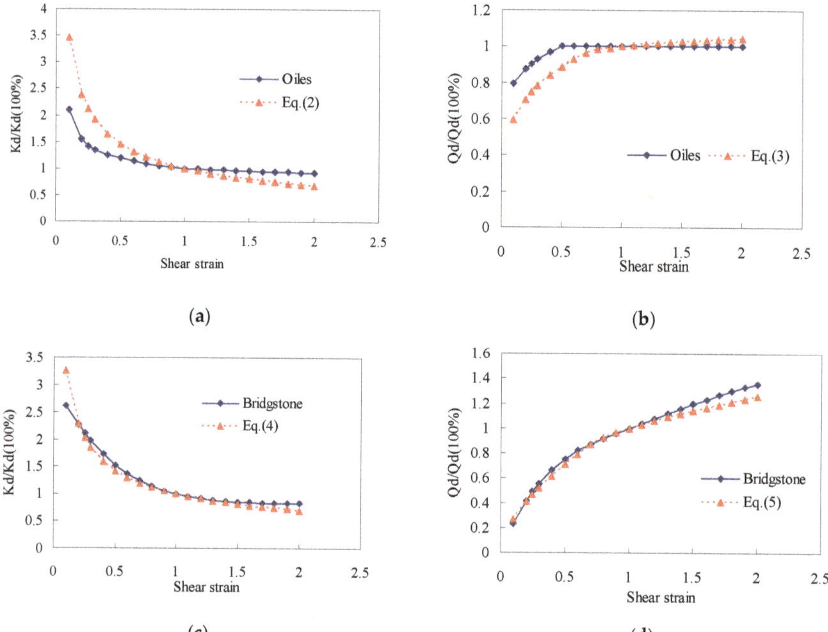

Figure 10. The difference of fitted curve of LRB and SHDR. (**a**) K_d for LRB; (**b**) Q_d for LRB; (**c**) K_d for SHDR; (**d**) Q_d for SHDR.

For LRB, Figure 10a,b shows the difference between the test values taken from Olies (i.e., Equations (6) and (7)) and Equations (2) and (3). Little difference can be found at large shear strain stages, whether for K_d or Q_d. However, at small shear strain stages, there is an obvious difference; moreover, with the shear strain decreasing, the difference increases. For example, at the 25% shear strain, the normalized value of K_d from Olies is about 1.41, while that from Equation (2) is about 1.92, showing a 26% reduction. Under the same shear strain stage, the normalized value of Q_d from Olies is about 0.90, while that of Equation (3) is about 0.75, showing an increase of around 20%.

For SHDR, Figure 10c,d shows the difference between the test values from Bridgestone (i.e., Equations (8) and (9)) and Equations (4) and (5). Compared to LRB, little difference is found between them, except that there is a little difference at a certain individual point for K_d. If this point cannot be considered, the maximum variation is about 8%. The G value of the specimens from Bridgestone is about 0.62 MPa; however, those in this paper are 0.8, 1.0, and 1.2 MPa. With this in view, we can conclude that as mentioned earlier, the G value has little effect on the dependency of the shear strain of SHDR.

4. Discussion on Special Experimental Results

4.1. Comparing Slope of Q_d for LRB under Low and High Temperatures

In Figure 5a, a significant phenomenon for LRB can be found, namely that the slope of the shear strain dependency parameter of characteristic strength Q_d is far smaller when the temperature is at 16 °C than at 40 °C. To explore the reason for this, hysteretic loops at different shear strains at 16 °C and 40 °C are plotted in Figure 11a,b. In Figure 11a, from shear strains of 25% to 200%, there is little change for Q_d, but in Figure 11b, there is an almost linear increase with the shear strain increasing. The reason may lie in the characteristic strength of lead and also the interaction between the lead and the inner rubber of the bearing. When the temperature is low, the lead has high characteristic strength, and when the temperature is high, the characteristic strength of the lead decreases much more. From 16 to 40 °C, the Q_d value at the latter temperature may be half of that at the former. Although the stiffness of the inner rubber may decrease, the change is very small; according to the literature [31], it is around 10%. From this, we can conclude that because the restraining function of the inner rubber on the lead under low temperature is far less than that under high temperature, the abovementioned phenomenon appears.

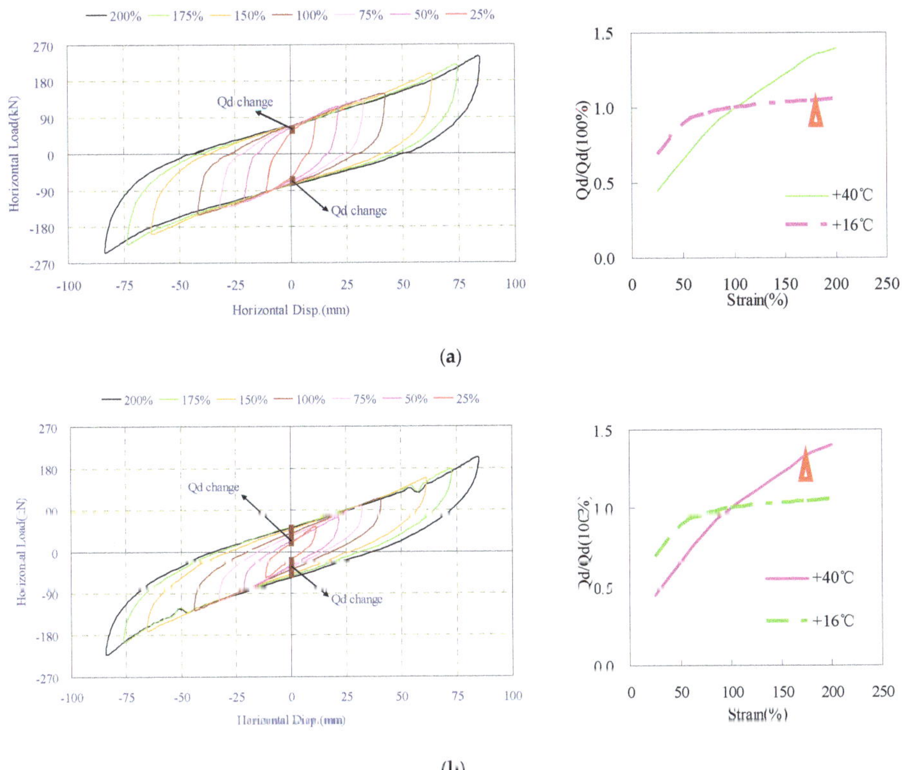

Figure 11. The different shear strain hysteresis loops of LRB at different temperature: (**a**) LRB at 16 °C; (**b**) LRB at 40 °C.

4.2. Normalized Value of K_d of LRB at Low Strain under Low Temperature Being Greather Than That under High Temperature

For LRB, in Figure 4a, one significant phenomenon can be found. At 16 °C, the normalized value of K_d at the 25% strain versus that at 100% is about 2.697, but at 40 °C, the value is about 1.658, meaning that the difference is very large. To explore the reason, the hysteretic loops of the horizontal force–displacement at the two different temperatures are plotted in Figure 12a,b.

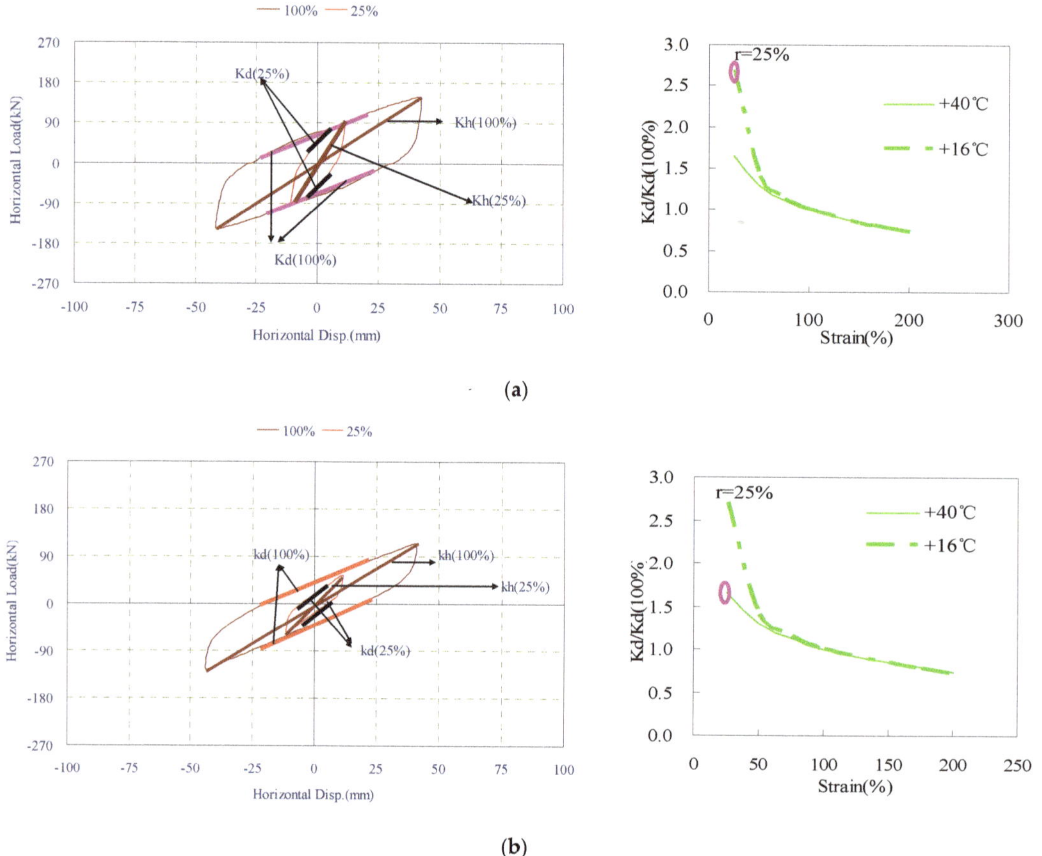

(a)

(b)

Figure 12. The 25% and 100% shear strain hysteresis loops of LRB at different temperatures: (**a**) LRB at 16 °C; (**b**) LRB at 40 °C.

According to the literature [27,32], for LRB, K_d is made up of two parts: one is the inner rubber, and the other is lead; the equation is in the following [27,32]:

$$K_d(\gamma) = C_{kd}(\gamma) \times (K_{r100\%} + K_{p100\%}) = C_{kd}(\gamma) \times K_{r100\%} + C_{kd}(\gamma) \times K_{p100\%} \tag{10}$$

where $K_{r100\%}$ is the stiffness provided by the inner rubber at $\gamma = 100\%$ and $K_{p100\%}$ is the stiffness provided by lead at $\gamma = 100\%$. $K_{r100\%}$ is calculated as:

$$K_{r100\%} = G\frac{A_r}{T_r}$$

where G is the shear modulus of the inner rubber. $K_{p100\%}$ is calculated as

$$K_{p100\%} = \alpha \frac{A_p}{T_r}$$

where α is the shear modulus of lead.

According to the literature [31], for LNR, there is little effect for the shear strain on the horizontal stiffness (i.e., K_r); from shear strains of 100% to 25%, only a 5% increase was found, and so Equation (10) can be slightly transformed to the following equation:

$$K_d(\gamma) = K_{r100\%} + C_p(\gamma) \times K_{p100\%} \tag{11}$$

where $C_p(\gamma)$ can be considered as the amplification factor of stiffness strengthening provided by the lead, and it can be determined by the function of lead restraint on the inner rubber in the bearings.

In Figure 12a,b, an interesting phenomenon can be observed. The shape of the loop curve at 16 °C resembles a spear, whereas at 40 °C it resembles a rectangle. The former is likely as a curve of a steel-reinforced concrete element yielded incompletely, while the latter is a curve of an element yielded completely. At the same time, at a temperature of 16 °C and at a 25% shear strain, the value of Q_d is about 48 kN; at 100% shear strain, the value is about 69.5 kN, meaning that the former is around 70% of the latter. At a temperature of 40 °C and at a 25% shear strain, Q_d is about 17.1 kN; at 100% shear strain, the value is about 38.2 kN, meaning that the former is around 45% of the latter. The change under high temperature is more than that under low temperature. At 16 °C and at a 25% shear strain, although the value of Q_d is slightly less than that at 100%, its incompletely yield state provides more restraint on the inner rubber, thus causing $C_p(\gamma)$ to be larger than that at the 100% shear strain. At 40 °C, on the one hand, the value of Q_d at 25% is far less than that at 100%, while on the other hand, the completely yielded state at 25% causes $C_p(\gamma)$ to increase in a limit. These could be the reasons that the normalized value of K_d at a low shear strain at 40 °C is less than that at 16 °C.

5. Conclusions

To investigate different factors' effect on the dependency of the shear strain of LRB and SHDR, three LRBs and three SHDR bearings were adopted. Both types of bearings have the same inner structure; only the G values were different. The shear strains used were 25%, 50%, 100%, 150%, 175%, and 200%. The factors of vertical pressure, loading frequency, G value, temperature, and loading sequence were studied. Tests on the mechanical properties, i.e., the post-yield stiffness, the characteristic strength, the area of a single cycle of the hysteretic loop, the equivalent stiffness, and the equivalent damping ratio, were conducted. Based on the test results, the following conclusions can be drawn:

1. In most of the conditions, the temperature may be the most significant factor in the shear strain dependency of LRB or SHDR.
2. The G value and pressure may be the least impactful factors, especially the G value, whose effect can be neglected, whether for LRB or SHDR.
3. For the loading frequency and loading sequence, there are certain impacts; in most of the conditions, their level of effect is found between the temperature and G value.
4. For LRB, excluding the temperature, the fitted curves of the post-yielded stiffness and the characteristic strength were given. By comparing them with those given by Olies, a little difference is found at small shear strains.
5. For SHDR, the corresponding fitted curves fit well with those given by Bridgestone.
6. For LRB, during all the shear strains, the change of the characteristic strength is very little under low temperature. However, the slope of characteristic strength versus the shear strain under high temperature is large.
7. For LRB, the post-yielded stiffness at 25% is larger than that at 100% under low temperature; however, the increase shows a significant reduction under high temperature.

The reason is caused by the high yield strength of lead at low temperature and its incompletely yielded state.

In this paper, most of results are based on the test of small size bearings. For the future, more research about large size bearings can be carried out, especially the effect of shear strain dependency of LRBs and SHDRs on seismic isolation, which needs to be more extensively investigated.

Author Contributions: Conceptualization, C.-Y.S. and X.-Y.H.; investigation, C.-Y.S. and Y.-Y.C.; writing—original draft preparation, C.-Y.S. and X.-Y.H.; writing—review and editing, Y.-H.M.; funding acquisition, X.-Y.H. and Y.-H.M. All authors have read and agreed to the published version of the manuscript.

Funding: This work was supported by the Major Program of National Nature Science Foundation of China (Grant No. 51991393), the National Key Research and Development Plan of China (Grant No. 2017YFC1500705), the National Key R&D Program of China (Grant No. 2019YFE0112500), the Natural Science Foundation of Guangdong Province (Grant No. 2016A030313544), the National Nature Science Foundation of China (Grant No. 51578168).

Institutional Review Board Statement: Not applicable.

Informed Consent Statement: Not applicable.

Data Availability Statement: The raw/processed data required to reproduce these findings cannot be shared at this time as the data also forms part of an ongoing study.

Conflicts of Interest: The authors declared no potential conflicts of interest with respect to the research, authorship, and/or publication of this article.

References

1. Zhou, F.L. *Earthquake Energy Absorbing Control on Engineering Structure*; Seismological Press: Beijing, China, 1997.
2. Li, A.Q. Investigation and consideration of seismic isolation and energy dissipation structures in Tohoku earthquake. *Eng. Mech.* **2012**, *29*, 69–77.
3. Yutaka, N.; Tetsuya, H.; Masanobu, H. Report on the effects of seismic isolation methods from the 2011 Tohoku-Pacific Earthquake. *Seism. Isol. Prot. Syst.* **2011**, *2*, 57–74.
4. China Ministry of Transport. *High Damper Rubber Bearings for Bridges on Road, JT/T 842-2012*; China Communications Press: Beijing, China, 2012.
5. Roeder, C.; Stanton, J.; Feller, T. Low-temperature performance of elastomeric bearings. *J. Cold Reg. Eng.* **1990**, *4*, 113–132. [CrossRef]
6. Nakano, O.; Nishi, H.; Shirono, T. Temperature-dependency of base-isolated bearings. In Proceedings of the Second U.S-Japan Workshop on Earthquake Protective Systems for Bridges, Technical Memorandum, Tsukuba Science City, Japan, 7–8 December 1992; Public Works Research Institute: Tsukuba Science City, Japan, 1993.
7. Kim, D.K.; Mander, J.; Chen, S. Temperature and strain rate effects on the seismic performance of elastomeric and lead-rubber bearings. In Proceedings of the Fourth World Congress on Joint Sealants and Bearing Systems for Concrete Structures, Sacramento, CA, USA, 29 September–3 October 1996.
8. Liu, W.G.; Zhuang, X.Z.; Zhou, F.L.; Feng, D.T.; Kato, Y. Dependence and durability properties of Chinese lead plug rubber bearings. *Earthq. Eng. Eng. Vib.* **2002**, *22*, 114–120.
9. Yakut, A.; Yura, J.A. Evaluation of Elastomeric Bearing Performance at Low Temperatures. *J. Struct. Eng.* **2002**, *128*, 995–1002. [CrossRef]
10. Yakut, A.; Yura, J.A. Parameters Influencing Performance of Elastomeric Bearings at Low Temperatures. *J. Struct. Eng.* **2002**, *128*, 986–994. [CrossRef]
11. Li, H.; Du, Y.F.; Di, S.K.; Tu, J.M.; Yang, W.X.; Li, Q.F. Cyclic test of laminated rubber damping bearing at low temperature and calculation of equivalent viscous damping ratio. *J. Lanzhou Univ. Technol.* **2006**, *32*, 116–119.
12. Li, L.; Ye, K.; Jiang, Y.C. Thermal effect on the mechanical behavior of Lead-Rubber Bearing. *J. Hua Zhong Univ. Sci. Technol.* **2009**, *26*, 1–3.
13. Fuller, K.; Gough, J.; Thomas, A.G. The effect of low temperature crystallization on the mechanical behavior of rubber. *J. Polym. Sci. Part B Polym. Phys.* **2010**, *42*, 2181–2190. [CrossRef]
14. Shirazi, A. Thermal degradation of the performance of elastomeric bearings for seismic isolation. Ph.D. Thesis, University of California, San Diego, CA, USA, 2010.
15. Cardone, D.; Gesualdi, G.; Nigro, D. Effects of air temperature on the cyclic behavior of Elastomeric seismic isolators. *Bull. Earthq. Eng.* **2011**, *9*, 1227–1255. [CrossRef]

16. Shen, C.Y.; Zhou, F.L.; Cui, J.; Huang, X.; Zhuang, X.; Ma, Y. Dependency test research of mechanical performance of HDR and its parametric value analysis. *Earthq. Eng. Eng. Vib.* **2012**, *32*, 95–103.
17. Shen, C.Y.; Tan, P.; Cui, J.; Ma, Y.H.; Huang, X.Y.; Wang, C. Experimental study on dependency of the horizontal mechanical property of elastomeric isolators with ultralow hardness. *Earthq. Eng. Eng. Vib.* **2014**, *34*, 204–216.
18. Basit, Q. Low temperature performance of elastomeric bearings in a full size field experience bridge. Master's Thesis, State University of New York at Buffalo, Buffalo, NY, USA, 2016.
19. Wang, J.Q.; Xin, W.; Li, Z.; Zhao, Z. Experimental study on vertical pressure dependency about shear properties of lead rubber bearing. *Earthq. Eng. Eng. Vib.* **2016**, *36*, 200–206.
20. Wang, J.Q.; Zhang, Z.Y.; Li, Z. Experimental research on the vertical pressure dependency about shear properties of High Damping Rubber Bearing. *J. Railw. Eng. Soc.* **2017**, *220*, 47–117.
21. Rohola, R.; Rovert, J.T. Numerical evaluation of steel-rubber isolator with single and multiple rubber cores. *Eng. Struct.* **2019**, *198*, 1–15.
22. Zhang, R.J.; Li, A.Q. Experimental study on temperature dependence of mechanical properties of scaled high-performance rubber bearings. *Compos. Part B* **2020**, *190*, 107932. [CrossRef]
23. Rohola, R.; Helder, D.C.; Rbecc, N. Static and dynamic stability analysis of a steel-rubber isolator with rubber cores. *Structures* **2020**, *26*, 441–455.
24. Javad, S.; Mojtaba, F. Natural Rubber Bearing Incorporated with Steel Ring Damper. *Int. J. Steel Struct.* **2020**, *20*, 23–34.
25. Radkia, S.; Rahnavard, R.; Tuwair, H.; Gandomkar, F.A.; Napolitano, R. Investigating the effects of seismic isolators on steel asymmetric structures considering soil-structure interaction. *Structures* **2020**, *27*, 1029–1040. [CrossRef]
26. Sheikhi, J.; Fathi, M.; Rahnavard, R.; Napolitano, R. Numerical analysis of natural rubber bearing equipped with steel and shape memory alloys dampers. *Structures* **2021**, *32*, 1839–1855. [CrossRef]
27. Masahiko, H.; Shin, O. *Response Control. and Seismic Isolation of Buildings*; Taylor & Francis Press: London, UK; New York, NY, USA, 2005.
28. Bridgestone Corporation Seismic Isolation Products & Engineering Development. In *Technical Report: High.-Damping Rubber Bearings for Base Isolation*; Product Code: HDR-X0.6R; Yokohama, Japan, 2010.
29. International Standardization Organization. *Elastomeric Seismic-Protection Isolators-Part. 1: Test. Methods. ISO 22762-1*; ISO: Geneva, Switzerland, 2005
30. The Society of Architecture of Japan. *Recommendation for the Design of Base Isolated Buildings*; Gihodo Shuppan Co., Ltd.: Tokyo, Japan, 2001.
31. Shen, C.Y.; Ma, Y.H.; Zhuang, X.Z.; Chen, Y.Y.; Lin, J.; Wang, S.B. Experimental Research on dependencies of mechanical property of elastomeric isolator with lower hardness and parameters analysis. *Appl. Mech. Mater.* **2012**, *166–169*, 2982–2994. [CrossRef]
32. International Standardization Organization. *Elastomeric Seismic-Protection Isolators-Part 3: Applications for Buildings—Specifications. ISO 22762-3*; ISO: Geneva, Switzerland, 2005.

Article

Road Roughness Estimation Based on the Vehicle Frequency Response Function

Qingxia Zhang [1,*], Jilin Hou [2], Zhongdong Duan [3], Łukasz Jankowski [4] and Xiaoyang Hu [5]

[1] Department of Civil Engineering, Dalian Minzu University, Dalian 116600, China
[2] Department of Civil Engineering, Dalian University of Technology, Dalian 116023, China; houjilin@dlut.edu.cn
[3] Department of Civil and Environmental Engineering, Harbin Institute of Technology, Shenzhen 518055, China; duanzd@hit.edu.cn
[4] Institute of Fundamental Technological Research, Polish Academy of Sciences, 02-106 Warsaw, Poland; ljank@ippt.pan.pl
[5] China Merchants Roadway Information Technology (Chongqing) CO. LTD, Chongqing 400067, China; zxhuxiaoyang@cmhk.com
* Correspondence: zhangqingxia@dlnu.edu.cn; Tel.: +86-41187557310

Abstract: Road roughness is an important factor in road network maintenance and ride quality. This paper proposes a road-roughness estimation method using the frequency response function (FRF) of a vehicle. First, based on the motion equation of the vehicle and the time shift property of the Fourier transform, the vehicle FRF with respect to the displacements of vehicle–road contact points, which describes the relationship between the measured response and road roughness, is deduced and simplified. The key to road roughness estimation is the vehicle FRF, which can be estimated directly using the measured response and the designed shape of the road based on the least-squares method. To eliminate the singular data in the estimated FRF, the shape function method was employed to improve the local curve of the FRF. Moreover, the road roughness can be estimated online by combining the estimated roughness in the overlapping time periods. Finally, a half-car model was used to numerically validate the proposed methods of road roughness estimation. Driving tests of a vehicle passing over a known-sized hump were designed to estimate the vehicle FRF, and the simulated vehicle accelerations were taken as the measured responses considering a 5% Gaussian white noise. Based on the directly estimated vehicle FRF and updated FRF, the road roughness estimation, which considers the influence of the sensors and quantity of measured data at different vehicle speeds, is discussed and compared. The results show that road roughness can be estimated using the proposed method with acceptable accuracy and robustness.

Keywords: structural health monitoring; road roughness; vehicle response; frequency response function; Fourier transform

Citation: Zhang, Q.; Hou, J.; Duan, Z.; Jankowski, Ł.; Hu, X. Road Roughness Estimation Based on the Vehicle Frequency Response Function. *Actuators* **2021**, *10*, 89. https://doi.org/10.3390/act10050089

Academic Editors: Zhao-Dong Xu, Siu-Siu Guo and Jinkoo Kim

Received: 25 February 2021
Accepted: 21 April 2021
Published: 26 April 2021

1. Introduction

Road surface conditions play an important role in road driving quality, comfort, and safety [1–3], and they are also essential for vehicle dynamics design and fatigue life [4–6]. Furthermore, they can provide valid data for road network maintenance [7,8] and durability applications [9]. Currently, artificial observation methods and accurate measurement technologies are commonly used for pavement condition evaluation [10,11]. The cost of the artificial observation method is low; however, its accuracy depends on observers with strong subjectivity. Automatic detection equipment is highly precise; however, it is expensive and not suitable for the frequent evaluation of ordinary roads. Therefore, the development of low-cost methods for accurate estimation of road roughness remains an important research topic. This work addresses the roughness estimation problem using measured vehicle accelerations and vehicle frequency response function in an approach that is easy to be performed and inexpensive.

For moving vehicles, the road roughness can be regarded as an external excitation; therefore, force construction methods based on a dynamic response [12–14] can usually be employed for road roughness identification. In recent years, road roughness identification methods based on vehicle responses have been widely studied because vehicle responses can be easily measured by low-cost and conventional sensors, such as displacement and acceleration sensors.

Using the measured vertical accelerations and displacements of vehicle wheels, and rotational movement of the vehicle body, Imine et al. [15] developed a method for road profile estimation based on sliding mode observers considering the full car model with known vehicle parameters. Ngwangwa et al. [16] reconstructed road surface profiles from measured vehicle accelerations through artificial neural networks (ANNs), which may eliminate the need for the characterization and calculation of systems through the utilization of supervised learning. Doumiati et al. [17] studied a real-time estimation method based on a Kalman filter using the measured dynamic responses of a vehicle. A known quarter-car model is considered, and the experimental results show the accuracy and potential of the proposed estimation process. Fauriat et al. [18] proposed a method for estimating road profiles via vehicle response using augmented Kalman filters in a stochastic framework, which offers a fast algorithm by combining information from different sensors through a simple linear quarter-car model of the vehicle with a priori knowledge of system parameters. Kang et al. [19] proposed a road-roughness estimation method based on a discrete Kalman filter with unknown input. Kim et al. [20] presented an improved Kalman filter that can simultaneously estimate the state variables and road roughness without any prior information about the vehicle suspension control system. Jiang et al. [21] proposed an inverse algorithm to construct road profiles in time using one iteration to update the wheel forces, which were then used to identify the road roughness. The proposed algorithm was evaluated for different types of road roughness profiles. Jeong et al. [22] proposed a deep learning estimation method utilizing the international roughness index (IRI) with the goal of using anonymous vehicles and their responses measured by a smartphone. The above methods were carried out in the time domain. Several other time-domain methods and algorithms, such as eigen perturbation techniques [23,24], are amenable for real-time structural inspection and can be adapted for the detection of road cracks and surface irregularities.

Compared with time-domain methods, frequency-domain methods are more efficient and less sensitive to noise. Therefore, the power spectral density (PSD) of road profiles provides a convenient way to assess and classify road roughness [25]. Liu et al. [26] proposed a construction method for road roughness in the left and right wheel paths based on the PSD and coherence function. In this method, the road roughness is divided into original and perturbed parts, and the perturbed parts of the two parallel wheels are considered to be stochastic and independent. González et al. [27] presented a method for estimating the PSD of a road profile from the PSD of the axle or body accelerations measured over the road profile considering a half-car model that requires prior knowledge of the vehicle transform function. Qin et al. [28] developed a method to estimate road roughness by measuring and calculating the PSD of unsprung mass accelerations using a two degrees-of-freedom (DOFs) quarter-car model through a transform function related to the vehicle parameters. Huseyin et al. [29] studied the estimator accuracy in road profile identification and derived a Cramer–Rao lower bound on the variance of all unbiased waviness parameter estimators. Turkay et al. [30] studied the modeling of road roughness from the power spectrum and coherence measurements of parallel tracks based on full-car models. Turkay et al. [31] utilized two methods to construct the road roughness model for the right and left tracks, of which one method is the Welch method and the other is a multi-input/multi-output subspace-based identification algorithm. Zhao et al. [11,32,33] evaluated the IRI using the dynamic responses of ordinary vehicles in the frequency domain.

The current methods of road profile estimation using the vehicle's response present different levels of complexity, precision, and computer intensity. However, most of them require the characteristics of the vehicle parameters to be known or identified in advance. This paper presents an estimation method for road profiles that uses the measured accelerations of a vehicle and is based on the vehicle frequency response function (FRF) with respect to the displacements of vehicle-road contact points in the frequency domain. The related formulations are deduced and expressed in a discrete system, which is convenient for use in practice. A half-car model, which can be relatively easily expanded to a complex full-vehicle model, is used to illustrate the proposed method. The time-shifting property of the Fourier transform is employed to build the relation between road profiles regarding the front and rear wheel contact points, such that it provides a road profile estimation with high efficiency by solving a linear equation.

This paper is structured as follows. First, the vehicle FRF is derived with regard to the vehicle-road contact points by analyzing the motion equation of the vehicle. Then, estimation methods of the vehicle FRF are discussed using the measured vehicle responses. Finally, a numerical example of the road roughness estimation is used to verify the proposed methods using a half-car model with four degrees of freedom.

2. Road Roughness Estimation

2.1. Vehicle Motion Equation

In this study, a half-car model is taken as an example to illustrate the theoretical derivation, and the theory can be easily expanded into a complex full-vehicle model. The half-car model is shown in Figure 1 with a four-DOF suspension system, which can reproduce the bouncing, pitching, and axle modes of the vehicle. The sprung body mass of vehicle m_1 has vertical body displacement $u_1(t)$ and rotation $u_2(t)$, and the body mass moment of inertia J is denoted as m_2. The two unsprung masses corresponding to the rear and front axles, that is, m_3 and m_4, respectively, have vertical axle displacements $u_3(t)$ and $u_4(t)$. In addition, the tire stiffness is modeled as a linear spring with constant values of k_3 and k_4 for the rear and front wheels, respectively, and the suspension system is also modeled as a linear spring with constant values of k_1 and k_2 for the rear and front axles in parallel with dampers c_1 and c_2, respectively. The horizontal distances from the centroid of the vehicle to the rear and front axles are denoted as e_1 and e_2, respectively.

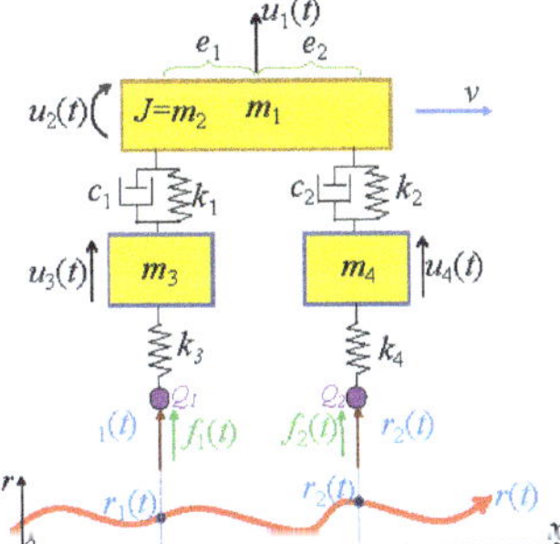

Figure 1. Half-car model of the vehicle.

As a car carries a small number of passengers, the coupling vibration between the wheels and the road is small, and thus can be ignored. Via a dynamic analysis of a vehicle moving on a road surface, the motion equation of the vehicle is shown in Equation (1), where matrices $\mathbf{M}$, $\mathbf{C}$, and $\mathbf{K}$ are the system mass, damping, and stiffness of the vehicle, respectively. Vector $\mathbf{F}$ represents the excitations applied to the vehicle, which are caused by road roughness $\mathbf{r}$, as shown in Equation (2), and is related to the stiffness $\mathbf{K}_t$ of the wheel. The component elements of the system matrices are shown in Equation (3).

$$\mathbf{M\ddot{u}}(t) + \mathbf{C\dot{u}}(t) + \mathbf{Ku}(t) = \mathbf{F}(t) \tag{1}$$

$$\mathbf{F}(t) = \mathbf{K}_t \mathbf{r}(t) \tag{2}$$

$$\left\{ \begin{array}{l} M = \begin{bmatrix} m_1 & & & \\ & m_2 & & \\ & & m_3 & \\ & & & m_4 \end{bmatrix}, \; K = \begin{bmatrix} k_1 + k_2 & k_1 e_1 - k_2 e_2 & -k_1 & -k_2 \\ k_1 e_1 - k_2 e_2 & k_1 e_1^2 + k_2 e_2^2 & -k_1 e_1 & k_2 e_2 \\ -k_1 & -k_1 e_1 & k_1 + k_3 & 0 \\ -k_2 & k_2 e_2 & 0 & k_2 + k_4 \end{bmatrix} \\[2em] C = \begin{bmatrix} c_1 + c_2 & c_1 e_1 - c_2 e_2 & -c_1 & -c_2 \\ c_1 e_1 - c_2 e_2 & c_1 e_1^2 + c_2 e_2^2 & -c_1 e_1 & c_2 e_2 \\ -c_1 & -c_1 e_1 & c_1 & 0 \\ -c_2 & c_2 e_2 & 0 & c_2 \end{bmatrix}, K_t = \begin{bmatrix} 0 & 0 \\ 0 & 0 \\ k_3 & 0 \\ 0 & k_4 \end{bmatrix}, r = \begin{bmatrix} r_1 \\ r_2 \end{bmatrix}, u = \begin{bmatrix} u_1 \\ u_2 \\ u_3 \\ u_4 \end{bmatrix} \end{array} \right. \tag{3}$$

2.2. Theoretical Reduction

By performing Fourier transform on the two sides of Equations (1) and (2) the vehicle motion equation is obtained in the frequency domain as shown in Equation (4). The expression for the vehicle frequency response $U(\omega)$ is shown in Equation (5), where matrix $H_{uu}(\omega)$ is the discrete vehicle FRF with its expression in Equation (6). Let C_0 be the observation matrix of the sensor locations on the vehicle, and Equation (7) shows the measured vehicle response $Y(\omega)$ in the frequency domain, where n represents the types of measured vehicle responses. The vehicle responses may be displacements, velocities, and accelerations, and correspondingly $n = 0, 1, 2$.

$$-\omega^2 M U(\omega) + \omega j C U(\omega) + K U(\omega) = K_t R(\omega) \tag{4}$$

$$U(\omega) = H_{uu}(\omega) K_t R(\omega), \tag{5}$$

$$H_{uu}(\omega) = \left(-\omega^2 M + \omega j C + K \right)^{-1} \tag{6}$$

$$Y(\omega) = (\omega j)^n C_0 U(\omega), \tag{7}$$

Based on the above reduction, the measured vehicle responses can be expressed by the product of the related frequency response function H_{yr} and road roughness $R(\omega)$ corresponding to the displacements of the vehicle–road contact points. Matrix H_{yr} is the FRF of the measured vehicle responses with respect to the displacements of the contact points, which can be estimated by Equation (9). Matrix $H_{ur}(\omega)$ is the FRF of the vehicle responses regarding the displacements of contact points, which can be estimated using Equation (10). Then, it can be seen that the frequency response function H_{yr} can be estimated from the FRF of vehicle $H_{uu}(\omega)$, which is related to the vehicle system parameters shown in Equation (6). With the estimated vehicle FRF H_{yr} and the measured responses $Y(\omega)$, the road roughness $R(\omega)$ can be obtained by solving the linear equation shown in Equation (8) and is expressed in Equation (11), where the matrix $H_{yr}^+(\omega)$ denotes the generalized inverse of matrix $H_{yr}(\omega)$.

$$Y(\omega) = H_{yr}(\omega) R(\omega), \tag{8}$$

$$H_{yr}(\omega) = (\omega j)^n C_0 H_{ur}(\omega), \tag{9}$$

$$H_{ur}(\omega) = H_{uu}(\omega) K_t \tag{10}$$

$$\left\{ \begin{array}{l} R(\omega) = H_{yr}^+(\omega) Y(\omega) \\ H_{yr}^+(\omega) = \left(H_{yr}^T(\omega) H_{yr}(\omega) \right)^{-1} H_{yr}^T(\omega) \end{array} \right. \tag{11}$$

2.3. Simplification of the Estimation Using Time Shift Property of Fourier Transform

Generally, vehicles run almost along a straight line; therefore, it can be assumed that the road roughness values corresponding to the front and rear wheels are almost the same with only a time difference. Thus, Equation (11) can be further simplified using the time shift property of the Fourier transform.

The distance between the front and rear wheels is $e_1 + e_2$, and if the vehicle velocity is v, then the time difference between the front wheel and the rear wheel passing through the same position is $t_0 = (e_1 + e_2)/v$. If $r_2(t)$ denotes the road roughness at the location of the front wheels at time t, then the road roughness with respect to the rear wheel $r_1(t)$ satisfies Equation (12) because the time that the rear wheel passes that position is $t + t_0$.

$$r_1(t + t_0) = r_2(t) \tag{12}$$

$R_2(\omega)$ and $R_1(\omega)$ denote the Fourier transforms of road roughness $r_2(t)$ and $r_1(t)$, respectively, and according to the time shift property of the Fourier transform, $R_2(\omega)$ and $R_1(\omega)$ satisfy the relation shown in Equation (13). Therefore, Equation (8) can be simplified to Equation (15), which is used to calculate the road roughness of rear wheel $R_1(\omega)$. Thus, only one variable must be solved. Then, the road roughness $r_1(t)$ in the time domain can be obtained by the inverse Fourier transform shown in Equation (17).

$$R_2(\omega) = e^{j\omega t_0} R_1(\omega) \tag{13}$$

$$\boldsymbol{R} = \begin{bmatrix} R_1 \\ R_2 \end{bmatrix} = \begin{bmatrix} 1 \\ e^{j\omega t_0} \end{bmatrix} R_1(\omega) \tag{14}$$

$$\boldsymbol{Y}(\omega) = \boldsymbol{H}_{yr}(\omega) \begin{bmatrix} 1 \\ e^{j\omega t_0} \end{bmatrix} R_1(\omega) = \boldsymbol{H}_{yr1}(\omega, v) R_1(\omega) \tag{15}$$

$$R_1(\omega) = \boldsymbol{H}_{yr1}^{+}(\omega, v) \boldsymbol{Y}(\omega) \tag{16}$$

$$r_1(t) = IFFT(R_1(\omega)) \tag{17}$$

3. Estimation of the Vehicle FRF with Regard to Road Roughness

Equation (16) shows that if the FRF of the measured vehicle response with respect to the displacements of contact points, that is, road roughness $\boldsymbol{H}_{yr}$, is obtained, the road roughness can be easily estimated using the measured vehicle response. However, it is worth noting that in Equation (16) a potential singularity exists in the calculation of the inverse matrix $\boldsymbol{H}_{yr1}^{+}(\omega, v)$. Thanks to the popular truncated singular value decomposition (TSVD) [34] or Tikhonov regularization method [35], the singularity can be eliminated. Furthermore, the estimation of vehicle FRF $\boldsymbol{H}_{yr}$ is investigated here via two approaches: the direct estimation and the updated estimation based on the shape function method.

3.1. Direct Estimation of the Vehicle FRF

To estimate the FRF $\boldsymbol{H}_{yr}$ of the measured vehicle response with respect to the road roughness, a driving test was designed based on Equation (8) using acceleration sensors to measure the vehicle response $\boldsymbol{Y}(\omega)$. A known hump is designed using the cosine wave expressed in Equation (18). Let a car drive over the hump with a constant velocity, as shown in Figure 2. The shape of the hump surface represents the displacement of the wheel–road contact point, which in the frequency domain is taken as the road roughness $R(\omega)$ in Equation (8).

$$r(z) = \frac{h\left(1 - \cos\left(\frac{2\pi z}{l}\right)\right)}{2}, \; 0 \le z \le l \tag{18}$$

Figure 2. Diagram of a car driving over a hump.

For the half-car model shown in Figure 1, the number of wheel–road contact points is two, and the dimension of the frequency transfer function matrix $H_{yr}(\omega)$ in Equation (8) is $q \times 2$, where q is the number of sensors located on the vehicle. The dimensions of the contact displacement $R(\omega)$ are 2×1. To accurately estimate the function $H_{yr}(\omega)$, it is advisable to perform multiple-group tests with different driving speeds to obtain effective data with different frequency bands. Although the shape of the road surface is the same in the tests, the time histories of the contact displacements $r(t)$ are different owing to different vehicle driving speeds. $r_i(t)$ and $y_i(t)$ denote the displacements of the contact points and measured vehicle responses in the ith test, respectively. Accordingly, $R_i(\omega)$ and $Y_i(\omega)$ are their Fourier transforms in the frequency domain.

Based on Equation (8), the measured responses $Y_i(\omega)$ and the corresponding contact displacements $R_i(\omega)$ from the tests are assembled and expressed in Equation (19).

The least-squares method can then be used to estimate the FRF $\overline{H}_{yr}(\omega)$ as shown in Equation (20) which represents the relation between the measured vehicle responses and the road roughness.

$$\begin{cases} \overline{Y}(\omega) = \overline{H}_{yr}(\omega)\overline{R}(\omega) \\ \overline{Y}(\omega) = [Y_1(\omega), Y_2(\omega), \cdots, Y_k(\omega)] \\ \overline{R}(\omega) = [R_1(\omega), R_2(\omega), \cdots, R_k(\omega)] \end{cases} \tag{19}$$

$$\overline{H}_{yr}(\omega) = \overline{Y}(\omega)\overline{R}^T(\omega)(\overline{R}(\omega)\overline{R}^T(\omega))^{-1} \tag{20}$$

3.2. Updating the Estimated FRF Based on the Shape Function Method

The direct estimation of the structural FRF using Equation (20) may result in errors due to noise and the frequency band range of the hump excitation. The noise mainly includes test and environmental noises. The frequency band range of the hump excitation depends on the driving velocity, and in the frequency band range corresponding to the excitation with a small amplitude, there will be a relatively large FRF estimation error, which may even cause singular data in those local frequency ranges.

The frequency response is complex and consists of real and imaginary parts. For the FRF of a car, the real and imaginary parts are generally continuous and smooth curves considering vehicle damping. In this way, the shape function method [36] is employed here to fit the real and imaginary parts of the estimated FRF in its local frequency band range with a data singularity to reduce the estimation error. In the shape function method, a continuous curve is compared to the bending deformation of a beam, and the curve is approximated by interpolation. In Figure 3 it can be seen that the curve is divided into several segments. Each segment is considered as a beam element, and the node of each segment is equivalent to the endpoint of the beam element. Using the property of the shape function in the finite element, the value at any point of the curve (the deformation of the beam) can be expressed by the rotation angle and displacement of the node of the segment.

As shown in Figure 3a, the curve is divided into n segments (elements) with a total of $n + 1$ nodes, and the number of shape functions is $2n + 2$. The frequency coordinate corresponding to the ith node is denoted by ω_i as shown in Figure 3b, and $N_{2i-1}(\omega)$ represents the shape function corresponding to its unit vertical deformation, the expression of which is shown in Equation (21). The shape function corresponding to its unit rotational angle is defined as $N_{2i}(\omega)$, the expression of which is shown in Equation (22).

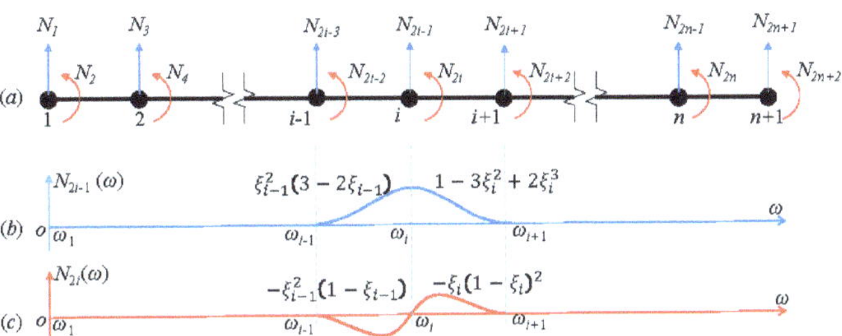

Figure 3. Basic principle of the shape function method. (**a**) Division of the curve; (**b**) The shape function corresponding to the unit vertical deformation; (**c**) The shape function corresponding to the unit rotational angle.

$$\begin{cases} N_{2i-1}(\omega) = 0 \; \omega < \omega_{i-1} \text{ or } \omega > \omega_{i+1} \\ N_{2i-1}(\xi_{i-1}) = \xi_{i-1}^2(3 - 2\xi_{i-1}); \; \xi_{i-1} = \frac{\omega - \omega_{i-1}}{\omega_i - \omega_{i-1}}, \omega_{i-1} < \omega < \omega_i \\ N_{2i-1}(\xi_i) = 1 - 3\xi_i^2 + 2\xi_i^3; \; \xi_i = \frac{\omega - \omega_i}{\omega_{i+1} - \omega_i}, \omega_i < \omega < \omega_{i+1} \end{cases} \tag{21}$$

$$\begin{cases} N_{2i}(\omega) = 0; \; \omega < \omega_{i-1} \text{ or } \omega > \omega_{i+1} \\ N_{2i}(\xi_{i-1}) = -\xi_{i-1}^2(1 - \xi_{i-1}); \; \xi_{i-1} = \frac{\omega - \omega_{i-1}}{\omega_i - \omega_{i-1}}, \omega_{i-1} < \omega < \omega_i \\ N_{2i-1}(\xi_i) = -\xi_i(1 - \xi_i)^2; \; \xi_i = \frac{\omega - \omega_i}{\omega_{i+1} - \omega_i}, \omega_i < \omega < \omega_{i+1} \end{cases} \tag{22}$$

Taking the shape functions $N_j(\omega)$ ($j = 1,2, \ldots ,2n + 2$) as a set of bases, $h(\omega)$ is assumed to be the real or imaginary part of the frequency response that needs to be updated, and the curve of $h(\omega)$ can be approximately expressed by these bases and is shown in Equation (23):

$$h(\omega) = \sum_{j=1}^{2n+2} N_j(\omega)\alpha_j, \tag{23}$$

where α_j is the coefficient of the jth shape function. All the directly estimated FRF data $h(\omega_a)$ ($a = 1,2, \ldots$) are assembled into a column vector h that corresponds to the frequency coordinates ω_a ($a = 1,2, \ldots$), and all the coefficients are assembled into vector $\boldsymbol{\alpha} = [\alpha_1, \alpha_2, \ldots, \alpha_{2n+2}]^{\mathrm{T}}$; then, Equation (23) can be rewritten as a system of linear equations as shown in Equation (24).

$$h = N\boldsymbol{\alpha}, \tag{24}$$

where N is a matrix that collects the shape functions $N_j(\omega_a)$ ($j = 1,2, \ldots ,2n + 2; a = 1,2, \ldots$). The coefficient $\boldsymbol{\alpha}$ can be calculated by the least-squares method, as shown in Equation (25).

$$\boldsymbol{\alpha} = \left(N^{\mathrm{T}}N\right)^{-1}N^{\mathrm{T}}h, \tag{25}$$

The singular data in the directly estimated FRF are certain to cause large errors in the calculation of the coefficient $\boldsymbol{\alpha}$. In this study, singular data are eliminated by setting the threshold value, and an iterative solution is adopted to calculate the coefficient $\boldsymbol{\alpha}$, as expressed in Equation (26):

$$\boldsymbol{\alpha}^{b+1} = \left(N^{\mathrm{T}}Q^b N\right)^{-1}N^{\mathrm{T}}Q^b h, \tag{26}$$

where $\boldsymbol{\alpha}^{b+1}$ is the calculated coefficient in the bth iteration. Herein, the initial value $\boldsymbol{\alpha}^1 = 0$. Q^b represents the weight matrix of the bth iteration, which is a diagonal matrix constituted by the weights $q^b(\omega_a)$ ($a = 1,2, \ldots$). The expression of $q^b(\omega_a)$ is given by Equation (27).

$$q^b(\omega_a) = \begin{cases} 0, & \left| h(\omega_a) - \sum_{j=1}^{2n+2} N_j(\omega_a)\alpha_j^b \right| > \beta^b \\ 1, & \left| h(\omega_a) - \sum_{j=1}^{2n+2} N_j(\omega_a)\alpha_j^b \right| \le \beta^b \end{cases}, \tag{27}$$

where β^b is the threshold for judging whether or not the data are singular. With an increase in iterations, the threshold β^b decreases. The iteration stops when the function converges. By substituting the coefficient α^{b+1} from the last iteration into Equation (24), the singular data in the frequency response curve can be eliminated. Then, using the obtained vehicle FRF, the road roughness can be estimated by Equation (16) using the measured vehicle responses.

3.3. On-Line Estimation of Road Roughness

Via the vehicle FRF calibrated in advance, the road roughness can be estimated using the measured vehicle responses. The process is performed in the frequency domain, which generally requires the responses to be measured in a certain time period to perform Fourier transform and the estimation. In this case, the following method of segmented data acquisition and calculation is proposed in this paper to achieve on-line identification of road roughness:

(1) Denote by T the time interval for on-line estimation. For the ith time period $t \in (t_i, t_i + 4T)$, $t_i = (i-1)T$, the responses measured in the time period $t \in (t_i, t_i + 4T)$ are used for road roughness estimation, defined as $\hat{r}_i(t), t \in (t_i, t_i + 4T)$.

(2) Since the initial state of the vehicle is usually unknown, estimation errors can appear in the initial time period. Considering the time coincidence between the ith and the $(i-1)$th estimation periods, the estimation accuracy can be improved by combining the two estimated roughness profiles in the overlapping time period. Therefore, the on-line road roughness estimation is performed as shown in Equation (28)

$$
\begin{cases}
r(t) = \left(\frac{t_i + 2T - t}{T}\right)\hat{r}_{i-1}(t) + \left(\frac{t - t_i - T}{T}\right)\hat{r}_i(t) & \text{for } t \in (t_i + T, t_i + 2T), \\
r(t) = \hat{r}_i(t) & \text{for } t \in (t_i + 2T, t_i + 4T),
\end{cases}
\tag{28}
$$

where $\left(\frac{t_i + 2T - t}{T}\right)$ and $\left(\frac{t - t_i - T}{T}\right)$ are the weighting coefficients of the two overlapping time periods in order to make the estimated road roughness curve continuous.

(3) Let $i = i + 1$, and repeat the above steps.

4. Numerical Simulation

To numerically demonstrate the proposed methods, a half-car model with four DOFs is used. The vehicle model parameters are listed in Table 1, which are selected based on reference [11].

Table 1. Parameters values of the vehicle model.

m_1 (kg)	m_2 (kg·m^2)	m_3 (kg)	m_4 (kg)	$k_1 = k_2$ (N/m)	$k_3 = k_4$ (N/m)	$c_1 = c_2$ (N·s /m)	e_1 (m)	e_2 (m)
1000	4000	100	150	20,000	300,000	4000	1.6	1.6

4.1. Characteristic Analysis of the Vehicle FRF

The four natural frequencies are shown in Table 2 and correspond to the bouncing, pitching, front axle, and rear axle-top modes of the vehicle.

Table 2. Four natural frequencies of the half-car model (Hz).

Order	1	2	3	4
natural frequency	0.779	0.974	7.355	9.006

Vehicle accelerations, which are generally easy to obtain via sensors, were measured in this study. It was assumed that three sensors were located along the DOFs of the vehicle body u_1, rear wheel u_3, and front wheel u_4, denoted by S_1, S_2, and S_3, respectively. The vehicle vertical acceleration responses are measured, and the observation matrix C_0 is shown in Equation (29).

$$C_0 = \begin{bmatrix} 1 & 0 & 0 & 0 \\ 0 & 0 & 1 & 0 \\ 0 & 0 & 0 & 1 \end{bmatrix} \tag{29}$$

By applying a unit impulse on the rear wheel along DOF u_3, the acceleration frequency responses of the vehicle at the measurement points can be expressed as $H_{su}(\omega) = -\omega^2 C_0 H_{uu}(\omega)B_{u3}$, where $B_{u3} = \begin{bmatrix} 0 & 0 & 1 & 0 \end{bmatrix}^T$, and the amplitude of the corresponding frequency responses are shown in Figure 4. Taking the unit vertical displacement of the rear wheel contact point, that is, the road roughness, as the excitation, the observed acceleration frequency responses of the vehicle can be expressed as $H_{sr}(\omega) = -\omega^2 C_0 H_{ur}(\omega)B_{r1}$, where $B_{r1} = \begin{bmatrix} 1 & 0 \end{bmatrix}^T$. Their amplitudes are shown in Figure 5. Although the amplitudes of the two types of frequency responses shown in Figures 4 and 5 are different, they have similar change regularities. Comparatively speaking, in the calculation of responses $H_{sr}(\omega)$, the stiffness and damping of the vehicle wheel are added; therefore, their amplitude is larger than the frequency responses $H_{su}(\omega)$.

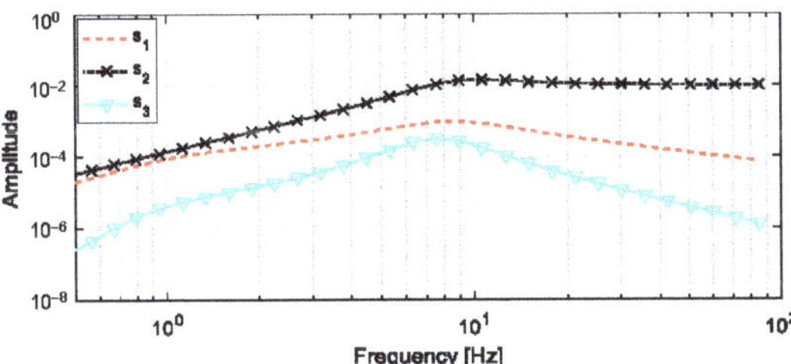

Figure 4. Amplitudes of the acceleration frequency responses of the vehicle at the measurement points to the unit impulse applied on the rear wheel along degrees of freedom (DOF) u_3.

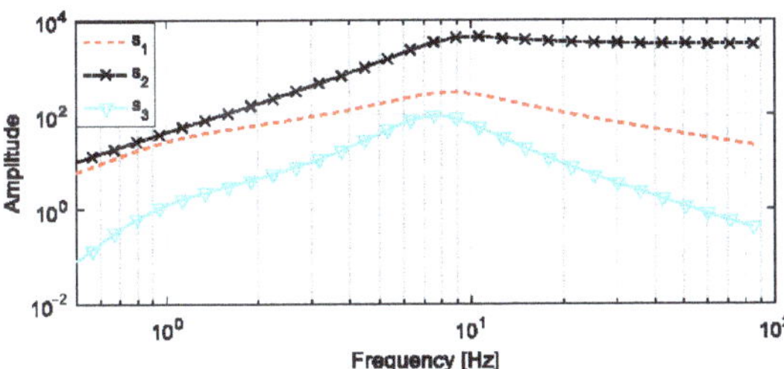

Figure 5. Amplitudes of the acceleration frequency responses of the vehicle at the measurement points to the unit vertical displacement of the rear wheel contact point.

The distance between the front and rear wheels is 3.2 m; therefore, with a driving velocity of 10 m/s, the time shift between the front and rear wheels is $t_0 = 0.32$ s. Assuming that the front and rear wheels drive along a straight line, the measured acceleration frequency responses of the vehicle with regard to the unit displacement of the rear wheel contact point $H_{yr1}(\omega, v)$ can be calculated using Equation (15). Figure 6 shows the amplitude of $H_{yr1}(\omega, v)$, and its real and imaginary parts are shown in Figure 7. Because both the front and rear wheels are excited in order, the frequency responses of the vehicle with respect to the rear wheel in Figures 6 and 7 are different from those shown in Figures 4 and 5. There exists a time shift of $t_0 = 0.32$ s between the excitations applied successively on the front and rear wheels, and there is one more item $e^{j\omega t_0}$ in the corresponding frequency responses; therefore, the frequency response curve has a periodic change rule, i.e., $1/t_0 = 3.12$ Hz, which can be seen as "S_3" in Figure 7. It can be seen that there is a phase difference of $\pi/2$ between the real and imaginary parts of the acceleration frequency responses of the vehicle.

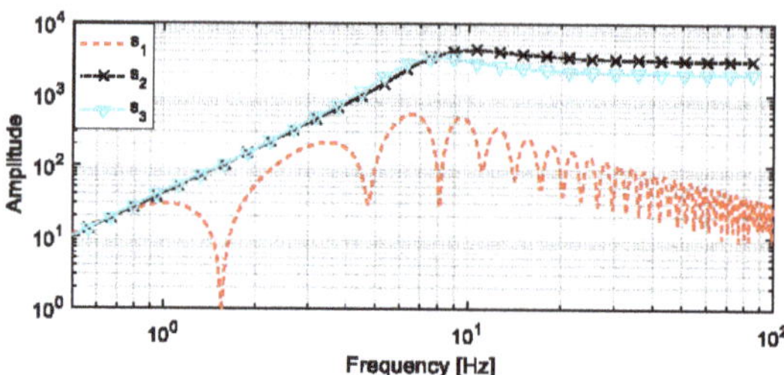

Figure 6. Amplitudes of the acceleration frequency responses of the vehicle at the measurement points to the excitation applied successively on the rear wheel along DOF u_3 and the front wheel along DOF u_4.

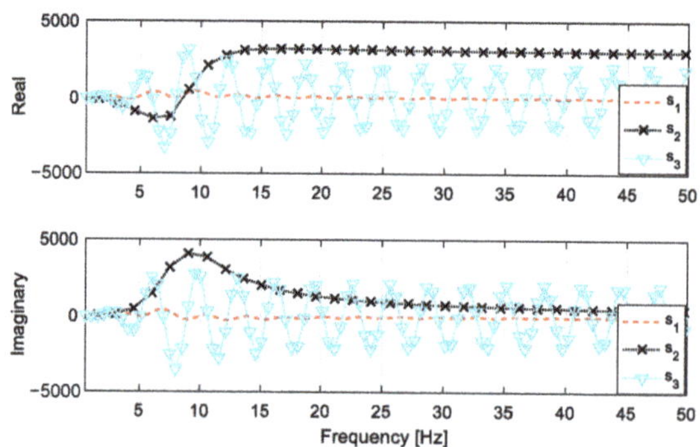

Figure 7. Real and imaginary parts of the acceleration frequency responses of the vehicle at the measurement points to the excitation applied successively on the rear wheel along DOF u_3 and the front wheel along DOF u_4.

4.2. Simulation of the Measured Vehicle Accelerations

Let the vehicle drive past a well-designed hump shown in Figure 2 with a height $h = 0.02$ m and length $l = 0.5$ m. Eight groups of driving tests were performed with velocities of [20, 17, 15, 13, 11, 7, 5, −3] m/s, where a negative value indicates that the car moves backward. In practice, the initial position of the vehicle needs to be at a certain distance from the hump in order to allow the vehicle to accelerate and then run at a constant speed, but in the numerical example, the initial position can be right in front of the hump and the influence of the vehicle wheel shape is also neglected. The measured vehicle accelerations are simulated via the equation of motion of the vehicle, i.e., Equation (1), using the Newmark-β method. Additionally, a 5% Gaussian noise is considered. Figure 8 shows the accelerations of the vehicle body measured by S_1 with velocities of [20, 15, −3] m/s, while Figure 9 shows the accelerations of the rear wheel measured by S_2 with velocities of [20, 15, −3] m/s.

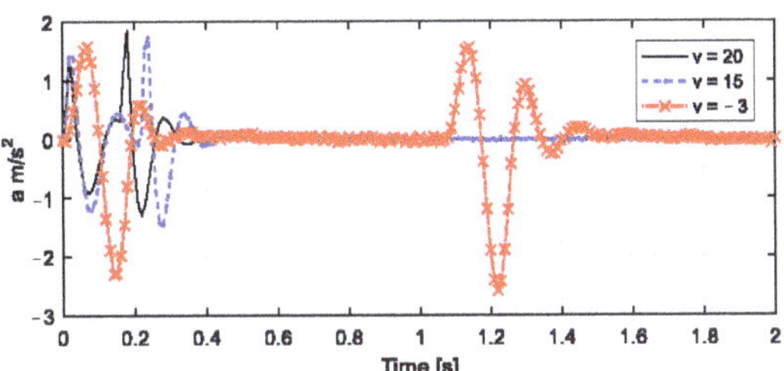

Figure 8. The accelerations of the vehicle body measured by S_1 with velocities of $[20, 15, -3]$ m/s.

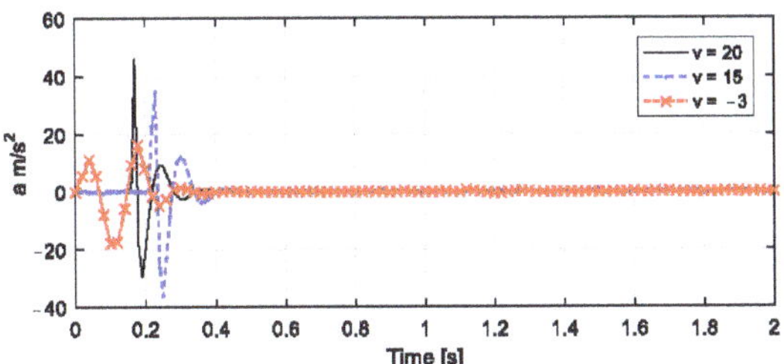

Figure 9. The accelerations of the rear wheel measured by S_2 with velocities of $[20, 15, -3]$ m/s.

4.3. Estimation of Vehicle FRF

To identify the road roughness via Equations (16) and (17), the vehicle FRF estimation is first discussed and verified using the direct estimation. Furthermore, the shape function method is used to deal with singular points.

4.3.1. Direct Estimation Using Measured Responses

Although the hump profile is the same in the different tests, the time histories of the hump profile, that is, the road roughness, differ with respect to the different driving velocities. The frequency spectra of the hump with respect to different velocities are shown in Figure 10, and the frequency spectra of the vehicle accelerations along DOF u_1 are shown in Figure 11.

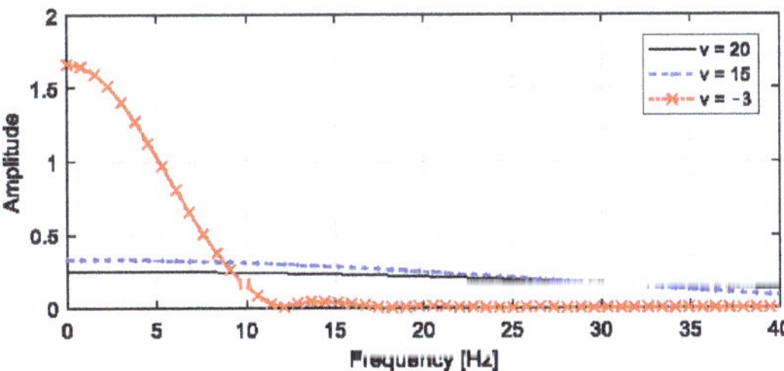

Figure 10. Frequency spectrum of the hump with velocities of $[20, 15, -3]$ m/s.

To investigate the influence of the measured data on the FRF estimation, the measured responses were combined into five cases, as shown in Table 3 which were employed to estimate the vehicle FRF. In different cases, the amplitudes of the vehicle acceleration FRF along DOF u_1 are estimated and compared in Figures 12–16 and denoted as "Direct". It can be seen that the accuracy is the highest in Case 1 which uses all the measured responses at different speeds. It also shows that the driving speed and measured data volume may influence the estimation results.

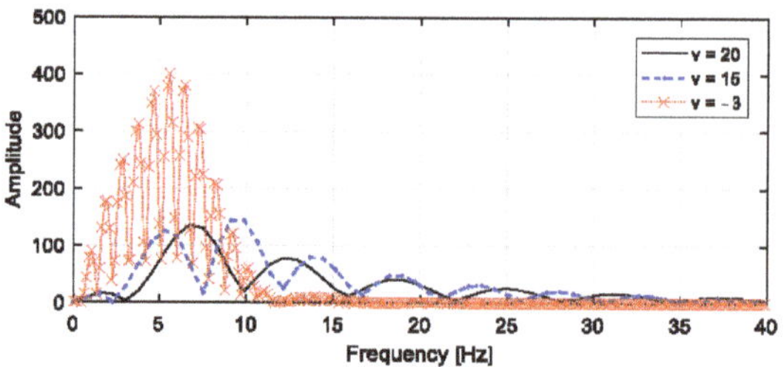

Figure 11. Frequency spectrum of the vehicle acceleration along DOF u_1 with velocities of [20, 15, −3] m/s.

Table 3. Combination of measured responses in different cases.

Case	Case 1	Case 2	Case 3	Case 4	Case 5
Sets of velocity (m/s)	20, 17, 15, 13, 11, 7, 5, −3	20, 15, 11, 5, −3	20, 15, 11, 5	20,15, 1	15, 11

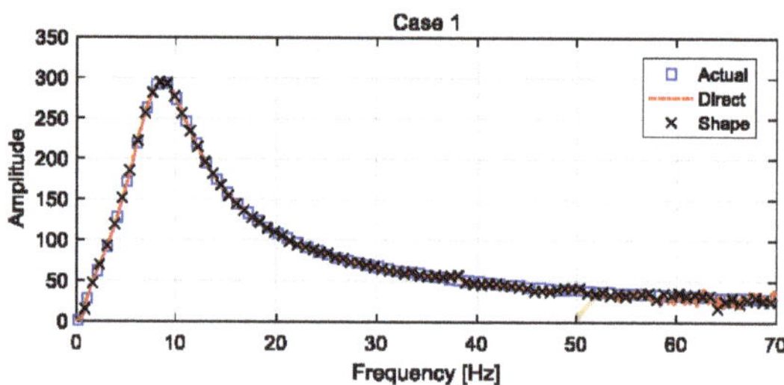

Figure 12. Comparison of the amplitudes of the vehicle acceleration frequency response function (FRF) along DOF u_1 in Case 1.

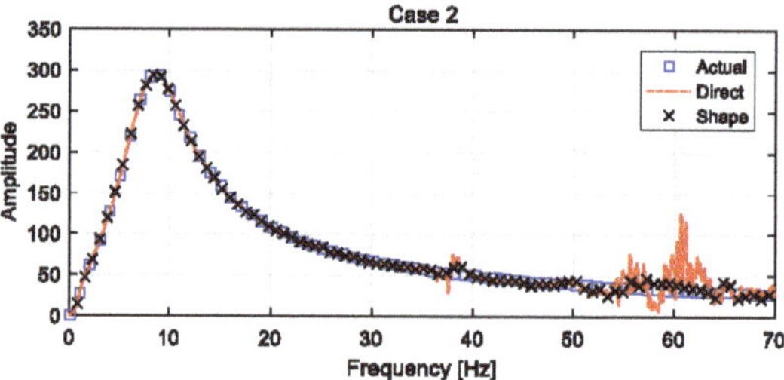

Figure 13. Comparison of the amplitudes of the vehicle acceleration FRF along DOF u_1 in Case 2.

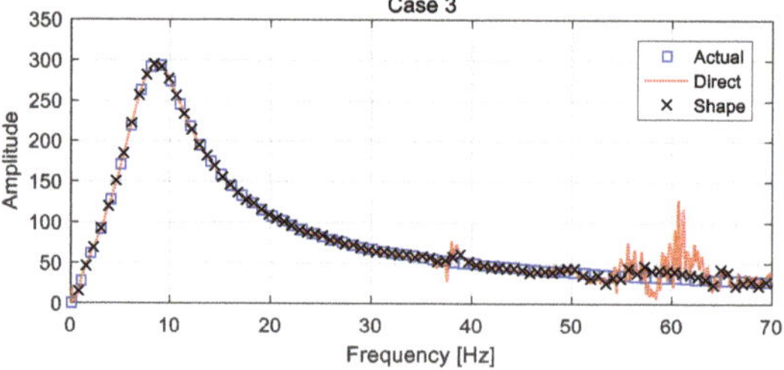

Figure 14. Comparison of the amplitudes of the vehicle acceleration FRF along DOF u_1 in Case 3.

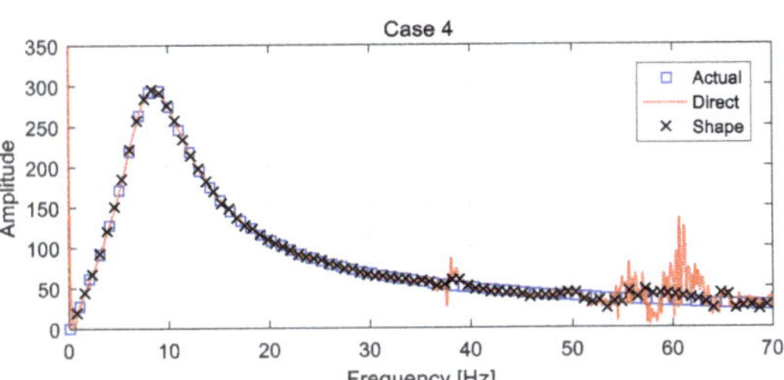

Figure 15. Comparison of the amplitudes of the vehicle acceleration FRF along DOF u_1 in Case 4.

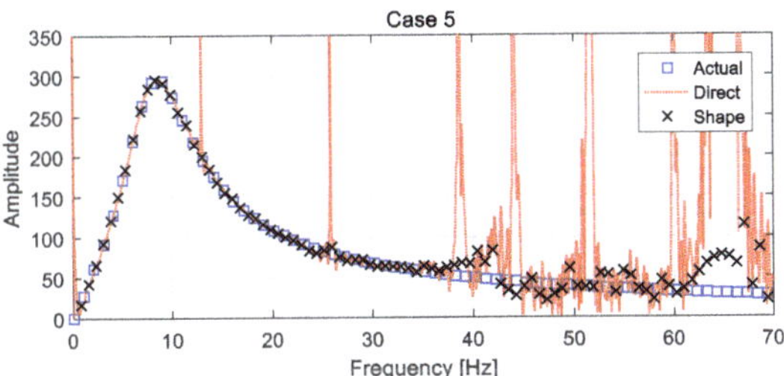

Figure 16. Comparison of the amplitudes of the vehicle acceleration FRF along DOF u_1 in Case 5.

It can be seen that the estimated FRF in Case 1 using eight groups of measured data is smoother than those estimated in the remaining cases, and singular data appear in the frequency band range greater than 50 Hz. In Case 2, which uses five groups of measured data, singular data in the FRF appeared around the frequency band range between 19 Hz and 38 Hz and greater than 50 Hz. More singular data appear in Cases 3 and 4, and there is a large quantity of singular data in Case 5. It is worth noting that the estimation error is obvious in the frequency band range larger than 40 Hz because the excitations in that range are close to zero. However, this part of the FRF has little influence on related problems. For brevity, only the related FRFs of the vehicle acceleration along DOF u_1 are shown in this paper, and the estimated FRFs of the rear and front wheels have similar accuracy.

4.3.2. Updating the Vehicle FRF

The vehicle FRF curve is divided into 80 segments, to which the imaginary and real parts are fitted locally via Equation (24), so as to reduce the FRF estimation error. The maximum value of the imaginary or real part is taken as the initial threshold value, and the threshold value is halved in the next iteration. A smooth frequency response curve can be obtained in four iterations.

For the five cases listed in Table 3, the amplitudes of the updated FRFs of the vehicle along DOF u_1 are shown in Figures 12–16 and denoted as "Shape." It can be seen that the influence of data singularities on FRF estimation can be effectively reduced by the shape function method, and even in Case 5, which has serious data singularities, the updated estimation is acceptable. As a result, the robustness of the method to noise and measured data is improved, which provides the advantage of a road roughness estimation.

4.4. Road Roughness Estimation

4.4.1. Road Roughness and Vehicle Response

Currently, the trigonometric series method, which uses a special triangular series to approximate the road surface irregularity curve, is commonly used for road roughness calculations. The expression for the road roughness $r(x)$ is as follows:

$$r(x) = \sum_{k=1}^{N_T} \alpha_k \cos(2\pi n_k x + \varphi_k) \tag{30}$$

$$\alpha_k^2 = 4G_d(n_k)\Delta n, \ \Delta n = \frac{n_u - n_l}{N_T},$$ (31)

$$G_d(n_k) = G_d(n_0)\left(\frac{n_k}{n_0}\right)^{-2}, \ n_k = n_l + k\Delta n,$$ (32)

where α_k is the coefficient of the triangular series, depending on the roughness degree of the pavement. $G_d(n_k)$ is the displacement power spectral density of the pavement calculated using the equation provided in [37]. $G_d(n_0)$ is defined as the coefficient of unevenness and depends on the degree of roughness of the pavement. n_0 is the reference special frequency ($n_0 = 0.1$ circle/m), and n_k is the special frequency. n_l and n_u are the lower and upper limits of the spatial frequency used to calculate the displacement power spectral density $G_d(n_k)$. φ_k is a uniformly distributed random phase angle in the range of [0 2π]. N_T is the number of trigonometric functions used to construct the road roughness.

In this study, the road surface grade was set to A, for which the irregularity coefficient $G_d(n_0)$ was 16×10^{-6} m^3, $n_l = 0.0221$ m^{-1}, and $n_u = 1.4142$ m^{-1}. The considered length of the road surface was 1600 m. The vehicle drove on the road with a velocity of 10 m/s, and the corresponding road roughness is shown in Figure 17.

Three acceleration sensors are located on the vehicle: S_1 measures the vertical acceleration response of the vehicle body along DOF u_1, S_2 measures the rear wheel acceleration response along DOF u_2, and S_3 measures the front wheel acceleration response along DOF u_3. The sampling frequency was 400 Hz. The time histories of the measured acceleration responses are shown in Figure 18.

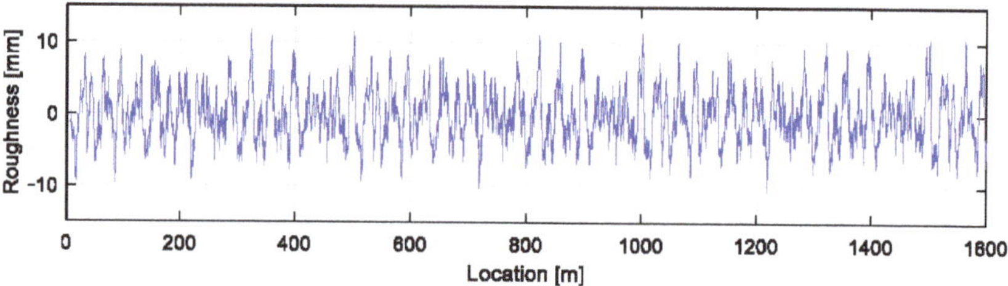

Figure 17. Time histories of the simulated road roughness.

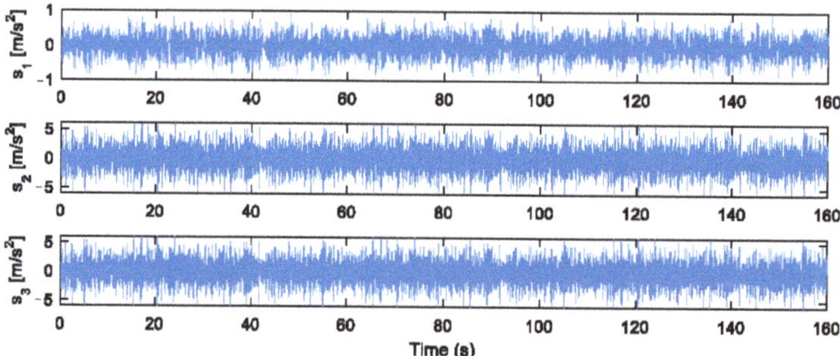

Figure 18. Time histories of the acceleration responses measured by sensors S_1–S_3.

4.4.2. Different Case Estimations

To check the influence of the measured responses, four cases are considered in the road roughness estimation and are listed as follows:

Case A: Estimation is performed using the responses measured by all the sensors, that is, S_1, S_2, and S_3.

Case B: Estimation is performed using the responses measured by sensors S_1 and S_3.

Case C: Estimation is performed using the responses measured by sensors S_1 and S_2.

Case D: Estimation is performed using the responses measured by sensor S_1.

In each case, the frequency spectrum of road roughness $R_1(\omega)$ was calculated via Equation (11) for the estimated vehicle.

First, the estimation for Case A is taken as an example to validate the proposed method. The vehicle FRF is estimated using the eight groups of measured responses in Case 1 listed in Table 3, and the estimated frequency spectra of road roughness $R_1(\omega)$ are shown in Figure 19, where "Actual" refers to the actual road roughness. "Direct" refers to the results estimated using the vehicle FRF by the direct method, while "Shape" refers to the results estimated using the vehicle FRF updated by the shape function method, and "FEM" refers to the results estimated using the vehicle FRF computed with the actual vehicle parameters. To compare the estimation accuracy, Figure 20 shows the corresponding absolute errors of the estimated road roughness, which refers to the difference between the estimated road roughness and the actual road roughness. It can be seen that the errors are quite small, which proves that the road roughness is estimated accurately using the above methods. The power spectral densities of the estimated road roughness are computed and shown in Figure 21 and are almost identical to the actual values.

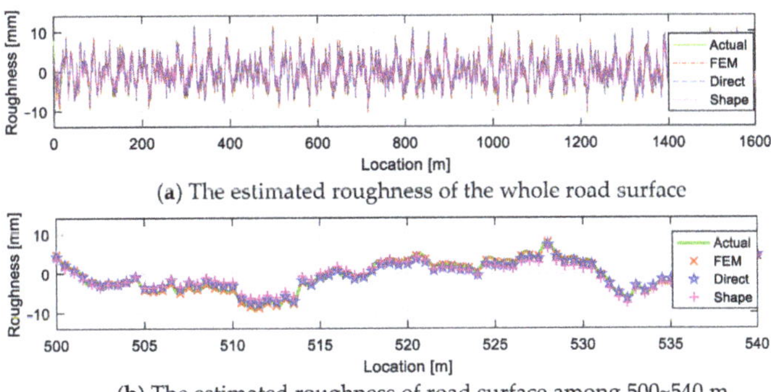

(a) The estimated roughness of the whole road surface

(b) The estimated roughness of road surface among 500~540 m

Figure 19. Comparison of the estimated road roughness using sensors S_1–S_3.

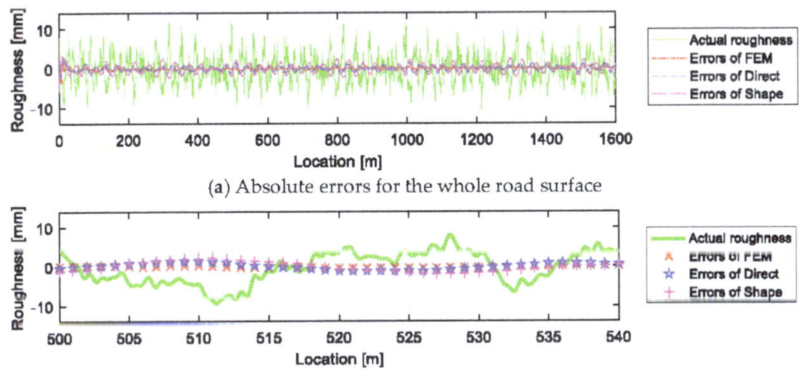

(a) Absolute errors for the whole road surface

(b) Absolute errors for road surface among 500~540 m

Figure 20. Absolute errors of the estimated road roughness for Case A using sensors S_1–S_3.

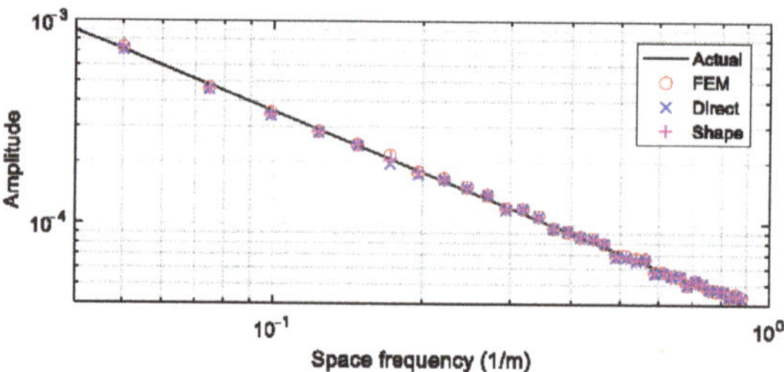

Figure 21. Estimated power spectral densities of the road roughness using sensors S_1–S_3 for Case A via the measured responses in Case 1.

In Cases A–D, the road roughness is estimated via the vehicle FRF using direct estimation and the updated method. The corresponding absolute errors are shown respectively in Figures 22 and 23. It can be seen that the estimation accuracy is good in Cases A–C, which have at least two sensors, while in Case D with only one sensor, the error is quite large.

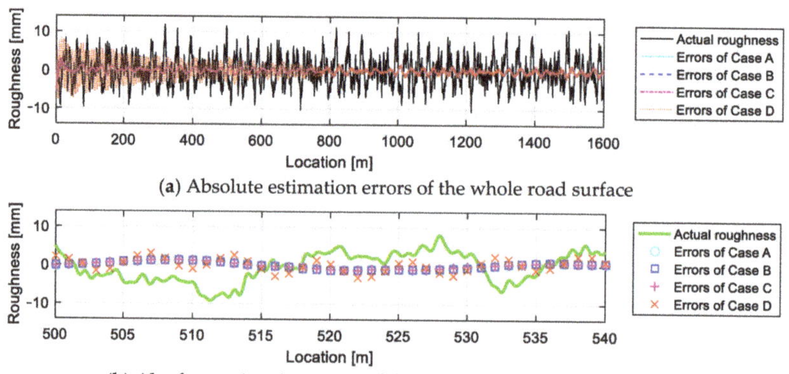

(**a**) Absolute estimation errors of the whole road surface

(**b**) Absolute estimation errors of the road surface among 500~540 m

Figure 22. Absolute estimation errors of road roughness in Case A–Case D using direct estimation.

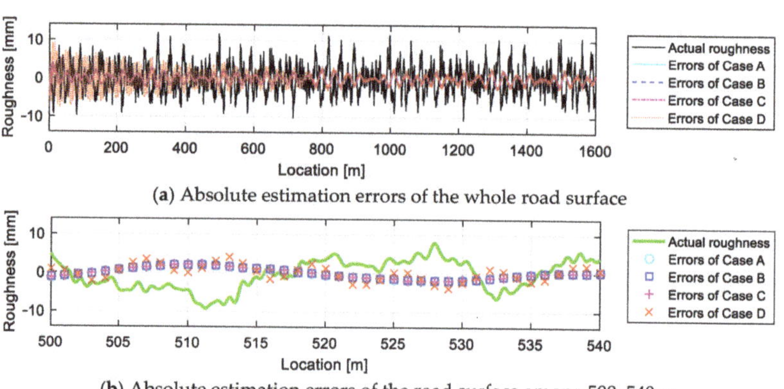

(**a**) Absolute estimation errors of the whole road surface

(**b**) Absolute estimation errors of the road surface among 500~540 m

Figure 23. Absolute errors of road roughness in Case A–Case D using the updated FRF.

4.4.3. Error Analysis

Considering that the vehicle FRF is estimated using the test of driving over the hump in Cases 1–5, which are listed in Table 3, the road roughness is estimated using the sensors shown in Cases A–D to analyze the influence of the frequency response estimation and the sensors. The estimation errors here are quantitatively calculated using Equation (33) in terms of the relative error:

$$\varepsilon = \|r_{id}(t) - r(t)\| / \|r(t)\|, \tag{33}$$

where $r_{id}(t)$ refers to the estimated road roughness, and $r(t)$ refers to the actual road roughness. The errors corresponding to the direct FRF estimation, the updated FRF, and the FRF obtained via the FEM are shown in Tables 4–6. In case D, where only one sensor S_1 is employed, the estimation errors are about 30% even with the accurate estimated vehicle FRF in Case 1. On the other hand, the estimation errors in Case 4 and Case 5 are also quite high, which means that the poor estimation of vehicle FRF definitely influences the road roughness estimation even with all the three sensors used in Case A.

Table 4. Errors of the estimated road roughness using the directly estimated FRF.

	Case A	Case B	Case C	Case D
Case 1	17.02%	16.82%	16.07%	31.10%
Case 2	24.67%	22.67%	20.17%	47.71%
Case 3	23.21%	23.16%	16.73%	38.03%
Case 4	54.33%	35.60%	53.53%	48.09%
Case 5	54.85%	42.51%	53.49%	52.85%

Table 5. Errors of the estimated road roughness using the updated FRF.

	Case A	Case B	Case C	Case D
Case 1	10.26%	10.43%	10.50%	28.76%
Case 2	11.67%	11.63%	11.57%	45.66%
Case 3	15.29%	12.68%	13.72%	33.36%
Case 4	59.89%	41.05%	57.57%	36.70%
Case 5	61.70%	46.58%	60.14%	42.95%

Table 6. Errors of the estimated road roughness using the FRF of the FEM.

	Case A	Case B	Case C	Case D
Errors	8.55%	8.75%	8.88%	16.39%

By observing the errors listed in Tables 4–6, the following conclusions can be drawn:

(1) Owing to the noise influence, the error still exists, approximately 8% in Cases A–C, even when using the FRF obtained by the vehicle FEM. The shape function method can improve the estimation accuracy of the vehicle FRF, and correspondingly, the estimation accuracy of the road roughness is increased using the updated FRF, which is approximately 10% in Case 1 and Cases A–C in Table 5.

(2) The more groups of driving tests with different speeds over a hump are performed, the more accurate the estimated frequency response is. Good results can be obtained by using the four groups of the driving tests shown in Case 3 and Cases A–C in Table 5.

(3) The location of two or three sensors in Cases A–C can provide a more accurate road roughness estimation, while the result via sensor S_1 in case D is relatively poor.

4.4.4. On-Line Estimation of Road Roughness

In this work, the road surface grade was set to A as mentioned in Section 4.4.1. The considered length of the road surface was 4000 m. The vehicle drove on the road with a velocity of 10 m/s, and the corresponding road roughness is shown in Figure 21.

Three acceleration sensors are employed with the same placements as those in Case A in Section 4.4.2. The sampling frequency was 400 Hz. The road roughness estimation is performed on-line with a time interval of 2.56 s. Assume that the vehicle runs for

400 s, and therefore it requires 154 times of on-line estimation step by step as described in Equation (28). The time histories of the estimated roughness of the whole road surface are shown in Figures 24 and 25 show the estimated results of the road surface between 2000 and 2050 m. It can be seen that the estimation accuracy is good.

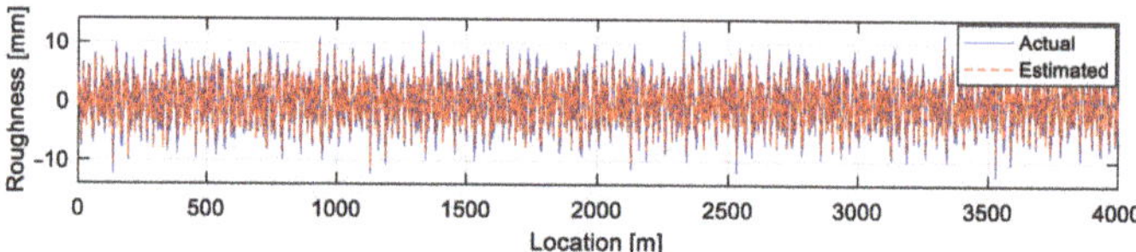

Figure 24. Time histories of the simulated road roughness.

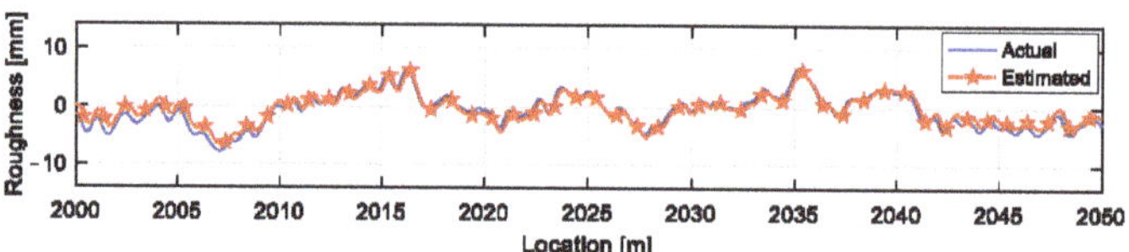

Figure 25. Time histories of the estimated roughness of the road surface among 2000~2050 m.

5. Conclusions

A road roughness estimation method is proposed in the frequency domain based on the vehicle FRF via the measured vehicle accelerations. A numerical simulation of the road roughness estimation was used to verify the effectiveness of the proposed methods. The main conclusions are as follows:

The formula for the vehicle response, road roughness, and vehicle FRF is deduced and set up in a linear equation system; hence, the road roughness can be estimated using the vehicle FRF and the measured vehicle responses.

The vehicle FRF is estimated by designing multiple groups of driving tests over a known-size hump at different driving speeds. It obviates the need for an updated finite element model of the vehicle with known vehicle parameters and vehicle modeling.

The vehicle FRF can be calculated by a direct estimation of the measured vehicle accelerations using the least-squares method. Moreover, the shape function method can be used to eliminate the singular and noisy parts of the estimated FRF and to improve the accuracy of the estimated road roughness profile. The road roughness can be estimated online with a few seconds time delay.

Author Contributions: Conceptualization, Q.Z. and J.H.; methodology, Q.Z., J.H., Z.D. and Ł.J.; software, Q.Z. and X.H.; validation, Q.Z.; writing—original draft preparation, Q.Z., J.H. and X.H.; writing—review and editing, Z.D. and Ł.J. All authors have read and agreed to the published version of the manuscript.

Funding: This research was funded by National Natural Science Foundation of China (NSFC) (51878118), Liaoning Provincial Natural Science Foundation of China (20180551205), the Fundamental Research Funds for the Central Universities (DUT19LK11), and the project 2018/31/B/ST8/03152 of the National Science Centre, Poland.

Institutional Review Board Statement: Not applicable.

Informed Consent Statement: Not applicable.

Data Availability Statement: Not applicable.

Conflicts of Interest: The authors declare no conflict of interest.

References

1. Ihs, A. The influence of road surface condition on traffic safety and ride comfort. In Proceedings of the 6th International Conference on Managing Pavements, Brisbane, Australian, 19–24 October 2004.
2. Hanafi, D.; Huq, M.S.; Suid, M.S.; Rahmat, M. A Quarter Car ARX Model Identification Based on Real Car Test Data. *J. Telecommun. Electron. Comput. Eng.* **2017**, *9*, 135–138.
3. Otremba, F.; Navarrete, J.A.R.; Guzmán, A.A.L. Modelling of a partially loaded road tanker during a braking-in-a-turn maneuver. *Actuators* **2018**, *7*, 45. [CrossRef]
4. Qinghua, M.; Huifeng, Z.; Fengjun, L. Fatigue failure fault prediction of truck rear axle housing excited by random road roughness. *Int. J. Phys. Sci.* **2011**, *6*, 1563–1568. [CrossRef]
5. Reza-Kashyzadeh, K.; Ostad-Ahmad-Ghorabi, M.J.; Arghavan, A. Investigating the effect of road roughness on automotive component. *Eng. Fail. Anal.* **2014**, *41*, 96–107. [CrossRef]
6. Shi, J.; Gao, Y.; Long, X.; Wang, Y. Optimizing rail profiles to improve metro vehicle-rail dynamic performance considering worn wheel profiles and curved tracks. *Struct. Multidiscip. Optim.* **2021**, *63*, 419–438. [CrossRef]
7. Ngwangwa, H.M. Calculation of road profiles by reversing the solution of the vertical ride dynamics forward problem. *Cogent Eng.* **2020**, *7*, 1833819. [CrossRef]
8. Yang, Y.; Cheng, Q.; Zhu, Y.; Wang, L.; Jin, R. Feasibility study of tractor-test vehicle technique for practical structural condition assessment of beam-like bridge deck. *Remote Sens.* **2020**, *12*, 114. [CrossRef]
9. Burger, M. Calculating road input data for vehicle simulation. *Multibody Syst. Dyn.* **2014**, *31*, 93–110. [CrossRef]
10. Kumar, P.; Lewis, P.; Mcelhinney, C.P.; Rahman, A.A. An algorithm for automated estimation of road roughness from mobile laser scanning data. *Photogramm. Rec.* **2015**, *30*, 30–45. [CrossRef]
11. Zhao, B.; Nagayama, T.; Xue, K. Road profile estimation, and its numerical and experimental validation, by smartphone measurement of the dynamic responses of an ordinary vehicle. *J. Sound Vib.* **2019**, *457*, 92–117. [CrossRef]
12. Zhang, Q.; Jankowski, Ł.; Duan, Z. Identification of coexistent load and damage. *Struct. Multidiscip. Optim.* **2010**, *41*, 243–253. [CrossRef]
13. Xu, Z.D.; Huang, X.H.; Xu, F.H.; Yuan, J. Parameters optimization of vibration isolation and mitigation system for precision platforms using non-dominated sorting genetic algorithm. *Mech. Syst. Signal Process.* **2019**, *128*, 191–201. [CrossRef]
14. Guo, S.S.; Shi, Q. Transient influence of correlation between excitations on system responses. *Commun. Nonlinear Sci. Numer. Simul.* **2020**, *80*. [CrossRef]
15. Imine, H.; Delanne, Y.; M'Sirdi, N.K. Road profile input estimation in vehicle dynamics simulation. *Veh. Syst. Dyn.* **2006**, *44*, 285–303. [CrossRef]
16. Ngwangwa, H.M.; Heyns, P.S.; Labuschagne, F.J.J.; Kululanga, G.K. Reconstruction of road defects and road roughness classification using vehicle responses with artificial neural networks simulation. *J. Terramech.* **2010**, *47*, 97–111. [CrossRef]
17. Doumiati, M.; Victorino, A.; Charara, A.; Lechner, D. Estimation of road profile for vehicle dynamics motion: Experimental validation. In Proceedings of the 2011 American Control Conference, San Francisco, CA, USA, 29 June–1 July 2011.
18. Fauriat, W.; Mattrand, C.; Gayton, N.; Beakou, A.; Cembrzynski, T. Estimation of road profile variability from measured vehicle responses. *Veh. Syst. Dyn.* **2016**, *54*, 585–605. [CrossRef]
19. Kang, S.W.; Kim, J.S.; Kim, G.W. Road roughness estimation based on discrete Kalman filter with unknown input. *Veh. Syst. Dyn.* **2019**, *57*, 1530–1544. [CrossRef]
20. Kim, G.W.; Kang, S.W.; Kim, J.S.; Oh, J.S. Simultaneous estimation of state and unknown road roughness input for vehicle suspension control system based on discrete Kalman filter. *Proc. Inst. Mech. Eng. Part D J. Automob. Eng.* **2020**, *234*, 1610–1622. [CrossRef]
21. Jiang, J.; Seaid, M.; Mohamed, M.S.; Li, H. Inverse algorithm for real-time road roughness estimation for autonomous vehicles. *Arch. Appl. Mech.* **2020**, *90*, 1333–1348. [CrossRef]
22. Jeong, J.H.; Jo, H.; Ditzler, G. Convolutional neural networks for pavement roughness assessment using calibration-free vehicle dynamics. *Comput. Civ. Infrastruct. Eng.* **2020**, *35*, 1209–1229. [CrossRef]
23. Bhowmik, B.; Tripura, T.; Hazra, B.; Pakrashi, V. First Order Eigen-Perturbation Techniques for Real-Time Damage Detection of Vibrating Systems: Theory and Applications. *Appl. Mech. Rev.* **2019**, *71*. [CrossRef]
24. Panda, S.; Tripura, T.; Hazra, B. First-Order Error Adapted Eigen Perturbation for Real Time Modal Identification of Vibrating Structures. *J. Vib. Acoust. Trans. ASME* **2021**, *143*. [CrossRef]
25. Andrén, P. Power spectral density approximations of longitudinal road profiles. *Int. J. Veh. Des.* **2006**, *40*, 2–14. [CrossRef]
26. Liu, X.; Wang, H.; Shan, Y.; He, T. Construction of road roughness in left and right wheel paths based on PSD and coherence function. *Mech. Syst. Signal Process.* **2015**, *60*, 668–677. [CrossRef]
27. González, A.; O'Brien, E.J.; Li, Y.Y.; Cashell, K. The use of vehicle acceleration measurements to estimate road roughness. *Veh. Syst. Dyn.* **2008**, *46*, 483–499. [CrossRef]
28. Qin, Y.; Guan, J.; Gu, L. The research of road profile estimation based on acceleration measurement. *Appl. Mech. Mater.* **2012**, *226–228*, 1614–1617. [CrossRef]
29. Akcay, H.; Turkay, S. Cramer-Rao bounds for road profile estimation. In Proceedings of the 2017 IEEE 3rd Colombian Conference on Automatic Control (CCAC), Cartagena, Colombia, 18 October 2017.

30. Türkay, S.; Akçay, H. Modelling of road roughness for full-car models: A spectral factorization approach. In Proceedings of the 20th International Conference on System Theory, Control and Computing (ICSTCC), Sinaia, Romania, 13–15 October 2016.
31. Turkay, S.; Akcay, H. Road roughness modelling by using spectral factorization methods. In Proceedings of the 2016 16th International Conference on Control, Automation and Systems (ICCAS), Gyeongju, Korea, 16–19 October 2016.
32. Zhao, B.; Nagayama, T. IRI Estimation by the Frequency Domain Analysis of Vehicle Dynamic Responses. *Procedia Eng.* **2017**, *188*, 9–16. [CrossRef]
33. Zhao, B.; Nagayama, T.; Toyoda, M.; Makihata, N.; Takahashi, M.; Ieiri, M. Vehicle model calibration in the frequency domain and its application to large-scale IRI estimation. *J. Disaster Res.* **2017**, *12*, 446–455. [CrossRef]
34. Hansen, P.C. *Rank-Deficient and Discrete Ill-Posed Problems: Numerical Aspects of Linear Inversion*; SIAM: Philadelphia, PA, USA, 1998; pp. 175–206.
35. Pan, C.D.; Yu, L.; Liu, H.L. Identification of moving vehicle forces on bridge structures via moving average Tikhonov regularization. *Smart Mater. Struct.* **2017**, *26*, 085041. [CrossRef]
36. Zhang, Q.; Dduan, Z.; Jankowski, Ł. Experimental validation of a fast dynamic load identification method based on load shape function. *J. Vib. Shock* **2011**, *9*, 98–102. [CrossRef]
37. GB/T7031-2005/ISO 8608. *Mechanical Vibration-Road Surface Profiles-Reporting of Measured Data of National Standard of the People's Republic of China*; International Organization for Standardization: Geneva, Switzerland, 1995.

Article

Investigation of Mechanical and Damping Performances of Cylindrical Viscoelastic Dampers in Wide Frequency Range

Teng Ge [1], Xing-Huai Huang [1,*], Ying-Qing Guo [2,3], Ze-Feng He [4] and Zhong-Wei Hu [1]

1 Key Laboratory of C&PC Structures of the Ministry of Education, Southeast University, Nanjing 211189, China; 230179091@seu.edu.cn (T.G.); 220181055@seu.edu.cn (Z.-W.H.)
2 Nanjing Dongrui Damping Control Technology Co., Ltd., Nanjing 210033, China; gyingqing@njfu.edu.cn
3 College of Mechanical and Electronic Engineering, Nanjing Forestry University, Nanjing 210037, China
4 College of Civil Engineering, Xi'an University of Architecture and Technology, Xi'an 710055, China; hzf2017@xauat.edu.cn
* Correspondence: huangxh@seu.edu.cn

Abstract: This paper aims to develop viscoelastic dampers, which can effectively suppress vibration in a wide frequency range. First, several viscoelastic materials for damping performance were selected, and different batches of cylindrical viscoelastic dampers were fabricated by overall vulcanization. Second, the dynamic mechanical properties of the cylindrical viscoelastic dampers under different amplitudes and frequencies are tested, and the hysteretic curves under different loading conditions are obtained. Finally, by calculating the dynamic mechanical properties of the cylindrical viscoelastic dampers, the energy dissipation performance of these different batches of viscoelastic dampers is compared and analyzed. The experimental results show that the cylindrical viscoelastic damper presents a full hysteretic curve in a wide frequency range, in which the maximum loss factor can reach 0.57. Besides, the equivalent stiffness, storage modulus, loss factor, and energy consumption per cycle of the viscoelastic damper raise with the frequency increasing, while the equivalent damping decreases with the increase of frequency. When the displacement increases, the energy consumption per cycle of the viscoelastic damper rises rapidly, and the equivalent stiffness, equivalent damping, storage modulus, and loss factor change slightly.

Keywords: cylindrical viscoelastic dampers; wide frequency range; dynamic mechanical performance tests; energy consumption capacity

Citation: Ge, T.; Huang, X.-H.; Guo, Y.-Q.; He, Z.-F.; Hu, Z.-W. Investigation of Mechanical and Damping Performances of Cylindrical Viscoelastic Dampers in Wide Frequency Range. *Actuators* **2021**, *10*, 71. https://doi.org/10.3390/act10040071

Academic Editor: Ramin Sedaghati

Received: 6 March 2021
Accepted: 2 April 2021
Published: 4 April 2021

Publisher's Note: MDPI stays neutral with regard to jurisdictional claims in published maps and institutional affiliations.

1. Introduction

Viscoelastic dampers are widely used in the fields of civil engineering, machinery, and precision instruments due to its low cost, simple structure, and satisfactory shock absorption performance [1–8]. When the structure is subjected to external excitation, the viscoelastic damper will undergo shear deformation, which will then dissipate the energy of vibration. In the past few decades, researchers have been doing experimental and theoretical research on viscoelastic dampers [9–12].

Chang et al. [13] carried out experimental research on the performance of three kinds of dampers with different sizes and found that the stiffness and energy dissipation capacity of the three kinds of dampers decreased in varying degrees with the ambient temperature increasing. Soong et al. and Tsai et al. [14–16] conducted performance tests on viscoelastic dampers at different temperatures, different displacement amplitudes, and different frequencies, respectively. The results show that temperature, frequency, and displacement amplitudes are the main factors affecting the viscoelastic damper performance. Bergman et al. [17] also found that the mechanical performance of viscoelastic dampers is greatly affected by the ambient temperature and excitation frequency. The changing trend of the influencing factors on the mechanical properties of viscoelastic dampers has been con-

cluded in above studies. There are few studies devoted to improving the energy dissipation performance of viscoelastic dampers from the perspective of materials.

Cazenove et al. [18] have studied the temperature rise effect of a plate viscoelastic damper, which shows that the temperature rise is a factor to be considered in the design of viscoelastic damper. Xu et al. [10,19,20] carried out performance tests of viscoelastic dampers with different rubber substrates under different temperatures, frequencies, and strain amplitudes, and carried out experimental research on the fatigue performance and ultimate deformation of the plate dampers. A series of novel vibration isolation devices based on viscoelastic materials have been developed [21–24], and its sufficient horizontal performance and vertical performance in experimental research has been presented. All the damper types studied above are based on plate damper, which limits the mechanical properties of viscoelastic damper to a certain extent. The cylindrical viscoelastic damper, in which viscoelastic material is wrapped by a steel plate with satisfactory performance, strong integrity, and large shear area per unit volume, is worth studying.

Additionally, plenty of experimental research on the viscoelastic damping structure have been carried out. Chang and Soong et al. [16,25] conducted shaking table tests on a 2/5 scale five-story steel structure model with viscoelastic dampers and its full-scale model, respectively, and investigated the influence of temperature and frequency on the vibration reduction effect of viscoelastic dampers. Aiken et al. [26] and others conducted shaking table tests on a nine-story steel structure model with a scale of one-fourth of the pure frame, additional steel brace, and viscoelastic damper, and compared the vibration reduction effects of several schemes, indicating that the viscoelastic damper has a better vibration reduction effect. Min et al. [27] carried out a shaking table analysis on a five-story steel structure with viscoelastic damping, and the results show that it has a good damping effect. Rao [8] introduced the application of passive damping technology using viscoelastic materials to control noise and vibration in vehicles. It can be seen above that viscoelastic dampers for high-frequency machinery and low-frequency building applications have been studied, but it has not been found that a specific viscoelastic damper can present sufficient mechanical properties at both low and high frequencies.

In this paper, it is necessary to study the energy dissipation performance of cylindrical viscoelastic dampers in a wide frequency range. Several batches of cylindrical viscoelastic dampers are made for different viscoelastic materials, and their mechanical properties are tested under different loading conditions. The polymer before vulcanization is prepared through mechanical mixing of nitrile butadiene rubber and some additives, and three batches of eight viscoelastic dampers were produced by a plate vulcanizer. The performance under different amplitudes and frequencies of these dampers is tested by a hydraulic servo machine, and the test results of these dampers are analyzed and compared.

2. Methodology and Theory of Viscoelastic Dampers Performance Test

Viscoelastic dampers are often subjected to alternating loads. For the convenience of analysis and mathematical treatment, a series of sine function combinations can be used to describe various complex alternating loads. Therefore, the dynamic mechanical performance test of the viscoelastic damper in this paper is also carried out by applying a sinusoidal excitation. From the basic theory of viscoelasticity, we know that viscoelastic materials show a hysteresis phenomenon under a cyclic load, and the strain response of materials lags behind the change of stress. It is because of this hysteresis that viscoelastic materials can dissipate energy during deformation.

Suppose that a sinusoidal alternating strain is applied to the viscoelastic material. The strain is as follows:

$$c(t) = c_0 \sin \omega t \tag{1}$$

where ω is the angular frequency and ε_0 is the maximum strain of the viscoelastic material. As the stress response of viscoelastic material is ahead of the strain by a phase angle δ, the stress response is as follows.

$$\sigma(t,\omega) = \sigma_0 \sin(\omega t + \delta) \tag{2}$$

where σ_0 is the maximum stress of the viscoelastic material. The above formula is expanded as follows.

$$\begin{aligned}\sigma(t,\omega) &= \sigma_0 \sin \omega t \cos \delta + \sigma_0 \cos \omega t \sin \delta \\ &= \varepsilon_0 [E'(\omega) \sin \omega t + E''(\omega) \cos \omega t]\end{aligned} \tag{3}$$

where $E'(\omega) = (\sigma_0/\varepsilon_0) \cos \delta$ and $E''(\omega) = (\sigma_0/\varepsilon_0) \sin \delta$ are storage modulus and loss modulus of viscoelastic material, respectively. The storage modulus E'' represents the work done by the stress in phase with the strain, which is converted into energy and stored in the sample, and the energy of this part can make its elastic deformation recover. The loss modulus represents the energy lost by the transformation into heat during deformation. The ratio of loss modulus and storage modulus is the loss factor η of viscoelastic material.

$$\eta = \frac{E''}{E'} = \tan \delta \tag{4}$$

By introducing Equation (1) into Equation (3), the following equation can be obtained.

$$\cos \omega t = \frac{1}{\varepsilon_0 E''(\omega)} [\sigma(t,\omega) - E'(\omega)\varepsilon(t)] \tag{5}$$

The expression of $\sin \omega t$ can be obtained from Equation (1), combining the above equation with $\sin^2 \omega t + \cos^2 \omega t = 1$.

$$\left(\frac{\sigma(t,\omega) - E'(\omega)\varepsilon(t)}{\varepsilon_0 E''(\omega)} \right)^2 + \left(\frac{\varepsilon(t)}{\varepsilon_0} \right)^2 = 1 \tag{6}$$

According to the above analysis, the relationship between force and displacement of viscoelastic damper conforms to Equation (6), which is also in the form of the elliptic equation. The relationship between the force F and displacement u of viscoelastic damper is given as follows [12].

$$\left[\frac{F - K_{d1}u}{\eta K_{d1} u_0} \right]^2 + \left[\frac{u}{u_0} \right]^2 = 1 \tag{7}$$

where K_{d1} is the energy storage stiffness of the damper, and u_0 is the maximum displacement of the damper. F_0 is the maximum force of the damper, F_1 is the damping force at the maximum displacement u_0, and F_2 is the damping force at the zero displacement, as shown in Figure 1.

The loading control and data acquisition system of the test system are controlled by a computer. During the test, the loading displacement and loading frequency are controlled, and the corresponding displacement and load values are collected.

To further analyze the dynamic mechanical properties of viscoelastic damper, it is necessary to calculate the storage modulus, loss modulus, loss factor, equivalent damping, equivalent stiffness, and single cycle energy consumption through a hysteretic curve. Therefore, the testing principle of viscoelastic damper and the acquisition method of corresponding mechanical parameters are briefly introduced.

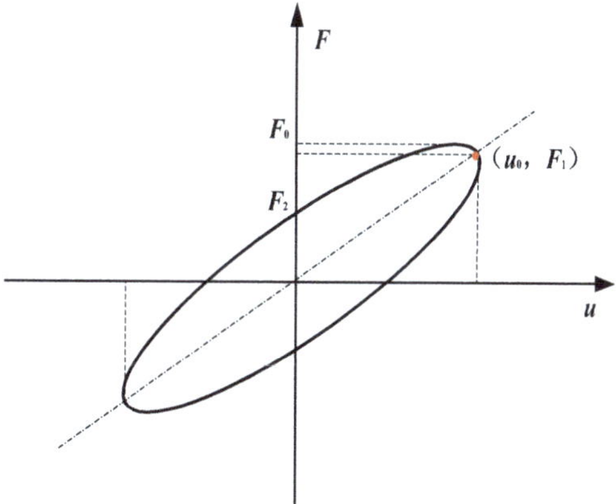

Figure 1. Force-displacement hysteresis curve.

The modulus of viscoelastic material can be expressed in the form of a complex modulus, as follows.

$$E^*(\omega) = E'(\omega) + iE''(\omega) \tag{8}$$

where is the form of viscoelastic material modulus in the frequency domain, which is also commonly used by us. Equation (8) usually represents the dynamic modulus of tension compression deformation, and the corresponding shear dynamic modulus can be expressed as.

$$G^*(\omega) = G_1(\omega) + iG_2(\omega) \tag{9}$$

where $G_1(\omega)$ and $G_2(\omega)$ are shear storage modulus and shear loss modulus, respectively.

The values of F_0, F_1, F_2, and u_0 can be obtained from the experimental hysteretic curves, and the viscoelastic layer number nn, shear area A_v, and shear thickness h_v can be obtained from the design parameters of viscoelastic damper, as shown in Table 1. The performance of viscoelastic damper is usually expressed by energy storage stiffness K_{d1}, equivalent damping C_e and single cycle energy consumption E_d, as follows [10].

$$K_{d1} = \frac{nn \cdot G_1 \cdot A_v}{h_v} \tag{10}$$

$$C_e = \frac{nnG_2A_v}{\omega h_v} \tag{11}$$

$$E_d = nn \cdot \pi \cdot G_2 \cdot A_v \cdot u_0^2/h_v \tag{12}$$

Table 1. Size of the viscoelastic dampers.

Viscoelastic Layer Number nn	Shear Area A_v	Shear Thickness h_v
1	6437 mm^2	5 mm

It can be seen from the above formula that the storage modulus G_1 and loss modulus G_2 of viscoelastic material are related to the equivalent stiffness K_{d1} and equivalent damping C_e of the viscoelastic damper, while the equivalent stiffness and equivalent damping of viscoelastic damper can be directly obtained from the test data, as follows [12].

$$K_{d1} = \frac{F_1}{u_0} \tag{13}$$

$$C_e = \frac{F_2}{\omega u_0} \tag{14}$$

Therefore, the relevant parameters of viscoelastic materials can be obtained through the above relationship from the test data as follows [19].

$$G_1 = \frac{K_{d1} \cdot h_v}{nn \cdot A_v} = \frac{F_1 \cdot h_v}{nn \cdot A_v \cdot u_0} \tag{15}$$

$$\eta = F_2/F_1 \tag{16}$$

$$G_2 = \eta G_1 \tag{17}$$

It can be seen from the above process that the mechanical parameters of viscoelastic materials and dampers for every single frequency can be calculated discretely based on the hysteretic curve obtained from the viscoelastic damper test and combined with Equations (13)–(17).

3. Production Process of Cylindrical Viscoelastic Dampers

Viscoelastic material directly determines the dynamic mechanical properties of viscoelastic damper. It is a kind of composite material that mainly includes the following four parts: matrix material, vulcanization system, reinforcement filling system, and anti-aging system.

The matrix material is a high molecular polymer with viscoelastic properties, which is the most important component in the formulation of viscoelastic materials. The type and amount of matrix play a decisive role in the properties of viscoelastic materials. As soon as the matrix material is specified, other mixing media as well as processing technology are determined. The loss factor of matrix polymer materials should be high, and the temperature range of the glass transition zone should be close to the service temperature range. Based on the study of the vibration reduction mechanism of viscoelastic materials [28], it can be known that the damping performance of matrix materials with more free network chains is higher, and the damping performance of matrix materials with a higher content of side groups of molecular chains is better. The viscoelastic material developed in this paper is high acrylonitrile butadiene rubber. In addition to its good damping performance, it also has the advantages of good processing performance, fast curing speed, and good mechanical properties. Usually, viscoelastic materials need to be vulcanized to form a cross-linking network structure between rubber molecular chains and improve their physical and mechanical properties to meet the application requirements. Therefore, the vulcanization system is a main content of viscoelastic material formulation design, which has a certain impact on the mechanical properties and damping properties of viscoelastic materials. Generally, the influence of the vulcanization system on the mechanical properties and damping properties of viscoelastic materials is contradictory, and the increase of the vulcanization degree will lead to the increase of mechanical properties, such as tensile strength and the decrease of damping properties such as loss factor. Because the base rubber in this paper is Nitrile rubber, the sulfur curing system is chosen. The reinforcement filling system is an important part of viscoelastic materials, and its effects on the properties of viscoelastic materials mainly include: (1) reinforcement, low strength of the pure polymer, and fillers, such as carbon black, which need to be added to improve the elastic modulus and tensile strength of materials and reduce the creep of materials. (2) In addition, changing the dynamic mechanical properties of materials, some fillers (such as mica) can significantly improve the internal friction energy of the materials. At the same time, the glass transition temperature of viscoelastic material can be changed by the filling system, and the temperature corresponding to the damping peak can be adjusted to match the temperature range. Carbon black and graphite powder are used in the filling and reinforcing system of viscoelastic materials developed in this paper. The protection system is mainly used to avoid or slow down the aging of viscoelastic materials in a harsh environment and improve the service life of materials. After adding the antioxidant, the viscoelastic material can maintain its performance for a long time in the process of use, and meet the use requirements.

The production process of viscoelastic materials includes three stages: plasticization, mixing, and vulcanization. The main purpose of plasticizing is to change the rheological properties of rubber, increase the plasticity, reduce the viscosity, and make it match with various additives in the mixing stage better. Mixing is the process of mixing raw rubber or plastic rubber and a compounding agent in the mixing machine until they are evenly dispersed. In the mixing process, the rubber and the compounding agent are mixed by the mechanical force of the rubber mixer, and evenly dispersed, which promotes the rubber and the compounding agent to have a certain chemical and physical action, thus, forming a more complex micro multi-phase structure and fundamentally changing the performance of the rubber mixture. Mixing is an important process that directly affects the mechanical properties of materials. The vulcanization process is carried out on the plate vulcanizing machine because the cylindrical viscoelastic damper in this paper adopts a vulcanization bonding integrated type, and the aluminum cylinder and viscoelastic material are filled into the mold before vulcanization. The corresponding adhesive is applied on the bonding contact surface in advance, and the determined vulcanization conditions (temperature: 160 °C, time: 15 mins, and pressure: 10 MPa) are set on the vulcanizing machinea Finally, the mold with a viscoelastic damper inside is put on the plate curing machine for curing, as shown in Figure 2.

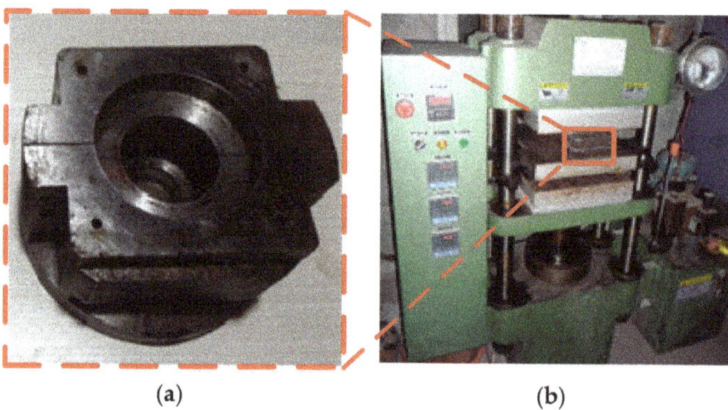

(a)   (b)

Figure 2. Vulcanization process of viscoelastic damper: (**a**) mold. (**b**) Flat vulcanizing machine.

As shown in Figure 3, the cylindrical viscoelastic damper is composed of an inner aluminum cylinder, an outer aluminum cylinder, and a viscoelastic layer bonded by vulcanization between the inner and outer cylinders. The viscoelastic layer is 50 mm in height, 5 mm in thickness, 46 mm in the outer diameter, 36 mm in the inner diameter, and 6437 mm^2 in a shear area. In practical application, the inner aluminum tube and the outer aluminum tube of the damper will be, respectively, installed between the base of the isolation structure and the damping components. The relative displacement between the inner and outer tubes will lead to the shear deformation of the viscoelastic layer, which plays the role of energy dissipation and vibration reduction.

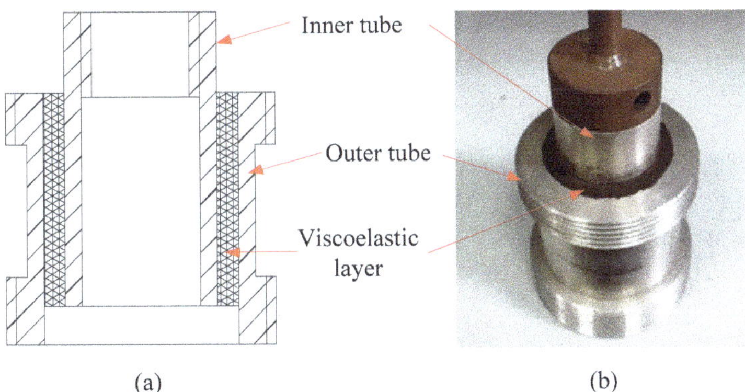

(a) (b)

Figure 3. Cylindrical viscoelastic damper. (**a**) cross Section; (**b**) physical map.

4. Performance Tests on Cylindrical Viscoelastic Dampers

4.1. Test Setup

The dynamic mechanical property tests on cylindrical viscoelastic dampers are carried out on a 10 kN hydraulic servo fatigue machine (Figure 4a). The loading frequency of the testing machine can reach 50 Hz, the control displacement accuracy can reach 5 μm, and the maximum loading force is 10 kN, which is suitable for the dynamic mechanical performance test of the damper under the conditions of high frequency and small amplitude.

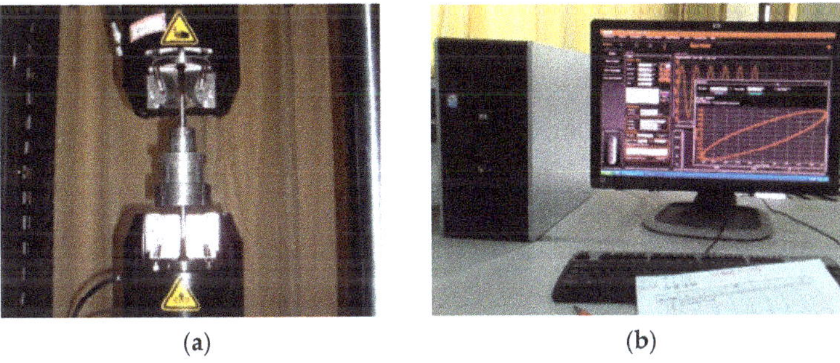

(a) (b)

Figure 4. Test setup of viscoelastic dampers: (**a**) loading device and (**b**) data acquisition system.

To obtain the hysteretic curves of a cylindrical viscoelastic damper, the mechanical properties of the viscoelastic damper under different working conditions (excitation frequency and displacement amplitude) were tested, and the effects of ambient temperature, loading frequency, and displacement amplitude on the mechanical properties and energy dissipation performance of viscoelastic damper were studied. In this paper, three batches of materials are composed of NBR, carbon black, graphite, sulfur, zinc oxide, stearic acid, and antioxidant, in which the damping agents are different including one damper in the first batch, marked as X01, three dampers in the second batch, marked as Y01, Y02, and Y03, four dampers in the third batch, marked as Y01-01 ~ 04. There are a total of eight damper specimens. Mica is added in the first batch of X01, three kinds of hindered phenols (AO60, AO80, and AO1098) are added in the second batch, and different fractions of AO60 are added in the third batch. The corresponding loading conditions are shown in Table 2.

Table 2. Loading procedure of property tests on the viscoelastic dampers.

Specimen Type	Specimen Number	Displacement Amplitude d (mm)	Frequency f (Hz)	Number of Cycles
Cylindrical VE damper	X01	0.15, 0.3	1, 5, 10, 15, 20, 25, 30, 35, 40, 45, 50	10
	Y01	0.1, 0.15, 0.3	1, 2, 5, 8, 10, 20, 30, 40, 50	
	Y02			
	Y03			
	Y01-01	0.1, 0.2, 0.5, 1.0	1, 2, 5, 8, 10, 20, 30, 40, 50	
	Y01-02		1, 2, 5, 8, 10, 20, 30, 40, 50	
	Y01-03		1, 2, 5, 8, 10, 20, 30, 40, 50	
	Y01-04		1, 2, 5, 8, 10, 20, 30, 40, 50	

In the process of each damper test, it should be checked that the connection between the chuck of the hydraulic servo machine and the test piece is firm and the viscoelastic layer has not degummed or torn to ensure that the collected test data are accurate and effective. The data acquisition system (Figure 4b) shows that the hysteretic curve is full under each working condition, which indicates that all parts of the specimen are well-connected. After every certain condition is loaded for 10 cycles, the loading frequency and amplitude need to be changed through the loading control software shown in Figure 4b for the next condition.

4.2. Analysis and Discussion of Test Results

Figures 5 and 6 show the force-displacement hysteretic curves of dampers X01 and Y02 at different frequencies. It can be seen that the hysteretic curves of the damper under different test conditions are standard ellipse shapes, which indicates that the damper has better energy dissipation capacity. At the same time, it can be seen that, with the increase of loading frequency, the tilt angle of the hysteretic curve increases, and the elliptic curve becomes fuller. At a low frequency, such as 0.1 Hz, the hysteresis curve is flat, while, at a high frequency, such as 10 Hz, the hysteresis curve is very full. This phenomenon shows that the stiffness and energy dissipation capacity of the viscoelastic damper increase with the rise in frequency.

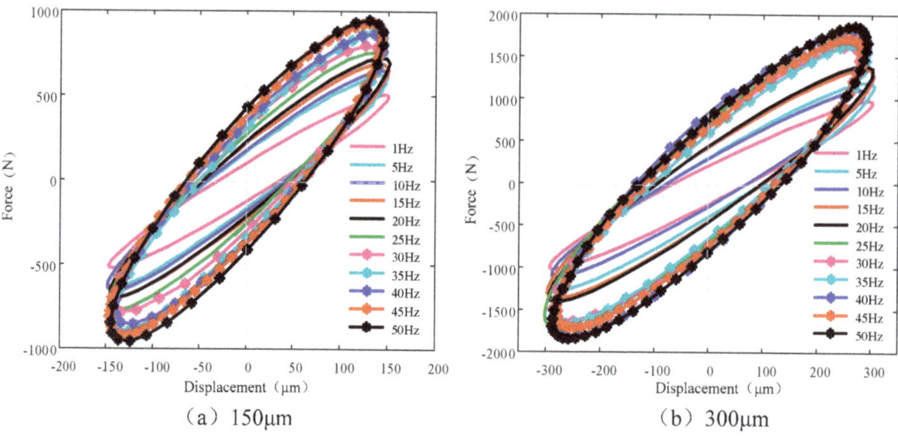

Figure 5. The hysteresis curves of specimen X01 at the same displacement amplitude with different frequencies: (**a**) 150 μm, (**b**) 300 μm.

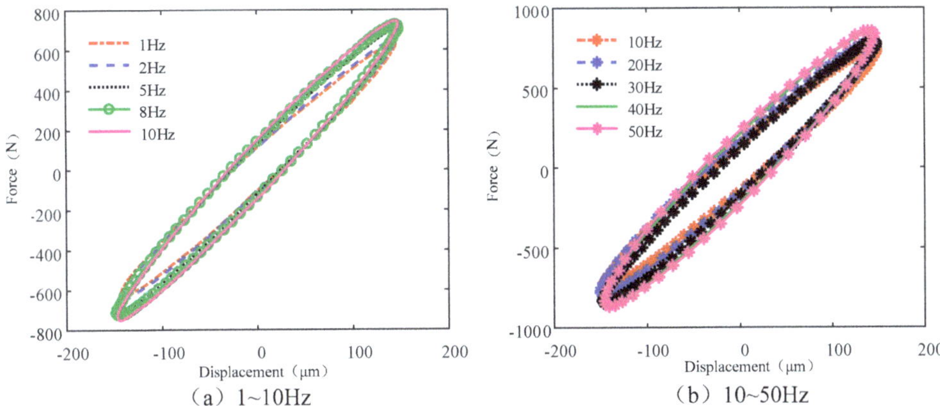

Figure 6. The hysteresis curves of specimen Y02 at the same displacement amplitude with different frequencies: (**a**) 1~10 Hz, (**b**) 10~50 Hz.

Figures 7 and 8 are force-displacement hysteretic curves of different displacement amplitudes of damper No. Y02 and Y01-03, respectively. It can be seen from the figure that the force-displacement hysteretic curves of the specimens under different working conditions are full ellipses. The results show that the viscoelastic damper has good energy dissipation capacity in a micro-vibration environment. At the same frequency and temperature, the envelope area of the curve increases with the rise of the displacement amplitude. This phenomenon shows that the energy dissipation of a single loop increases with the growth of the displacement amplitude. It can also be seen from the figure that the inclination angle of the hysteretic curve of the damper is only slightly reduced and basically remains unchanged, which indicates that the displacement amplitude has almost no effect on the stiffness of the damper.

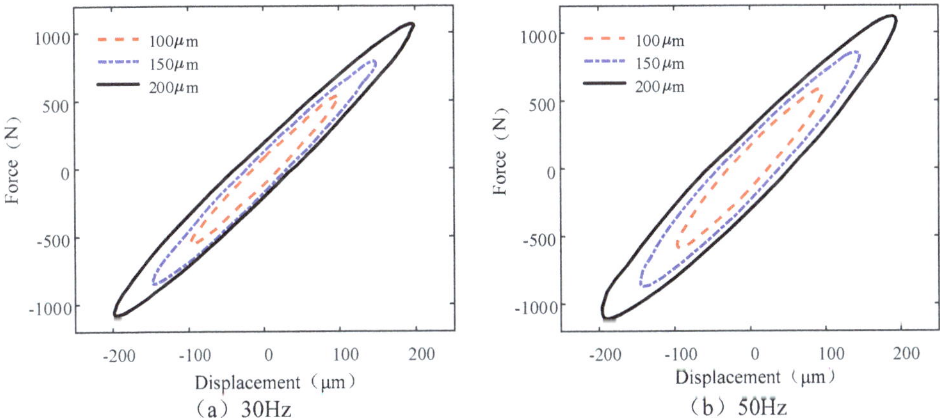

Figure 7. The hysteresis curves of specimen Y02 at the same frequency with different displacements: (**a**) 30 Hz, (**b**) 50 Hz.

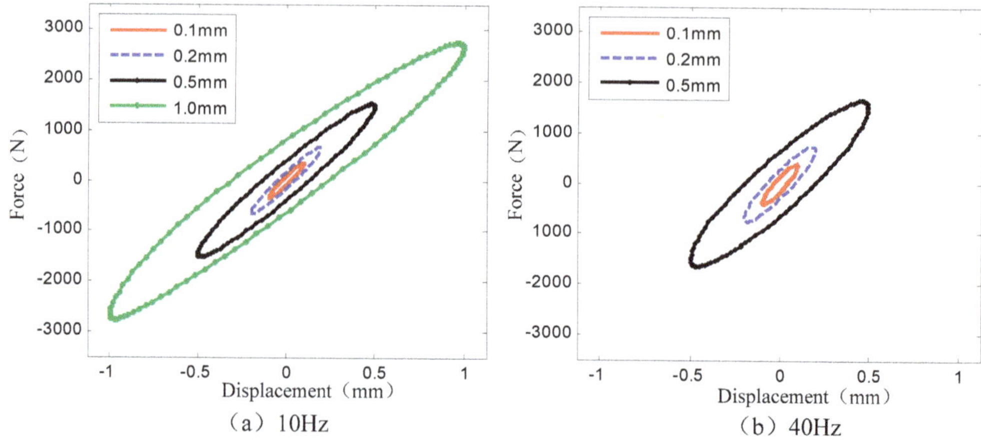

Figure 8. The hysteresis curves of specimen Y01-03 at the same frequency with different displacements: (**a**) 10 Hz, (**b**) 40 Hz.

Figure 9 shows the equivalent stiffness and equivalent damping of the viscoelastic damper specimen X01 at different frequencies. It can be seen that, under the same displacement amplitude, the equivalent stiffness of the viscoelastic damper increases with the rise of frequency. Under the same displacement amplitude, the equivalent damping decreases with the increase of frequency. In the lower frequency range, it can be seen that the equivalent damping decreases sharply, and, with the increase of frequency, the reduction tends to moderate, and the difference of equivalent stiffness and equivalent damping is small under different displacement amplitudes.

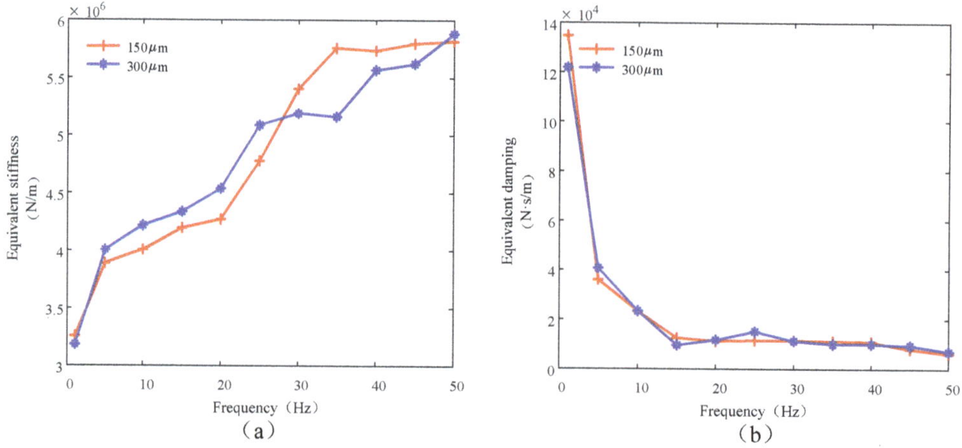

Figure 9. Dynamic parameters of specimen X01 vary with frequency: (**a**) equivalent stiffness, and (**b**) equivalent damping.

Figure 10 shows the storage modulus and loss factor of the viscoelastic damper specimen X01 at different frequencies. It can be seen that the storage modulus and loss factor of viscoelastic materials increase with the growth of frequency. The change of dynamic mechanical properties of viscoelastic materials with frequency is mainly related to the motion state of molecular chain and the relaxation time of the material. In the tested frequency range, with the increase of the excitation frequency, the external load action time becomes shorter, gradually approaching the relaxation time of the molecular chain. The motion of the chain segment intensifies, and the interaction between the chains strengthens, so the performance indexes are increased.

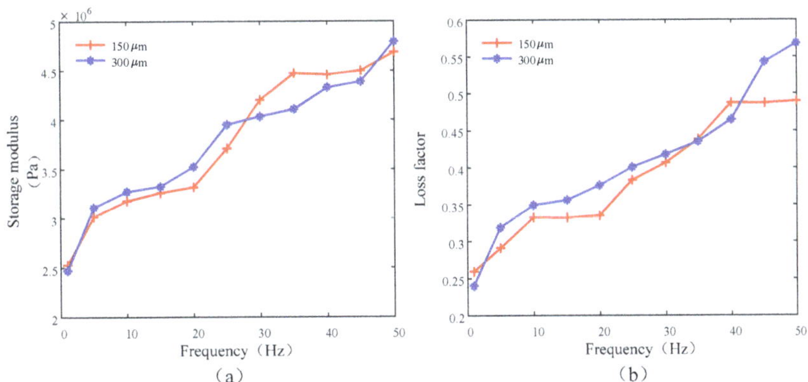

Figure 10. Dynamic parameters of specimen X01 vary with frequency: (**a**) storage modulus and (**b**) loss factor.

When the frequency is 1 Hz, the curves of equivalent stiffness and equivalent damping of viscoelastic damper specimen Y01 ~ 03 varying with amplitude have been drawn, as shown in Figure 11. The relative difference between the equivalent stiffness and the equivalent damping of specimen Y01 ~ 03 is less than 20%. The damping parameters of the three specimens are in good agreement, which indicates that the viscoelastic damper has a good consistency. Taking Y03 as an example, when the loading frequency is 1 Hz, the change rates of equivalent stiffness and equivalent damping are −1.26%, −2.40%, and −0.5%, −1.52%, respectively, when the displacement amplitude is from 0.1 to 0.15 mm and 0.15 to 0.2 mm. It can be seen that, when the displacement amplitude is very small, the equivalent stiffness and damping of viscoelastic damper slightly decrease with the increase of displacement, and the effect of displacement amplitude on the stiffness and damping of the viscoelastic damper is very small.

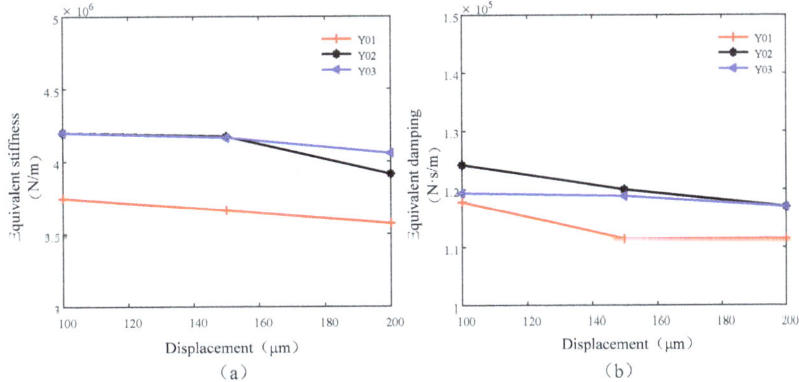

Figure 11. Dynamic parameters of specimens Y01~03 vary with displacement at 1 Hz: (**a**) equivalent stiffness and (**b**) equivalent damping.

Figure 12 shows the storage modulus and loss factor of viscoelastic material under different amplitudes of specimen Y01~03 at the frequency of 1 Hz. Taking Y03 as an example, when the loading frequency is 1 Hz and the displacement amplitude is from 0.1 to 0.15 mm and 0.15 to 0.2 mm, the change rates of storage modulus and loss factor are −0.79%, −2.56%, 0.39%, and 1.01%, respectively, and the growth rates of single cycle energy consumption are 124.26% and 75.14%. It can be seen that, with the increase of displacement amplitude, the energy consumption of a single loop increases greatly. When the displacement amplitude is very small, the displacement amplitude slightly affects the storage modulus and loss factor of the viscoelastic damper.

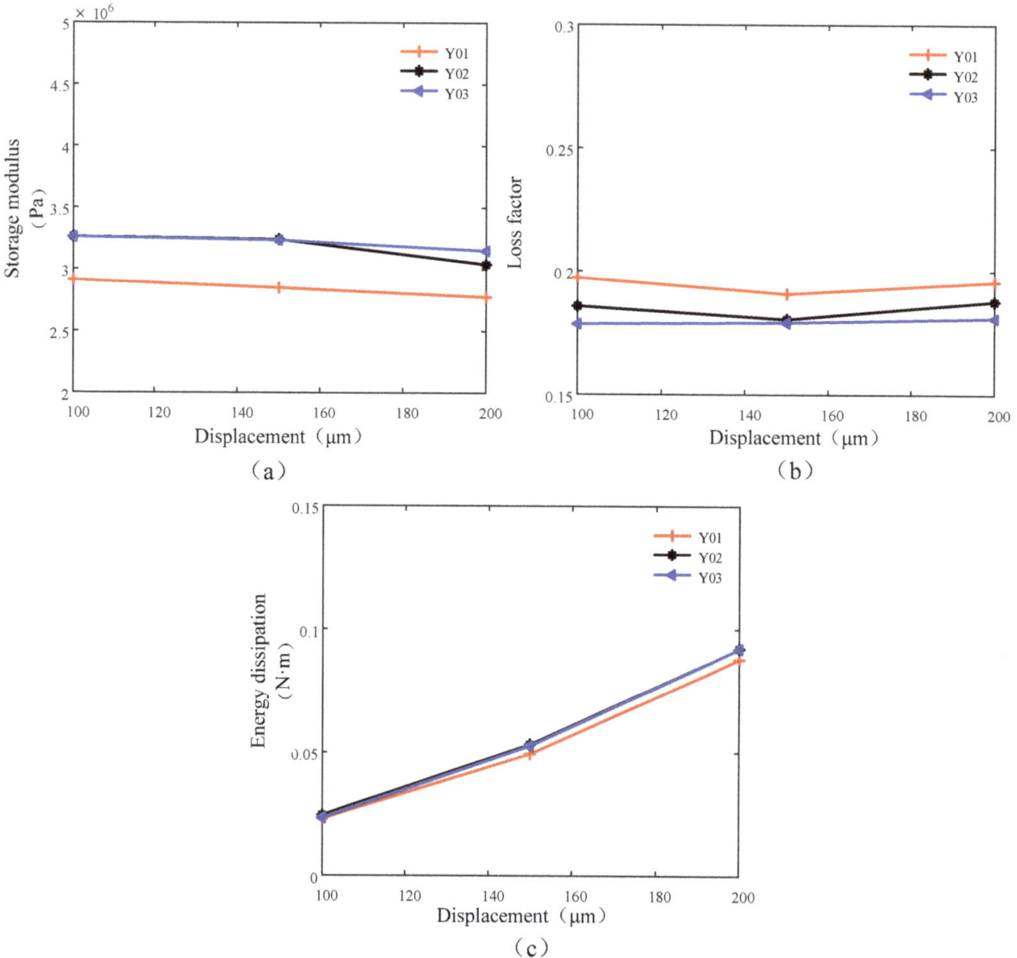

Figure 12. Dynamic parameters of specimens Y01~03 vary with displacement at 1 Hz: (**a**) storage modulus, (**b**) loss factor, and (**c**) energy dissipation.

When the displacement amplitude is 0.2 mm, the equivalent stiffness and damping curves of specimen Y01-01 ~ 04 at different frequencies have been plotted, as shown in Figure 13. Similar to the change law of mechanical properties of specimen X01, the stiffness of the viscoelastic damper increases slightly with the rise in frequency. When the frequency is small, the stiffness increases rapidly with the growth in frequency. When the frequency is large, the stiffness increases slowly and the curve tends to be smooth. The equivalent damping of the viscoelastic damper decreases with the increase of frequency. When the frequency is small, the decrease is faster. When the frequency is high, the decrease is slower and the curve tends to be smooth. It can be seen from the figure that the stiffness of the four damper specimens is slightly different from other specimens except for Y02-02. The damping parameters of the four specimens are in good agreement, which indicates that the viscoelastic damper has a good consistency.

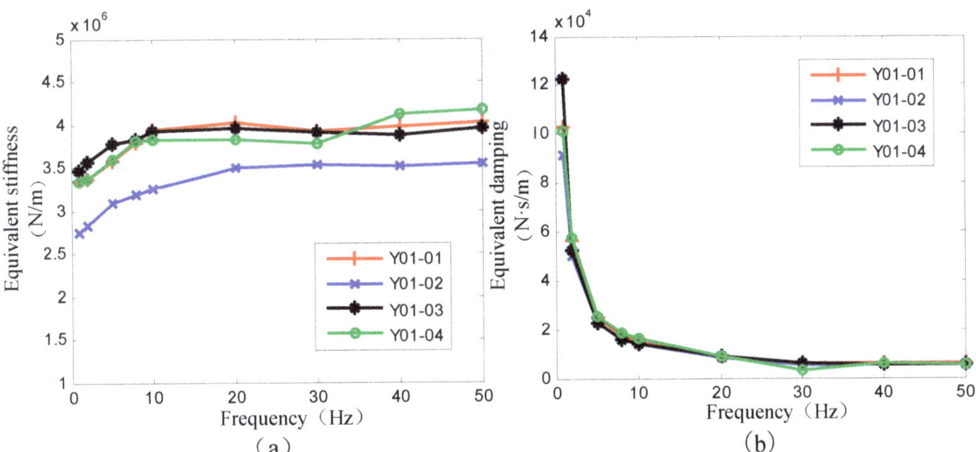

Figure 13. Dynamic parameters of specimens Y01-01~04 vary with frequency at 0.2 mm: (**a**) equivalent stiffness and (**b**) equivalent damping.

Figure 14 shows the equivalent stiffness and damping of specimen Y01-01 ~ 04 under different amplitudes at the frequency of 5 Hz. Taking Y01-01 as an example, when the displacement amplitude is from 0.1 to 0.2 mm, 0.2 to 0.5 mm, and 0.5 to 1 mm, the change rates of equivalent stiffness and damping are −5.66%, −7.39%, −10.24%, and −6.63%, 0.64%, 0.64%, and 17.94%, respectively. It can be seen that the stiffness and equivalent damping of viscoelastic damper decrease slightly with the increase of displacement, which is consistent with the stable slope and fullness of the hysteresis curve at different displacements in Figures 7 and 8.

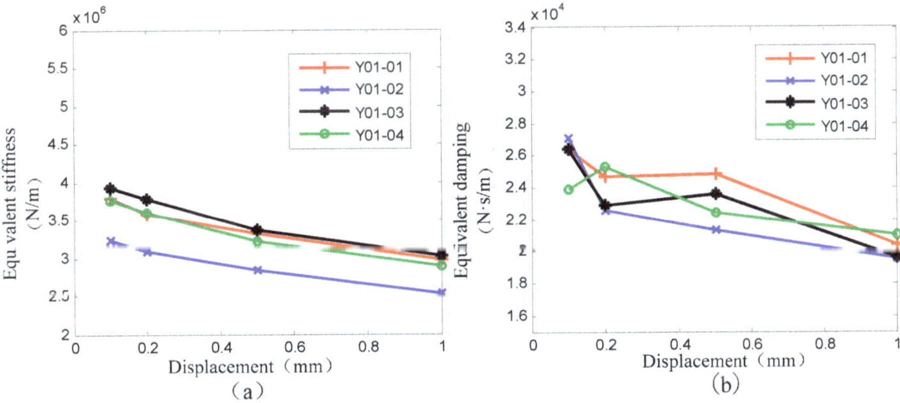

Figure 14. Dynamic parameters of specimens Y01-01~04 vary with displacement at 5 Hz: (**a**) equivalent stiffness and (**b**) equivalent damping.

Figure 15a,b, respectively, show the energy consumption of a single cycle under different frequencies and different displacements when the displacement amplitude of specimen Y01-01 ~ 04 is 0.2 mm and the frequency is 5 Hz. It can be seen from the curve that the damping parameters of the four specimens are in good agreement, which indicates that the viscoelastic damper has a good consistency. Taking the specimen Y01-01 as an example, when the displacement amplitude is 0.2 mm, and the frequency is from 1 to 10 Hz, 10 to 20 Hz, 20 to 30 Hz, 30 to 40 Hz, and 40 to 50 Hz, the increasing values of energy consumption of a single cycle are 0.034 N·m, 0.03 N·m, 0.028 N·m, 0.039 N·m, and 0.047 N·m, respectively. When the frequency is 5 Hz, the increasing values of energy consumption of a single cycle are 0.071 N·m, 0.605 N·m, and 1.397 N·m, respectively, when the displacement is from 0.1 to 0.2 mm, 0.2 to 0.5 mm, and 0.5 to 1 mm. It can be seen that the energy dissipation of a single loop of viscoelastic damper raises with the frequency and displacement increasing.

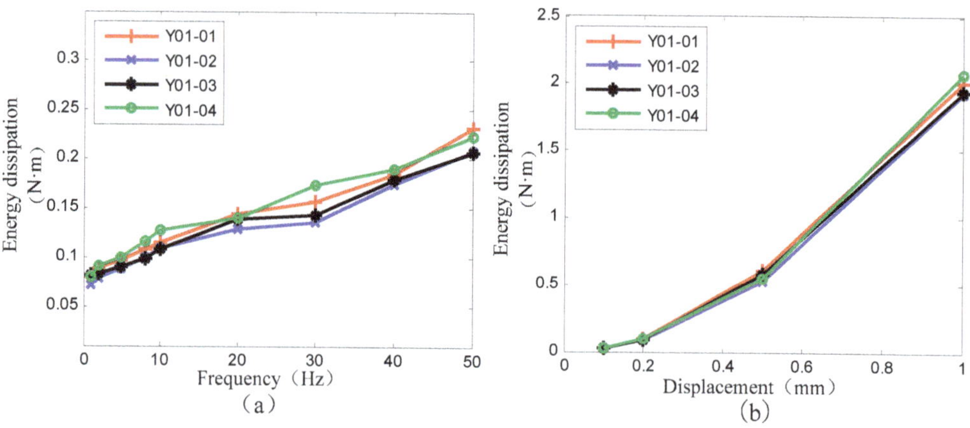

Figure 15. Energy dissipation of specimens Y01-01~04 vary with (**a**) frequency at 0.2 mm and (**b**) displacement at 5 Hz.

5. Conclusions

In this paper, several batches of cylindrical viscoelastic dampers are vulcanized for different viscoelastic materials, and the dynamic mechanical properties are tested at different frequencies and amplitudes. The main conclusions are as follows.

(1) The cylindrical viscoelastic dampers present a full hysteretic curve and excellent damping performance in a wide frequency range, which further promotes the application of viscoelastic damper under different frequency conditions.

(2) The equivalent stiffness, storage modulus, loss factor, and energy consumption per cycle of the viscoelastic damper increase with the rise in frequency, while the equivalent damping decreases with the increase in frequency.

(3) When the displacement increases, the energy consumption per cycle of the viscoelastic damper increases rapidly, and the equivalent stiffness, equivalent damping, storage modulus, and loss factor change slightly.

Author Contributions: Conceptualization, T.G. and X.-H.H.; methodology, Y.-Q.G.; software, T.G.; validation, T.G., Y.-Q.G. and Z.-F.H.; formal analysis, T.G.; investigation, T.G.; resources, X.-H.H.; data curation, T.G.; writing—original draft preparation, T.G.; writing—review and editing, Y.-Q.G.; visualization, Z.-W.H.; supervision, Y.-Q.G.; project administration, X.-H.H.; funding acquisition, Y.-Q.G. All authors have read and agreed to the published version of the manuscript.

Funding: This study was financially supported by the National Key Research and Development Program of China (2016YFE0200500), the National Key R&D Programs of China (Grant Number: 2019YFE0121900), Jiangsu International Science and Technology Cooperation Program (BZ2018058), National Natural Science Foundation of China (Grant Number: 56237845), Natural Science Foundation of Jiangsu Province (Grant Number: BK20170684), and the State Foundation for Studying Abroad, China.

Institutional Review Board Statement: Not applicable.

Informed Consent Statement: Informed consent was obtained from all subjects involved in the study.

Data Availability Statement: Not applicable.

Conflicts of Interest: The authors declare no conflict of interest.

References

1. Kim, J.; Ryu, J.; Chung, L. Seismic performance of structures connected by viscoelastic dampers. *Eng. Struct.* **2006**, *28*, 183–195. [CrossRef]
2. Rashid, A.; Nicolescu, C.M. Design and implementation of tuned viscoelastic dampers for vibration control in milling. *Int. J. Mach. Tools Manuf.* **2008**, *48*, 1036–1053. [CrossRef]
3. Paolacci, F. An energy-based design for seismic resistant structures with viscoelastic dampers. *Earthq. Struct.* **2013**, *4*, 2. [CrossRef]
4. Tsai, C.S.; Lee, H.H. Applications of viscoelastic dampers to high-rise buildings. *J. Struct. Eng.* **1993**, *119*, 1222–1233. [CrossRef]
5. Xu, Z.D. Earthquake mitigation study on viscoelastic dampers for reinforced concrete structures. *J. Vib. Control* **2007**, *13*, 29–43. [CrossRef]
6. Xu, Z.D.; Shen, Y.P.; Zhao, H.T. A synthetic optimization analysis method on structures with viscoelastic dampers. *Soil Dyn. Earthq. Eng.* **2003**, *23*, 683–689. [CrossRef]
7. Moliner, E.; Museros, P.; Martínez-Rodrigo, M.D. Retrofit of existing railway bridges of short to medium spans for high-speed traffic using viscoelastic dampers. *Eng. Struct.* **2012**, *40*, 519–528. [CrossRef]
8. Rao, M.D. Recent applications of viscoelastic damping for noise control in automobiles and commercial airplanes. *J. Sound Vib.* **2003**, *262*, 457–474. [CrossRef]
9. Lewandowski, R.; Pawlak, Z.Ł. Dynamic analysis of frames with viscoelastic dampers modelled by rheological models with fractionalderivatives. *J. Sound Vib.* **2011**, *330*, 923–936. [CrossRef]
10. Xu, Z.D.; Liao, Y.X.; Ge, T.; Xu, C. Experimental and theoretical study of viscoelastic dampers with different matrix rubbers. *J. Eng. Mech.* **2016**, *142*, 04016051. [CrossRef]
11. Chang, K.C.; Soong, T.T.; Lai, M.L.; Neilson, E.J. Viscoelastic dampers as energy dissipation devices for seismic applications. *Earthq. Spectra* **1993**, *9*, 371–387. [CrossRef]
12. Xu, Z.D.; Xu, C.; Hu, J. Equivalent fractional Kelvin model and experimental study on viscoelastic damper. *J. Vib. Control* **2015**, *21*, 2536–2552. [CrossRef]
13. Chang, K.C.; Lai, M.L.; Soong, T.T.; Hao, D.S.; Yeh, Y.C. *Seismic Behavior and Design Guidelines for Steel Frame Structures with Added Viscoelastic Dampers*; National Center for Earthquake Engineering Research: Buffalo, NY, USA, 1993.
14. Shen, K.L.; Soong, T.T.; Chang, K.C.; Lai, M.L. Seismic behaviour of reinforced concrete frame with added viscoelastic dampers. *Eng. Struct.* **1995**, *17*, 372–380. [CrossRef]
15. Tsai, C.S. Temperature effect of viscoelastic dampers during earthquakes. *J. Struct. Eng.* **1994**, *120*, 394–409. [CrossRef]
16. Chang, K.C.; Soong, T.T.; Oh, S.T.; Lai, M.L. Seismic behavior of steel frame with added viscoelastic dampers. *J. Struct. Eng.* **1995**, *121*, 1418–1426. [CrossRef]
17. Bergman, D.M.; Hanson, R.D. Viscoelastic mechanical damping devices tested at real earthquake displacements. *Earthq. Spectra* **1993**, *9*, 389–417. [CrossRef]
18. De Cazenove, J.; Rade, D.A.; De Lima, A.M.G.; Araújo, C.A. A numerical and experimental investigation on self-heating effects in viscoelastic dampers. *Mech. Syst. Signal Process.* **2012**, *27*, 433–445. [CrossRef]
19. Xu, Z.D.; Ge, T.; Liu, J. Experimental and theoretical study of high-energy dissipation-viscoelastic dampers based on acrylate-rubber matrix. *J. Eng. Mech.* **2020**, *146*, 04020057. [CrossRef]
20. Xu, Z.D.; Wang, D.X.; Shi, C.F. Model, tests and application design for viscoelastic dampers. *J. Vib. Control* **2011**, *17*, 1359–1370. [CrossRef]
21. Xu, Z.D.; Lu, L.H.; Shi, B.Q.; Huang, X.; Zeng, X. Experimental and numerical studies on vertical properties of a new multi-dimensional earthquake isolation and mitigation device. *Shock Vib.* **2013**, *20*, 401–410. [CrossRef]
22. Xu, Z.D.; Gai, P.P.; Zhao, H.Y.; Huang, X.H.; Lu, L.-Y. Experimental and theoretical study on a building structure controlled by multi-dimensional earthquake isolation and mitigation devices. *Nonlinear Dyn.* **2017**, *89*, 723–740. [CrossRef]
23. Xu, Z.D.; Xu, F.H.; Chen, X. Vibration suppression on a platform by using vibration isolation and mitigation devices. *Nonlinear Dyn.* **2016**, *83*, 1341–1353. [CrossRef]
24. Xu, Z.D.; Ge, T.; Miao, A. Experimental and theoretical study on a novel multi-dimensional vibration isolation and mitigation device for large-scale pipeline structure. *Mech. Syst. Signal Process.* **2019**, *129*, 546–567. [CrossRef]
25. Chang, K.C.; Lin, Y.L. Seismic response of full-scale structure with added viscoelastic dampers. *J. Struct. Eng.* **2004**, *130*, 600–608. [CrossRef]
26. Aiken, I.D.; Nims, D.K.; Whittaker, A.S.; Kelly, M.K. Testing of passive energy dissipation systems. *Earthq. Spectra* **1993**, *9*, 335–370. [CrossRef]
27. Min, K.; Kim, J.; Lee, S. Vibration tests of 5-storey steel frame with viscoelastic dampers. *Eng. Struct.* **2004**, *26*, 831–839. [CrossRef]
28. Lin, Y.Y. *Polymer Viscoelasticity: Basics, Molecular Theories, Experiments and Simulations*; World Scientific: Singapore, 2010.

Review

A Review of Piezoelectric Material-Based Structural Control and Health Monitoring Techniques for Engineering Structures: Challenges and Opportunities

Abdul Aabid [1,*], Bisma Parveez [2], Md Abdul Raheman [3], Yasser E. Ibrahim [1], Asraar Anjum [4], Meftah Hrairi [4], Nagma Parveen [5] and Jalal Mohammed Zayan [4]

[1] Department of Engineering Management, College of Engineering, Prince Sultan University, P.O. Box 66833, Riyadh 11586, Saudi Arabia; ymansour@psu.edu.sa
[2] Department of Manufacturing and Materials Engineering, Faculty of Engineering, International Islamic University Malaysia, P.O. Box 10, Kuala Lumpur 50728, Malaysia; mirbisma5555@gmail.com
[3] Department of Electrical and Electronic Engineering, NMAM Institute of Technology, Nitte, Karkala Taluk, Udupi 574110, India; mararkeri@nitte.edu.in
[4] Department of Mechanical Engineering, Faculty of Engineering, International Islamic University Malaysia, P.O. Box 10, Kuala Lumpur 50728, Malaysia; asraar.anjum@gmail.com (A.A.); meftah@iium.edu.my (M.H.); zayan_mohammed@yahoo.co.in (J.M.Z.)
[5] Department of Electrical and Computer Engineering, Faculty of Engineering, International Islamic University Malaysia, P.O. Box 10, Kuala Lumpur 50728, Malaysia; nagmaparveen1192@gmail.com
* Correspondence: aabidhussain.ae@gmail.com or aaabid@psu.edu.sa

Citation: Aabid, A.; Parveez, B.; Raheman, M.A.; Ibrahim, Y.E.; Anjum, A.; Hrairi, M.; Parveen, N.; Mohammed Zayan, J. A Review of Piezoelectric Material-Based Structural Control and Health Monitoring Techniques for Engineering Structures: Challenges and Opportunities. *Actuators* **2021**, *10*, 101. https://doi.org/10.3390/act10050101

Academic Editors: Zhao-Dong Xu, Siu-Siu Guo and Jinkoo Kim

Received: 5 April 2021
Accepted: 7 May 2021
Published: 10 May 2021

Publisher's Note: MDPI stays neutral with regard to jurisdictional claims in published maps and institutional affiliations.

Abstract: With the breadth of applications and analysis performed over the last few decades, it would not be an exaggeration to call piezoelectric materials "the top of the crop" of smart materials. Piezoelectric materials have emerged as the most researched materials for practical applications among the numerous smart materials. They owe it to a few main reasons, including low cost, high bandwidth of service, availability in a variety of formats, and ease of handling and execution. Several authors have used piezoelectric materials as sensors and actuators to effectively control structural vibrations, noise, and active control, as well as for structural health monitoring, over the last three decades. These studies cover a wide range of engineering disciplines, from vast space systems to aerospace, automotive, civil, and biomedical engineering. Therefore, in this review, a study has been reported on piezoelectric materials and their advantages in engineering fields with fundamental modeling and applications. Next, the new approaches and hypotheses suggested by different scholars are also explored for control/repair methods and the structural health monitoring of engineering structures. Lastly, the challenges and opportunities has been discussed based on the exhaustive literature studies for future work. As a result, this review can serve as a guideline for the researchers who want to use piezoelectric materials for engineering structures.

Keywords: piezoelectric material; vibration control; noise control; active control; damage structure; SHM

1. Introduction

In the direction of smart material applications in engineering systems, various efforts were made. These intelligent materials have certain properties that can be desirably altered by varying stress, temperature, and a magnetic or electric field, which serves as an external stimulus in a controlled environment. A *smart material* is mainly divided into two types, "piezoelectric transducer" and "shape memory alloys", and these types of materials are more frequently used in various fields. Piezoelectric material transducers have both sensors as well as actuator functionality [1]. The piezoelectric materials have become popular over the last three decades due to their electromechanical effects and versatile applications. The emphasis has shifted to piezoelectric-based methods due to their implications, which include operational cost reductions in preservation as well as an improvement in the structure's life cycle. The piezoelectric materials have been utilized for many purposes in engineering structures. In the event of the initiation of cracks/damages, any type of structure requires high maintenance of safety or can result in the replacement of the whole structure, and these damages are mostly due to fatigue/corrosion. Such cases have been solved by the application of piezoelectric materials [2–10]. Indeed, in the last decade, piezoelectric materials were utilized for the control/repair of structures such as aerospace, concretes, and photovoltaic solar panels.

On the other hand, the use of piezoelectric-based structural health monitoring (SHM) has assisted in the transformation of the industry for various engineering aspects, while the electromechanical process, a relatively recent non-destructive research tool, has been studied for more than two decades and there are still a number of issues that must be resolved before it can be extended to actual structures. The methodology requires the use of a single piezoelectric for exciting and detecting the host structure and can lead to the advancement of one of the most efficient SHM systems. Moreover, many researchers are investigating the electromechanical impedance (EMI) technique for SHM via experimental and computational standpoints. Structural disruption, sensor/actuator faults, and delamination can all be detected using the EMI technique. The sensor's self-detection is important because it can lead to a defective diagnostic, causing the device to malfunction. As a result, several researchers in the SHM group are focusing on sensor self-diagnosis. Although the EMI analysis is complicated, it can be carried out by breaking down the structural loss into actual and theoretical impedance components.

In this review, literature has been carried out based on the control of damaged structures (vibration, noise, and active), the SHM of engineering structures. The control of damaged structures using piezoelectric materials is a highly developed research concept in current industries, particularly in the aerospace industries. The SHM is the most useful character of piezoelectric material for monitoring any type of structure; hence, there are a number of studies that have reported on this. The next two sections are about piezoelectric materials and their modeling. Section 3 expresses the control of structures and Section 4 extracted some of the studies of SHM to cover the piezoelectric applications in recent years. Section 5 elucidates the challenges and opportunities in this field and is particularly related to the present review contents. Finally, a conclusion has been constructed based on the current review work.

2. Piezoelectric Materials

The direct piezoelectric effect is the capability of piezoelectric materials to create an electric field under the influence of mechanical stress. This property of the piezoelectric materials is utilized for the generation of electrical energy. The reciprocal of the direct piezoelectric effect is the inverse piezoelectric effect in which mechanical strain is developed in response to the electric field. These effects are strongly dependent on the crystal orientation with respect to the strain or electric field [11–16]. The direct effect makes it possible to use them as sensors, and the converse effect as actuators (Figure 1). The designed structures constructed using piezoelectric materials can be bent, expanded, or contracted upon the application of voltage, and they can be used for sensing and actuating [17] purposes and for easy control [18]. Piezoelectric patches (or films) are thin ceramic strips that are either intended to be bonded to the substructure surface or to be inserted within the structure. The stacks instead are built by piling up multiple piezoelectric layers of alternating polarity [19]. The piezoelectric materials are widely used in valves, micropumps, earphones and speakers, ultrasonic cleaners, emulsifiers, and sonic transducers.

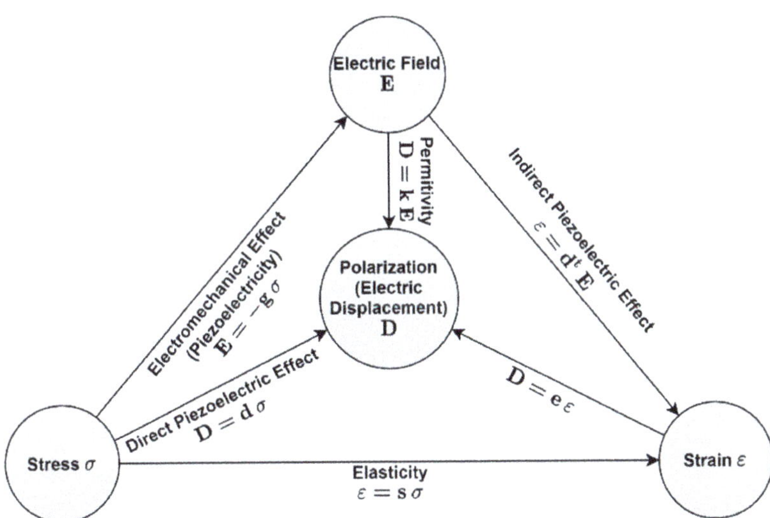

Figure 1. Effect of Piezoelectric material [15]. Reprinted under the Creative Commons (CC) License (CC BY 4.0).

In addition to piezoelectric ceramics, piezoelectric polymers are another group of piezoelectric materials that have found widespread use. Polyvinyl fluoride (PVDF) is versatile and light in weight in comparison to piezoelectric ceramics. Because of this, thin films of any desired form can be drawn into them, giving them an advantage over piezoceramics in various applications involving complex designs of sensors or actuators. Besides being versatile and lightweight, however, they have lower electromechanical coupling compared to piezoelectric ceramics, and the other characteristics that make the piezoelectric polymers attractive are their low electrical permittivity, low acoustic impedance, high voltage sensitivity, and relatively lower cost. An updated overview of the applications of piezoelectric polymers in touch devices, pyroelectric infrared sensors, property measurement with photopyroelectric spectroscopy, and shock sensors can be observed in the article by Lang and Muensit [20].

2.1. Piezoelectric–Mechanical Constitutive Equations

In piezoelectric ceramics, nonlinear dielectric, elastic and piezoelectric relationships were well established in piezoelectric constitution equations as well as Preisach-type models, which were employed to define the hysteretic path-dependent strain–field relationship in piezoelectric actuator models [21]. Generally, the relation for the constitutive equation for piezoelectric materials is written as:

$$S = c^E \cdot T + [d]^t \cdot E \tag{1}$$

$$D = d \cdot T + \varepsilon^T \cdot E \tag{2}$$

where T is the constant stress, E is the constant field, C is the stiffness coefficient, D is the constant electrical displacement, S is the constant strain, ε^T is the dielectric permittivity, and d is the piezoelectric constant matrices. The superscript E and t indicate a constant electric field and charting time, respectively, for the compliance matrix that is evaluated. These matrix relations are generally used for modeling finite element (FE) analysis. Only some of the relationships are typically useful for theoretical methods to simplify the problem further [22].

2.2. Reduced-Order Modal Equations

Reduced-ordered modal equations from the piezoelectric constitution equation can be governed for the modal coordination system with the mode of the superposition process. In addition, the FE nodal displacement vector can transfer the modal coordinate vector with the used modal matrix system. With this, there is a possibility to express the approximate relation of generalized nodal displacement vector $\{d\}$:

$$\{d\} \approx \sum_{i=1}^{r} \varnothing_i q_i = [\varnothing]\{q\} \tag{3}$$

where

$$[\varnothing] = [\varnothing_1, \ldots, \varnothing_n](n < r) \tag{4}$$

where $\{q\}$ is the coordinate vector (modal), in which n order is a time-dependent vector, and n variable represents the variation of the modal in a form of numbers for preserved/controlled.

After being introduced, the damping of the feedback control system for a reduced-order modal equation for a linear de-coupled form is as follows:

$$[\overline{M}]\{\ddot{q}\} + [\overline{C_d}]\{\dot{q}\} + [\overline{K}]\{q\} = \{\overline{F}\} + [\overline{K_A}]\{u_a\} \tag{5}$$

Here, $[\overline{M}]$ is the modal mass and it is expressed as $[\overline{M}] = [\varnothing]^T\{F\}$, $\{u_a\}$ is the applied voltage of piezoelectric actuators (input vector control), and $[\overline{K_A}]\{u_a\} = [\varnothing]^T\{F_P\}$, where $[\overline{K_A}]$ is the stiffness matrix of the modal actuator or control input effect matrix.

Concerning mass, normalize the modal matrix $[\varnothing]$ and a structural damping coefficient ζ_i ($i = 1, \ldots, r$) is assumed; then, the modal system becomes:

$$\{\ddot{q}\} + \text{diag}\,[2\zeta_i\omega_i]\{\dot{q}\} + \text{diag}\left[\omega_i^2\right]\{q\} = \{\overline{F}\} + [\overline{K_A}]\{u_a\} \tag{6}$$

where ω_i is the natural frequency and a mode form vector that corresponds to each mode is $\varnothing_i$ ($i = 1, \ldots, r$).

2.3. Piezoelectric Material Type-Based Investigation and Issues

Lee et al. [23] prepared specimens of lead meta niobate (LMN) [24] from the commercially available piezoelectric transducer, and these were polled by the manufacturer for use in the transducers [25]. The damping of structural vibration LMN was employed because that material tends to stick. It has been observed that the piezoelectric effect is mostly in the form of artificial piezoelectric material and advantageous features of generating electricity, and it can effectively repair the crack and can carry out SHM. Mohammad et al. [26] studied the ceramics containing 1% of SrTiO3 (SPN) with orthorhombic and tetragonal structures. These ceramics simultaneously exhibited a maximum piezoelectric constant. Because of their denser and similar composition to stoichiometry, microwave sintered ceramics showed a higher piezoelectric constant in comparison to traditional sintered ones. Arian et al. [27] studied their usage in multi-layer ceramic capacitors, lead-based perovskite materials (MLCCs). There are benefits to these materials over traditional materials such as barium titanite. Due to the diffuse phase transition, they show a very large dielectric. The best results were obtained by the stoichiometric lead magnesium niobate (PMN) [28] with a lower sintering temperature of about 900 °C and a relative dielectric constant of 10,000 at room temperature. The piezoelectric properties were defined in the constitutive equations, assuming that the total strain in the transducer is the sum of the mechanical strain caused by the mechanical stress and the controllable actuation strain induced by the electrical voltage applied. A study carried out by Pasquali et al. [29] presented the nonlinear piezoelectric plate model that was capable of accurately demonstrating the direct and convergent piezoelectric effect. The starting point of the nonlinear piezoelectric plate model for full electromechanical coupling may be an empirical expression of electrical potential. Some of the issues/requirements in piezoelectric materials that can be considered an important aspect while using the structures are as follows [30]:

- In comparison to the host structure, the PZT transducer should be non-reactive and have marginal stiffness and strength. It should also be protected from environmental factors such as humidity, precipitation, and temperature.
- The frequency spectrum of excitation determines the sensing region of the PZT transducer. A wider sensing domain is covered by frequencies below 30 kHz. High-frequency EMI models are only applicable to a small region.
- An arrangement has parts that are weak or essential that need more effort than others. High-stress fields, corrosive environment areas, and so on, must all be monitored closely. The length, distance, and thickness of the PZT actuation are all three directions. As a result, a detailed estimate of the PZT-sensing region must be determined using these three types of actuations, based on the geometry and material properties of the host structure.
- In the absence of damages, PZT transducers are effective at measuring the loading on a structure, or vice versa. The study on PZT-based EMI for load applications is largely limited to 1D structures, but [31] shows some work on 2D, 3D, and complex structures.
- To track any structure using EMI-based SHM, the current usual practice is to first acquire a baseline signature. This is then compared to later levels of the structure's EM admittance signatures to see if the structure has any flaws. It is very difficult to achieve the no-damage baseline signature for older current systems, making comparisons with later stage signatures almost impossible. As a result, any signature obtained from any structure at any point in time should provide overt or implied knowledge about the structure.
- As the embedded or surface bonded PZT transducer is excited, the 'structural responses' are extracted and expressed as conductance and susceptibility signatures. The structural reaction varies with the frequency of excitation. The effectiveness of any non-parametric index is determined by its ability to detect harm using all modified peaks.
- Ultrasonic technology, acoustic absorption, magnetic field analysis, global structural reaction analysis, and visual inspection techniques have also been proven to be useful at identifying damage early on. Regardless of their usefulness, both of these approaches should be used in conjunction with the EMI technique.

3. Structural Control Using Piezoelectric Material

Control of structures is a study in which the piezoelectric materials have been utilized for various purposes. Due to the versatile application, the PZT transducer can control metallic and non-metallic structural components, and this section reviews the piezoelectric materials as the main object to control structural component's vibration, noise, and activity. Furthermore, this section summarizes the methodologies that were used to achieve the goals.

3.1. Vibration Control

This concept was first presented by [32], proposing the use of piezoelectric transducers in combination with electrical components, called passive vibration control shunt circuits and flexible piezo patches [33]. A study was made via experimentally and numerically for vibration control with a wide range of operating temperatures [34]. The core principle consists of the transformation of the host structure's vibrant strain energy into electric energy. By using the direct piezoelectric effect, routing this energy through the shunt circuit where it can be partly absorbed was achieved. These are subsequently a superior class of energy conversion materials with associated mechanical and electrical features; shunt circuits coupled with such materials play an important role in the performance of wave propagation and/or vibration control in smart periodic structures [35].

With low and medium ranges of frequency, information can be collected for a damped thin plate with piezoelectric patches that are associated with the time-varying RL shunt circuits. The plate damping was considered in the lower range and five time-varying shunted patches were used [36]. As a result, the damping factor and frequency were executed via PZT effects in such a way as to either switch between given values, or sweep within certain ranges, to control the resonant response of targeted flexural modes of the plate. Figure 2 shows a smart panel made of a thin rectangular plate of aluminum and a series of five thin piezoelectric patches polarized in the transverse direction [36]. The physical and geometrical parameters of the plate and piezoelectric patches are summarized in [36]. The panel's dimensions and physical properties were selected to represent a part of an aircraft skin wall made up of two stringers and two rings, which can be easily modeled as a supported panel. The patches were chosen so that they would occupy a good part of the panel and have the same width as the panel. As a result, four patches were bonded on one side of the panel and the fifth patch was bonded on the opposite side at the panel's middle, as seen in Figure 2b. The patches are made of a traditional piezoceramic material that is isotropic in the x–y plane and therefore can effectively act on the two-dimensional hosting plate structure. The plate is subjected to a white noise rain-on-the-roof excitation, which is an idealized excitation made up of a uniform distribution of uncorrelated point forces that similarly excites the structure's natural modes. The rain-on-the-roof excitation has been approximated in this analysis with a 4-4 finite array of forces such that they are divided around a flexural wavelength at 1400 Hz and hence achieve the even modal excitation in the entire frequency band considered in the simulations, as seen in Figure 2a.

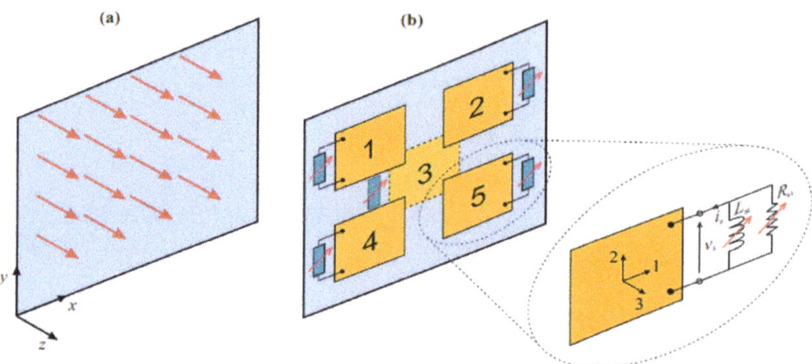

Figure 2. Vibration plate with shunted piezoelectric patches. Reprinted from [36], Copyright 2021, with permission from Elsevier.

With a parametric actuator and control theory in a distributed form, the active vibration damper of a cantilever beam was designed. A piezoelectric polymer, polyvinylidene fluoride (PVDF), was the distributed-parameter actuator and Lyapunov's second approach was used for distributed-parameter systems to design a damper control algorithm [37]. Honghao [38] has also conducted an experimental study of the AVC of a piezoelectric laminated paraboloidal shell by positive position feedback. The piezoelectric active damper was proposed by Bailey et al. [37] in 1985. The first modal damping ratio of the beam was increased to 4.5 times with distributed PVDF piezoelectric film layers laminated on one side of the flexible cantilever, and these PVDF patches were laminated inside and outside the shell, of which eight were used as sensors and eight as actuators to monitor the vibration of the first two natural modes. Through the frequency response feature review, Modal VIEW software [38] obtained lower natural frequencies and vibration modes of the paraboloidal shell. Hagood and Flotow [39] derived the mechanical impedance for the piezoelectric part shunted by an arbitrary circuit. It was found that the shunted piezoelectric has a frequency-dependent stiffness and loss factor that also depends on the shunting circuit.

An electrical resonance is introduced by shunting with a resistor and inductor, which can be optimally tuned to structural resonances in a way similar to a mechanical vibration absorber [39]. In the moving system, it is caught in its high stiffness state to store energy in actuators. The motion of the device is responsible for obtaining control back from the actuator, so the actuator is changed to a low state of stiffness, dispersing the energy [40].

It is necessary to know that the actuators for the structural portion of wing vibration used in aerospace applications are controlled by the actuator. In this case, highly flexible multi-functional wings with embedded piezoelectric material are used for adaptive vibration control and energy harvesting, according to recent studies by Natsuki [41]. The use of an actuator for an unmanned aerial vehicle, combined with a non-destructive health monitoring system based on vibration, was proposed. The indication was that someplace excitation and record acquisition occur simultaneously from the piezoelectric transducer against a truth expansion. This removes the need for roofing training with hundreds of monitoring sensors, as this concept uses a particular piezoelectric transducer to monitor a structure. By converging unmanned aerial vehicles, the expected handiwork creates new fields of inquiry [42]. A Simple-FSDT-based iso-geometric approach with the mathematical expression for piezoelectric functionally graded plates has been studied on vibration analysis [43]. Such studies have been found to shape the control of the antenna reflector to the desired shape—a closed-loop iteration based on the influence coefficient matrix and FE model [44]. The AVC method was also found in civil infrastructures in which the review has been made by considering different types of application of civil structures that have been controlling piezoelectric material [45]. For the case of suspension bridges, an investigation has been made for the active damping controlled by decentralized integral force feedback [46] and, similarly, model frame structures, [47] including smart model frames [48] and loop share buildings [49], are also controlled.

3.2. Noise Control

Controlling noise can be done through the smart piezoelectric transducer in any type of structural device. This section illustrates the previous work done by the researchers to control the noise using piezoelectric materials.

Aridogan and Basdogan [50] analyzed existing advanced systems of active vibration and noise control for plate structures with varying boundary conditions. Numerical and experimental techniques were reviewed to search various facets of control structural design. First, according to their designs, they identified the controls, then compared their vibration and noise reduction efficiency, and at last included recommendations for further development. Shivashankar and Gopalakrishnan [51] reviewed the use of piezoelectric materials for active vibration, flow control, and noise to seek analysis to outline the improvements achieved in both areas by focusing solely on the application of the piezoelectric material. Gripp and Rade [52] offered a comprehensive literature analysis of numerous piezoelectric shunt damping techniques industrialized for vibration and noise reduction in mechanical systems, an evaluation of the fundamental principles, as well as design procedures and computational simulation of piezoelectric shunt damping variance. Casadei et al. [53] described simulation and experimental examinations of the operation of a periodic series of piezoelectric forced RL patches for the reduction of broadband noise radiated in an enclosed cavity by a flexible layer. Frequency bandgaps described the reaction of the resultant periodic system where vibrations and related noise were highly attenuated.

Ang et al. [54] provided an analysis of current practices in various industries, such as automobile, maritime, aerospace, and defense used for cabin noise control. Nevertheless, the focus was put on cars and armored vehicles. In general, car cabins typically consist of thin structural plates, where the simple frequency usually drops below 200 Hz. Booming noise happens if a certain structural mode couples with a particular acoustic mode of the cabin. Lai et al. [55] investigated the effects of equivalent series resistance on the noise mitigation performance of piezoelectric shunt damping, developed an understanding of the impact on noise mitigation efficiency of the equivalent series resistance (ESR) of the piezoelectric damper in a piezoelectric shunt damping (PSD) device and showed that an improved ESR contributes to a substantial improvement in noise transmissibility due to a decrease in the mechanical damping of the device.

Salvador et al. [56] presented the possibility to extort such noise in the form of electricity into renewable energy using a combination of a piezoelectric transducer and a super-capacitor. A prototype was strategically built and constructed between the streets of Lerma and Nicanor Reyes usually congested by traffic, situated within the university belt area in Sampaloc, metro manila, Philippines. Araujo and Madeira [57] obtained an active control bonded with PZT sensors and actuators to optimal noise reduction solutions in laminated viscoelastic soft-core sandwich plates. An in-house finite element implementation of the active laminated sandwich plate was used to accomplish the frequency reaction of the sheets. Using the Rayleigh integral method, the sound propagation features of the panels were determined by computing their radiated sound power, since the structural/acoustic problem are loosely coupled. As an actuator and sensor based on a numerical solution method also called the generalized differential quadrature approach, intelligent control, and dynamic investigation of a reinforced composite graphene nanoplatelet (GPLRC) cylindrical shell surrounded by a piezoelectric layer were provided (GDQM). The strains and stresses were measured using the First-order Shear Deformable Theory (FSDT). The results showed that the PD controller's weight fraction, viscoelastic base, slenderness factor, external voltage, and graphene nanoplatelets (GPLs) have a major impact on the vibration and amplitude of the GPLRC cylindrical shell [58].

Li et al. [59] discussed several simple instruments and techniques widely used in various controlled objects for different components of the active control system. For reducing periodic noise produced in a high magnetic field, such as noise generated by magnetic resonance (MR) imaging devices (MR noise), an active noise control (ANC) method was suggested. Optical microphones and piezoelectric loudspeakers were used for the proposed ANC system, as specialized acoustic equipment was required to address the high-field issue and consisting of a structure mounted on the head to control noise near the user's ear and to recompense piezoelectric loudspeaker's low performance [60].

3.3. Active Control

The damaged structures, such as cracked and delaminated composite, which are damaged due to the external load or accident can also be controlled by the piezoelectric material application. In many cases, composite material is inexpensive and lightweight, but due to high rigidity, it can crack or delaminate easily as compared to other types of body kit materials [61], and such type of delamination can be controlled by a PZT actuator. As an active control, PZT actuators were used over the last two decades, during which researchers have investigated various subjects, such as the control of buckling, stresses, and cracks in the structures [62], and the use of the piezoelectric actuator to sense and operate the structure has been a point of interest [63–69].

The PZT actuators have adequate output in compromised systems. The active electromechanical coupling effect greatly alters the properties of the damaged structure due to its adjustable mechanical properties [70] for active control in damaged structures in which Rao et al. [71] investigated an aluminum plate with a central crack for repairing, by using four equal sizes (40×40 mm^2) patches made up of piezoelectric material polarized in the X-direction. The active repair of an aluminum center-cracked plate was carried out to minimize the stress concentration at the crack tip by bonding four piezoelectric patches to the host structure. The single-strap adhesive joint system has been analytically modeled to research the effect of piezoelectric patch surface bonding and stress distribution in the adhesive layer [72]. Investigation of the delamination control of composite plate with the PZT actuator by inducing low-speed impact using explicit FE code LS-DYNA was also carried out [73] in such studies.

Sohn et al. [74] developed a signal processing technique for composite plates, to predict delamination. This method, together with an active signal system, was used for the continuous monitoring of the composite structures under consideration. The parameters that influenced the delamination size detection were the distance between piezoelectric patches, actuating frequency, wavelength, and the size of the grids. They proposed a damage detection algorithm that easily detected the delamination under different boundary conditions and temperatures [4], to repair delamination to prevent beam fractures. Their findings showed that the voltage needed in beams to repair delamination depends solely on the delamination position. A related study was conducted using ANSYS software by Liu to restore delamination using piezoelectric materials. It has been indicated that the lower voltage is ideal and economical for safer operations, while the patch length, layer and thickness are highly influential [75].

The use of piezoelectric patches for repairing the delaminated beam under static loading conditions was also reported in [76]. The higher voltage suggested for a spring model to analyze the bonding adhesive effect on the output of piezoelectric patches (single and multi-layered patches as shown in Figure 3) used as active cantilever beam repair [6,76]. To decrease the stress intensity factor, the voltage applied to close the crack was higher [77,78]. Another study was documented by Wu and Wang using FE analysis during static loading conditions. To eliminate compressive and tensile forces around the delamination site, they concocted a discrete electrode from piezoelectric actuator patches to reduce stress singularity around the damaged part [79–84].

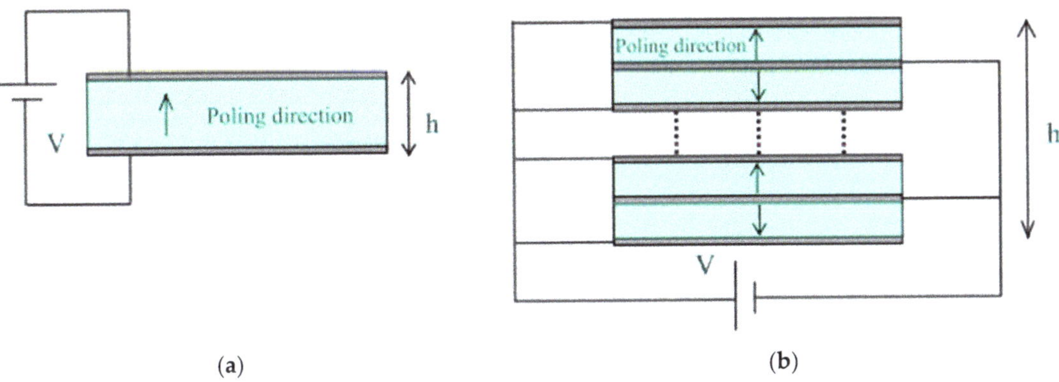

(a) (b)

Figure 3. Piezoelectric patches: (a) single-layer patch and (b) multi-layer patch. Reprinted from [6], Copyright 2021, with permission from Elsevier.

Furthermore, an example of this type of study has been found by Wang [85], in which the model with a simply supported beam undergoing axial compression was tested. At the center of each actuator's surface, a resistive strain gauge is connected. Each piezoelectric patch is polarized along the Z-axis as an actuator and applied through its thickness with a voltage. A cantilevered beam exposed to axial compression was the second design. The actuator pair positions were set to shift from the clamped end to the free end along the beam to find the optimum locations.

In recent investigations, Abuzaid et al. [86–91] utilized the piezoelectric actuator (PIC 151) to control the crack propagation in aluminum rectangular thin plates and they determined the fracture parameters, such as stress intensity and stress concentration factors. The ideas developed to produce stress (compression/tension) on the stress distribution around the hole and along with the width of the damaged plate by the piezoelectric actuator. The methodologies were adopted via experimental work for an edge-cracked plate [92] and edge- and center-cracked analytical modeling using the crack-closer method [91] and weight function method [92] and numerical simulation via ANSYS simulation [92,93]. To determine SIF for the center-cracked plate, concentrated on the location of the piezoelectric patch on a cracked plate with respect to the dimension by changing the patch location and thickness of adhesive layers. For the same reason, piezoelectric actuators additionally attached a composite patch for the repair of the cracked plate with the determination of SIF [94] and SCF [95].

4. Structural Health Monitoring

Generally, SHM used in wide applications with its advanced technologies and a number of research studies has been reported in the literature over the last two decades. For the sake of piezoelectric material application, this review has reported with some fundamentals/methodologies/overview used to perform SHM on a damaged structure and its enhancement. The enhancement of orthotropic and isotropic material for piezoelectric transducers can improve its properties significantly. The structural strength and stiffness of the material together make it a high-performance material. Delamination in composite structures plays a key role in reducing structural strength and rigidity, subsequently reducing device integrity and reliability so that the lamb-wave technique can be efficiently produced using piezoelectric transducers embedded within a composite plate for health monitoring [96]. The SHM is an innovative technique built from non-destructive testing (NDT), blends sophisticated sensor technologies with intellectual algorithms to cross-examine systemic health conditions [63]. Statistical model creation is concerned with the implementation of algorithms that use the extracted features to measure the extent of the damaged structure. These algorithms can be classified into two classes, as shown in Figure 4. To improve the damage detection process, all of these algorithms test statistical distributions of the measured or derived features. A broader and more comprehensive discussion can be found in [97,98], which are two fundamental texts for all people working on SHM. Moreover, supervised learning strategy for classification and regression tasks applied to aeronautical SHM problems was discussed in detail by Miorelli et al. [99].

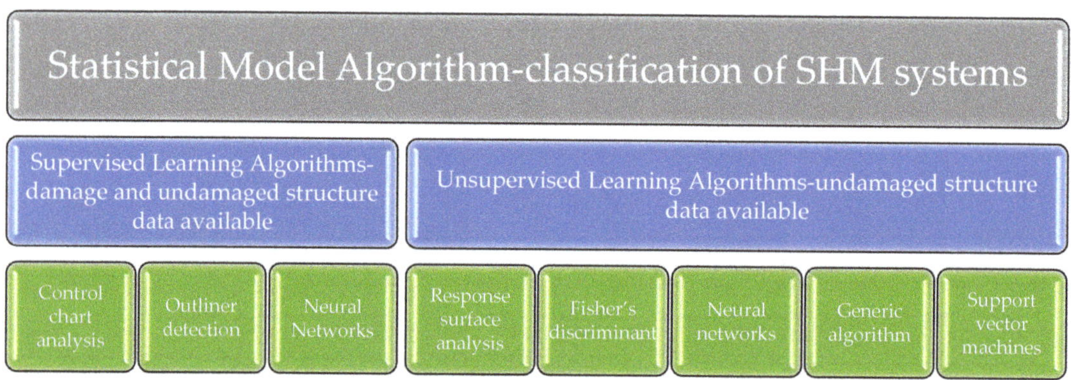

Figure 4. Algorithms classification for Statistical Model Development for SHM.

4.1. Aerospace Structures

For the identification of minor emerging vulnerabilities in engineering systems, several researchers are focusing on SHM techniques based on guided wave propagation. Low-velocity impacts on structures may cause these defects, which are often not apparent to the naked eye [100,101]. Instead of using traditional non-destructive methods, guided wave propagation techniques for SHM are often used in the aerospace industry [102]. Continuous SHM for aerospace systems during service is a difficult but promising technique [103]. As a result, several researchers are developing these techniques for the continuous SHM of aerospace systems when they are in use [30,104,105]. Because of the environmental consequences, sophisticated SHM methods are very difficult to adopt even in ideal circumstances [106].

Because of its ability to detect very small losses, guided waves are the most active area of study in SHM for aerospace, as demanded by the aircraft industry. The definition is straightforward, as seen in Figure 5: A PZT bonded/embedded into the structure emits a short ultrasonic pulse (the frequency used is a few hundred kHz) that propagates as an elastic wave through the plate and is absorbed by other PZTs, though warped. The signals received are saved and compared to signals received later in the structure's lifespan. Any new signal distortion must be the result of a structural change in the emitter–receiver PZT direction. In flat laminates, the idea works well, and minor delamination's produced by an effect can be observed and even found. The method also fits well for cylindrical tubing [107]. As the concept is applied to specific systems with boundaries, stiffeners, and thickness adjustments, the complexities increase. Elastic waves behave like all other waves, with reflection and refraction happening at either interface, complicating the signals obtained [108]. Furthermore, variations in thickness encourage mode switching, and all modes are dispersive, moving at varying speeds. As a result, signal processing and measurement were much more challenging. The obtained signals are often distorted by temperature and operating loads [109]. Modeling wave propagation and association with defects with structures with increasing geometrical complexity, such as stiffened structures, is currently a major focus [110].

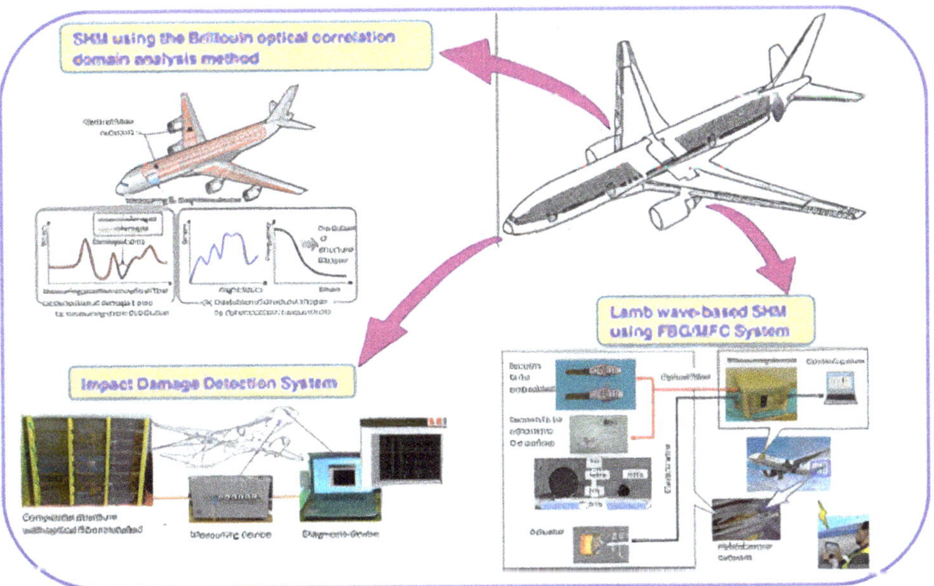

Figure 5. Damage detection alternatives with fiber optic sensors [99]. Reprinted under the Creative Commons (CC) License (CC BY 4.0).

While extensive SHM advancement has been accomplished, a great deal of work is still needed for more practical applications of SHM in composite materials [111]. Through a down-select procedure based on the control and health monitoring of structural design for the space vehicle, the tested SHM sensors and their sensing techniques were chosen [112]. To design online SHM systems for aerospace vehicles or aircraft, an extended fluid–structure interaction was proposed. It is a strongly coupled version of a standard FSI problem with a wave propagation problem coupled, in which the wave propagation problem is posed on the moving mesh which is automatically adopted from the FSI problem at each time stage [112].

For the damage/delamination studies, to identify skin/stiffener debonding and delamination cracks suitable for laminated composite structures, the SHM method was proposed [7]. An improvement of the SHM methods, the lamb-wave techniques for quasi-isotropic graphite/epoxy thin patches and sandwich beams containing representative damage modes, transverse ply cracks, delamination, and through-holes were studied. The detection of damage by measuring transmitted waves with piezoceramic sensors was experimentally optimized and given a technique capable of simple and precise determination [113], and it is recognized that piezoelectric materials constitute both the electrical and mechanical properties and are used in the field of SHM engineering [114]. Moreover, piezo composite [115] patches were applied to detect the defects using lamb-wave focusing [116]. At the time of design, the piezoelectric sensor was embedded and demonstrated in the structure and the bonded patch served as sensors for both the global dynamic technique and the EMI technique [117]. Assessments of SHM for fatigue cracks in metallic structures were made by using lamb waves guided by piezoelectric transducers [118]. The piezo ceramic transducer-based electromechanical impedance technique (EMI) and the digital image correlation (DIC) method that uses structural surface adjustments with monitoring were experientially studied [119]. The consignment of tiredness generally exacerbated the fissure if there were any defects in the structure. The adjacent active electrode multiple-crack monitors triggered numerous airplane defects that were caused by EMI output and the DIC system in the specimens over the weakness test [119]. Although the EMI analysis is complicated, it can be carried out by breaking down the structural loss into actual and theoretical impedance components. Figure 6 depicts the ultimate process for SHM depending on impedance.

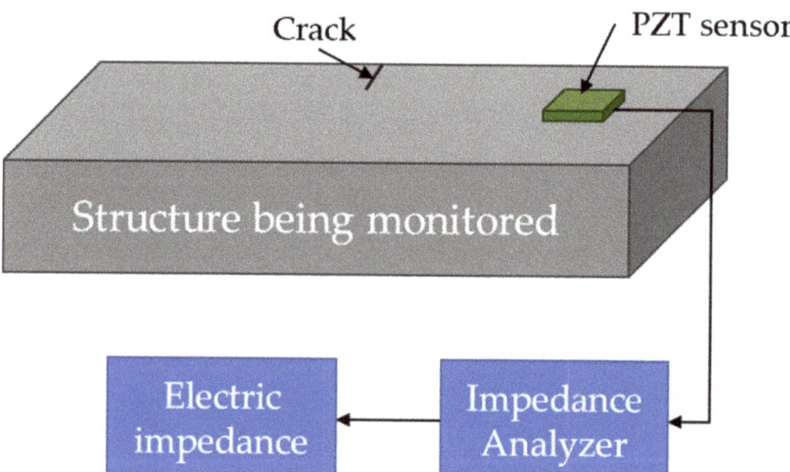

Figure 6. SHM of the cracked structure by impedance analyzer.

Sensor amplifiers and data acquisition units make up passive diagnostic hardware. Through a sensor network, the impact monitoring system collects stress wave signals produced by impact loads [120]. Figure 7 illustrates the process of the impact monitoring of aircraft structures.

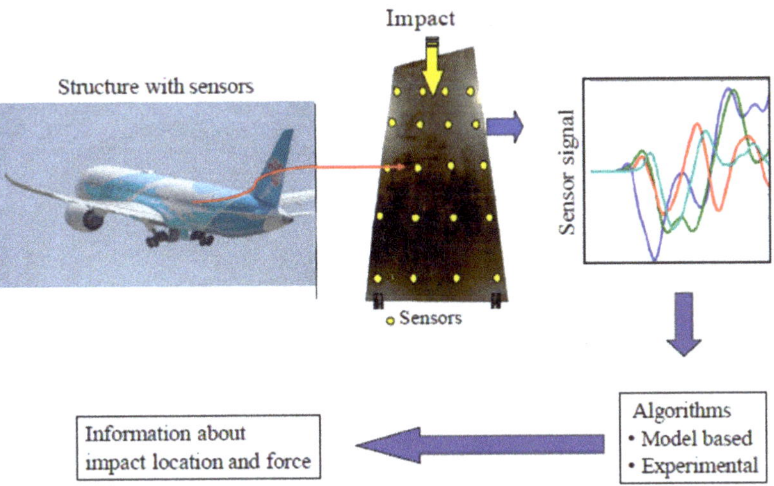

Figure 7. Impact monitoring of aircraft structures [120]. Reprinted under the Creative Commons (CC) License (CC BY 4.0).

Richard et al. [121] examined, by the analytical method, the vibrations of the piezo-electric composite plate through cylindrical bending and experimentally established the various vibrations for the bolted composite plate based on the SHM technique. This novel SHM approach combines vibration-based thermography with the idea of local defect resonance to create a novel SHM method. The use of hard shakers to apply high excitation and infrared cameras to observe thermal responses are also major challenges for face layer debonding detection in aerospace sandwich structures [122]. For an aircraft skin health monitoring system, a piezoelectric sensor network with shared signal transmission wires was proposed with multiple PZT using the design principle method [123]. On such cases, the PZT sensor network, which uses mutual signal communication cables, must be wired longitudinally and transversely, which is difficult for some real-world aircraft structures with several frames and ribs. Some suggestions has been made and reviewed based on the piezoelectric sensors for aerospace structures for SHM [124].

As an overview of this section based on the above studies using the SHM approach, some conclusions can be derived. For the case of composite materials, this method was found to be very useful to detect delamination, cracks, and minor damages. With the SHM technique, it is possible to improve the mechanical and electrical properties of intact and damaged materials. Hence, using piezoelectric transduces for SHM techniques will be an innovative concept in engineering research fields such as aerospace structures. The example is shown in Figure 8. By improving the material's properties, and particularly the material's toughness, the slope of the curve is decreased and the durability of the structure is increased [125].

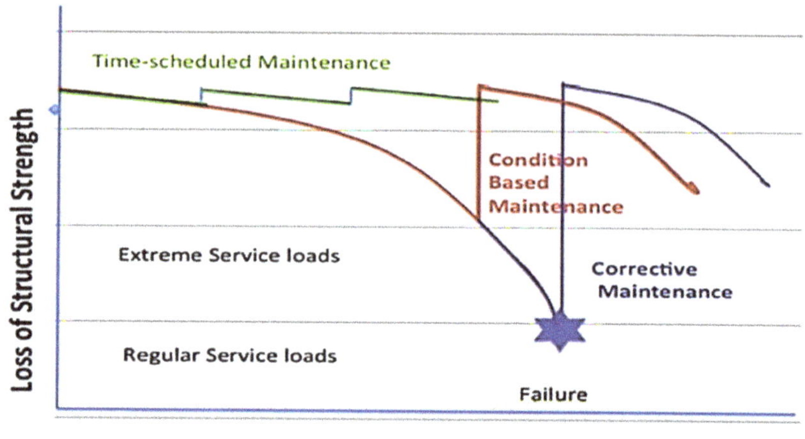

Figure 8. Maintenance strategies with/without SHM [125]. Reprinted under the Creative Commons (CC) License (CC BY 4.0).

In critical view, a piezoelectric wafer active sensor has been used to monitor the onset and progress of structural damage, such as fatigue cracks and corrosion, on current ageing aircraft structures. The state of the art in active SHM sensors and detection of damage was reviewed by the researchers [126]. In general, the efficiency of the piezoelectric sensor is the most versatile and low-cost sensor. For global and local-level damage and cracks, piezoelectric patches have been used as a sensor and they explored the possibility of an embedded piezoelectric sensor. As multi-purpose sensors for research, piezoelectric patches were also used, using various methods such as modal analysis, acoustic emission, lamb-wave, and strain-based methods simultaneously by adjusting driving frequencies and sampling rates [127]. There are relatively fewer sensors in the piezoelectric material and are therefore suitable for engineering applications such as aerospace, automotive, and civil structural health monitoring [128].

4.2. Concrete Structures

Due to the advancement of piezoelectric material, it has been used for civil concrete structures to detect and repair damages. Hence, this section reviewed the works related to civil engineering structures.

SHM field monitoring, which uses embedded sensors or real field testing to track the condition of existing or new civil engineering infrastructure, is an emerging technology. The use of SHM as a critical component of infrastructure design will be critical in the construction of the next generation of smart civil engineering structures. Intelligent sensing systems have four key components: (i) sensors and actuators that collect information and operate in a target environment; (ii) a network infrastructure for data and control signal transmission; (iii) data processing and visualization systems; and (iv) basic analysis and decision-making applications.

The presented model deals with the bonding layer as a mass-spring damping mechanism between the piezoelectric patch and the substrate structure in the corresponding electromechanical analysis. The effect of the bonding layer was therefore considered on the dynamic interaction between the piezoelectric sensor drive and the main structure [129]. The piezoceramic transducer was developed in the contemporary past as an anti-intelligent material commonly used in electromechanical impedance (EMI) and guided ultrasonic wave broadcast techniques. A piezoelectric transducer interrelates with the horde construction in the EMI technique to study exclusive health marks as an inverse action of structural impedance in the incidence of the sensitive region on the application of the high-frequency structural load [30].

Effective monitoring of rock components in civilian infrastructures such as tunnels and caves remains challenging. Yang et al. [130] introduced the use of smart fiber-optic and PZT impedance sensors for integrated rock condition monitoring, load profile monitoring/detection, and damage assessment. Rock samples were loaded periodically, and their condition was constantly monitored by fiber-optic and piezoelectric sensors. The surface of a fiber optic sensor was based on a multi-fiber Bragg grating sensor in combination with rock samples. The strain sensitivity was compared to that of a conventional electrical strain gauge. The EMI technique of the piezoelectric transducer, consisting of real and virtual parts, served as an indicator for predicting the state/integrity of the host structure. In practice, however, components such as panels, beams, and columns were constantly subjected to external loads. Experimental and statistical studies showing the impact of exposure to valid electronic results were studied [131]. Moreover, it was noted that an acceptance indicator was better than a conductivity indicator for locating pressure in situ in the main structure. This observation was further confirmed by statistical analysis.

Experiments were performed to examine the complex tribulations in the EMI technique's real-life implementations, seeking to moderate the space between suspicion and application. Experimental studies showed that the bonding thickness is expected to be much thinner than one-third of the scrap to duck any negative realization induced by the reminiscence of the piezoelectric direction on the permission signatures restoring the structural actions of the horde. In order to be thoroughly connected to the thickness of bonding, the hotness on the access signatures was created, as an intensification of hotness would ease the tautness of the bonding sheet, resulting in disturbing strain transfer [132]. The structural mechanical impedance extracted from the incoming piezoelectric electromagnetic signal is used as an error indicator. A comparative study of the sensitivity of the transmission of electromagnetic waves to damage concrete structures and the mechanical resistance of the structure was carried out. The results show that structural mechanical impedance is more susceptible to damage than EM, which is a better indicator of damage detection. Genetic algorithms were used in dynamic systems to find optimal values for unknown parameters. The experiments were carried out on a two-story concrete frame, subject to basic vibrations that simulate an earthquake. Several piezoelectric sensors were regularly assembled and connected to the frame structure to obtain the PZT-EM approval method. The relationship between noise and the disturbance distance of a PZT sensor was examined to cover the sensitivity and sensitive areas of the PZT sensor [133].

Xio and Jiang [134] proposed a method for detecting shear bolt failure in reinforced concrete composite beams based on EMI measurements of PZT ceramic sensors. Several piezoelectric patches were superficially attached to the top flange of a steel beam and concrete slab, and their EMI was measured with an impedance analyzer before and after loosening the connecting screws. Based on the impedance measurement, the impedance spectrum was estimated and compared for the presence of interconnecting volts using several general error statistics including standard deviation, absolute percent deviation, and correlation coefficient deviation. Five kinds of scratch requirements were considered to examine the impedance ethics at sundry frequency bands. Reliable regulations are originating by control and analysis. Equally, the core median pays off deviation and the correlation coefficient deviation smash up indices are adapted for detecting the structural damage. The mathematical and experimental studies verify that the EI structure can accurately detect changes in the quantity of break-in armored definite slabs. The smash up alphabetical listing changes evenly with the vastness of costs to the sensor [135]. Piezoelectric transducers in the structure of smart aggregates are embedded into the specimen during casting. Piezoceramic equipment can be as old as actuators to breed peak frequency vibrating waves, which promulgate in physical structures; meanwhile, they tin beside old sensors to reveal the waves [136].

Load-induced structural stress/compression stress and stress-induced injury were tested qualitatively by evaluating and contrasting the electromechanical input technique typescript with that of the non-stressed piezoelectric transducer. Quantification assessment of stress and harm through algebraic source shabby pay deviation indicator was also presented [137]. For damage detection of a 6.1 m extensive non-breakable physical attachment bent cap, piezoceramic transducers are used. At pre-determined spatial locations before casting, piezoceramic transducers are embedded in some designs. This delves into being able to be cautious as a prior work persistence, wherever four piezo ceramic transducers were embedded near one point of the bent cap in planar locations. This involves ten piezo ceramic patches in four separate cross-sections embedded at spatial locations for the investigation [138].

Talakokula et al. [139] studied the effects from a series of accelerated oxidation tests performed on rebars embedded in separate cubes in which measurements were ready to float up bonded on rebars using the electromechanical impedance technique via piezo ceramic patches. The comparable structural parameters derived from the entry signatures of the PZT information were measured against the deterioration objective based on which a new representation of the oxidation assessment was proposed. The experimental fallout suggested that in the practical detection and quantification of the decomposition, equal parameters were efficient. To continue its trend, realistic studies associated with the EMI method for the past decade have been reviewed. New ideas and dreams planned by a variety of writers were also discussed, and the tabloid ended with a discussion of the promising guidelines for imminent works [140]. In the next study, the PZT sensor was placed near the crack on the concrete sample surface, from which the impedance and admission were decided by a connected impedance analyzer to determine the structural conditions [141]. The wireless smart aggregate SHM device configuration and signal flow are depicted in Figure 9. A detected reinforced concrete (RC) structure, a signal excitation module, a signal data acquisition module, a wireless communication module, and a power module make up the established device. The smart aggregates are pre-embedded in the RC structure that was observed, and the detailed study can be found in [142].

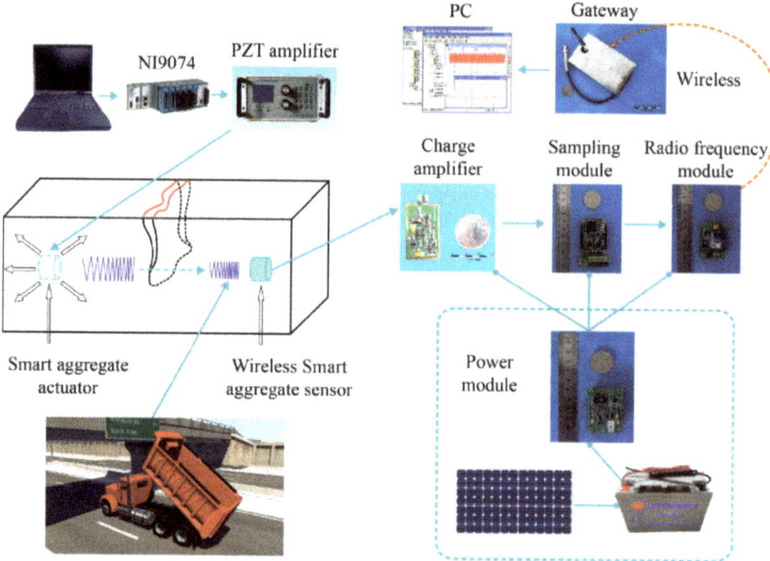

Figure 9. System setup and signal flow of the SHM system [142]. Reprinted under the Creative Commons (CC) License (CC BY 4.0).

4.3. Practical Issues

Numerous experiments have been conducted in this field since the advent of the EMI methodology, with many reporting positive findings. However, the majority of the experiments were conducted in laboratories or were mostly theoretical, raising concerns about their potential in real-world applications, especially in harsh environments [140]. While damage to the structure alters the impedance signature, other variables such as temperature and the reliability of PZT transducers may also alter the signature. Sun et al. [143] used the EMI technique in a temperature-varying environment and found that raising the temperature softened the total stiffness of the host system, shifting the resonance frequency spectrum and changing the peak amplitudes. By horizontally rotating the signature, the authors used cross-correlation to eliminate signature variance due to temperature change. Park et al. [144] observed that the actual part of impedance signatures could be preferable to the imaginary part since the real part of the free PZT patch's signature shifted just slightly with temperature fluctuations. Besides, the authors proposed another technique for compensating for signature variations caused by temperature changes; several other researchers have also looked into this subject.

Grisso and Inman [145] suggested using a frame structure to perform experiments under different temperatures to separate the temperature variance effect from sensor defects. The researchers discovered that the estimated susceptance slope had a linear relationship with temperature changes. Wandowski et al. [146] evaluated the suggested temperature correction algorithm using carbon-fiber-reinforced polymer samples. The two-step algorithm works by shifting the signature first in the horizontal direction using cross-correlation, then in the vertical direction using signal normalization with root mean square values. Although many techniques for compensating for the temperature effect of the EMI technique have been proposed, completely compensating for this effect is extremely difficult. Some peaks can change or increase/decrease in amplitude more than others, making impedance signature variations unpredictable [147].

5. Challenges and Opportunities

The critical analysis and discussion of the previous selected studies of this review have been categorized into SHM and control of structures. However, over the last two decades, there have been several studies reported based on the use of piezoelectric materials in engineering structures, and the challenges and opportunities of the review have been illustrated from recent studies.

In order to tackle real-life field implementations, a lot of work is being put into studying these deployment problems in SHM techniques. For more than two decades, the EMI technique has been used, and based on recent studies, numerous problems must be resolved before it is practical to structures. The technique, which likely played a part in the formation of a large amount of force in SHM systems, involves the wear and tear of a unique piezoelectric for sensing and exciting the mass structure. The study presents a modified model of the EMI of piezoelectric drives from many researchers. The presented model deals with the bonding layer between the piezoelectric patch and the familiar structure as a mass-spring damping system in the corresponding electromechanical analysis. In the contemporary past, the piezoceramic transducer has emerged as a useful smart material, which is generally employed in EMI and guided ultrasonic wave broadcast techniques. In the EMI technique, a piezoelectric transducer interrelates with the horde construction for findings in exclusive health signature, as an inverse behavior of structural impedance and at what time it is exposed to high-frequency structural excitations in the occurrence of the sensitive field. Effective tracking of rock components in civilian infrastructures such as caves and tunnels remains challenging.

SHM-based electromechanical impedance damage detection is rapidly evolving in structural design. In the electromagnetic impedance method, a piezoceramic transducer is mounted on the surface of the main structure to operate electrically. The EM technique of the piezoelectric transducer consists of real and virtual parts, serving as an indicator for predicting the state/integrity of the host structure. In practice, however, components such as panels, beams, and columns are constantly subjected to external loads. The EM approval mark obtained for permanently loaded structures is different from that acquired in the event of structural damage. In a new era in the subject of SHM focused on non-destructive assessment, the beginning of smart tools such as the piezo-impedance transducer and optical fiber has begun. Thus, the research on the EMI potential using a piezo-impedance transducer is often laboratory based and largely theoretical. The technique's real-life attention, especially in harsh environments, has typically been challenged. The repeatability of electrical input signatures obtained from the floating piezoelectric patches bonded to aluminum structures was initiated to be brilliant for up to one and a half years. The EMI based on piezoelectric ceramics for SHM has been successfully applied in a variety of technical systems. However, a basic study of the sensitivity of piezoelectric impedance sensors to fault detection is still needed. Traditional EMI methodology uses the EM tolerance of the piezoelectric (as opposed to impedance) as a failure indicator, making it difficult to determine the impact of damage on structural properties. The following are some open research areas for SHM for engineering structures and applications that may need focus.

- It is difficult to install the piezoelectric to the host structures and generate a high-frequency range; therefore, the piezoelectric is packaged in a way that makes installation easier [148].
- Modern civil infrastructure projects were designed to provide specialized functionality for multi-purpose applications in extreme weather situations, including earthquakes, hurricanes, and typhoons. These dynamic structural structures raise significant questions about their stability. Integrating a network of smart and embeddable sensors into civil infrastructure systems with local artificial intelligence (AI)/machine learning (ML) data-processing platforms is a promising solution to this problem, allowing next-generation smart, civil infrastructure systems to be installed on the skeleton of conventional systems [149].
- The lack of a computational platform on which to develop new techniques for realizing massively distributed smart sensors is one of the crucial issues listed in [150] and the possible development of the next generation of SHM systems.
- Although they have been the focus of intense study for many decades, computational approximation and simulation of fluid–structure interactions remain an undeniably difficult topic with many unsolved problems and concerns, and the "arbitrary Lagrangian-Eulerian" method is a standard framework for solving fluid–structure interaction problems, as we emphasize [112].
- SHM can be used to measure civil industries that are live loads on bridge systems as well as environmental loads and to spot any significant differences from the values assumed in bridge-building codes. Given raising questions about climate change and its possible effect on the stability and serviceability of concrete bridge systems due to increased wind loads, floods, thermal gradients, freeze–thaw cycles, deicing salt use, and other factors, this is becoming an important topic [151].

Strong suit monitoring is critical in the first part of the age of structures to determine the skill of the structures for operation. Strong point monitoring techniques based on piezoelectric avails an innovative experimental approach including performing material health monitoring at primitive ages. The piezoceramic transducer has emerged as a new tool for the health monitoring of large-scale structures expected to benefit from their low cost, work sensing, clever response, simplicity for implementation, and availability, not the same shapes and simplicity. As far as piezoelectric materials are concerned, Figure 10 provides the figures and the timeline for different fields of use. Figure 10 concentrates on the use of materials and the hope is that this number will keep growing in the near future within engineering structures. This expectation is due to fact that these materials have already built a reputation, and because of their advancement and growth, which were previously stumbling blocks for conventional approaches, they have encouraged new possibilities in resolving various engineering issues. The existing literature provides a clear picture of the progressive production of piezoelectric actuators in these fields in the areas of SHM. The study of effective repair, on the other hand, is still in its infancy. One important thing to be noted is that piezoelectric-based MFC actuators can be preferred over ceramic-based materials due to their high impact resistance and flexibility. The development of piezoelectric materials creates new possibilities for the repair and control of active damage techniques. Active control is a very challenging task, in both technological and design aspects. The implementation of research software to explain the repair process would be important.

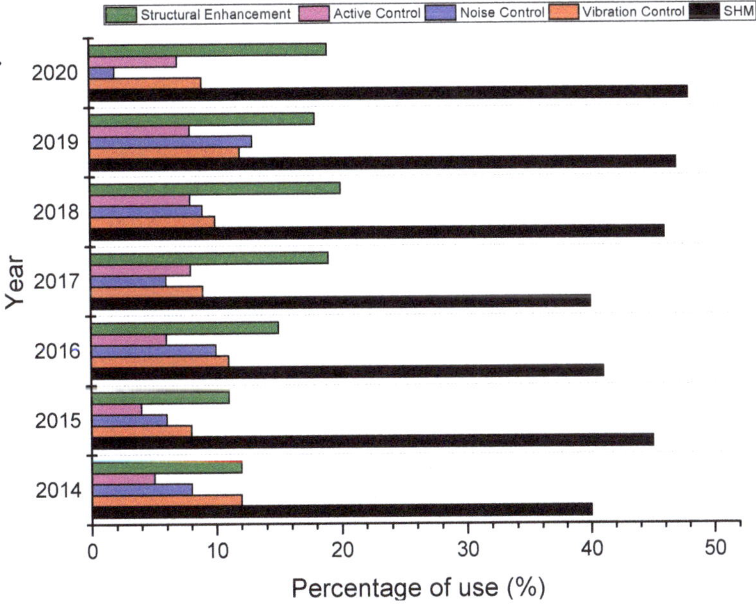

Figure 10. Trends in piezoelectric research.

Most of the previous works related to dynamic reactions of piezoelectric coupled systems with open-circuit electrical boundary conditions were performed using FE simulations. All previous repair techniques have been extended to thin beams and plate systems. There is no mathematical or theoretical model for the structural repair of piezoelectric materials that provides an active feedback control mechanism for multiple damages to various structures. The composite laminates were used in many structural applications, involving active repair and control by specific requirements. The stress concentration at the adhesive joints can be controlled by using the piezoelectric actuator and the induced surface moment. The value of the fracture tolerance of essential damage systems and their components needs further study and the creation of innovative design solutions, according to the researchers. They say the crack tip strain is exceeding infinity at the tip of the crack, so the previous solution is more realistic. Table 1 shows a summary of previous work with limitations and opportunities based on structural control/repair.

Table 1. Summary of previous work based on the structural control/repair.

Type of Structure	Technique Adopted	Number of Piezoelectric	Focused Parameters	Remark	Reference
Simply supported and cantilever beam	Multi-domain boundary integral formulation and spring model	2 (single and multi-layered)	Displacements, electric potential, and friction coefficient	Friction contact does not affect the repairing mechanism	[78]
Graphite/epoxy plate	wavelet-based signal processing	16	Piezoelectric patches and delamination size	Simple and needs minimum interaction	[74]
Cantilever and simply supported beam	Euler–Bernoulli beam theory and numerical simulation	2	Delamination locations, shear stress, and voltage	Piezoelectric materials are capable of repair of delaminated beam	[4]
Cantilever model	FE analysis using ANSYS	2 (single and multi-layered)	Crack location and length, repair voltage, patch thickness, and length	FEM analysis allows a detailed understanding of active repairs as well	[75]
Beam	FEM analysis using ABAQUS	2 (top and bottom)	Repair voltage, repair index,	The repair index depends upon the delamination location	[76]
Cantilever beam	Multi-domain boundary integral formulation and spring model	2 (single and multi-layered)	Shear stress, peal stress, crack displacement	The piezoelectric patch actuation capacity decreases due to shear stress transfer at the interface of structure and crack	[6]
Composite (drop ply) and cantilever beam	Springle model and multi-domain boundary integral formulation	2 (single and multi-layered)	Total ERR distribution, normal and tangential crack surface displacements	The optimal position for the patch is on the top of the skin, which eliminates crack opening.	[77]
Rectangular plate	Stiffness ratio and induce strain	2 (top/bottom)2 (left/right) around a circular hole	Reduction of stress concentration factor	SIF	[62]
Aluminum alloy 2024-T3 and 7075-T6 plate	J-integral using the FE method	1 composite patch (carbon/epoxy)	Crack propagation	Parametric study and SIF with experimentation give more ideas	[152]
Aluminum 2024-T3 plate	Von-mises stress, J-integral using FE analysis	1 compsite patch (boron/epoxy)	Fatigue life	SIF	[153]
GH2036 superalloy (novel model)	low and high cycle fatigue loading	No patch	crack closure and behavior of growth	Reproduce with bonded composite patches and piezoelectric actuators	[154]
Aluminum plate	Step heating thermography with FE modeling	1 (Composite patch)	delamination and disbond) with thermal heat transfer	SIF	[155]

6. Conclusions

Based on the present investigations, some conclusions have been made:

- The procedure of control and repair used to preserve the structural integrity of damaged components is distinctive. It is established on the converse piezoelectric effect, in which the local moment and force induced in the piezoelectric materials by an applied electric field would make it easier for the structure to prevent the development of high stress and strain levels because of external load and thus lessens the criticality of the damage.

- Structural health monitoring is also proving highly significant in avoiding the premature collapse of structures based on aerospace and civil industries such as offshore platforms, houses, bridges, and underground structures.

In this review, guidelines for scientists attempting to apply piezoelectric actuators/sensors in engineering structures are introduced. These guidelines include descriptions, findings, and analysis of the critical literature of piezoelectric material applications. A brief idea of the research areas of piezoelectric material can be presented in the classification. Furthermore, researchers may give transparent views and indices for their research areas through the challenges and opportunities. In short, these guidelines can help researchers to develop new ideas, particularly in the early stages of this research field.

Author Contributions: Conceptualization, A.A. (Abdul Aabid) and M.A.R.; methodology, A.A. (Abdul Aabid) and M.A.R.; formal analysis, Y.E.I. and M.H.; investigation, A.A. (Abdul Aabid), B.P., M.A.R., A.A. (Asraar Anjum), N.P. and J.M.Z.; resources, A.A. (Abdul Aabid); data curation, M.H.; writing—original draft preparation, A.A. (Abdul Aabid), M.A.R., B.P., A.A. (Asraar Anjum) and N.P.; writing—review and editing, B.P., Y.E.I., A.A. (Asraar Anjum), M.H. and J.M.Z.; supervision, Y.E.I. and M.H.; project administration, A.A. (Abdul Aabid), M.H. and Y.E.I.; funding acquisition, Y.E.I. All authors have read and agreed to the published version of the manuscript.

Funding: This research is supported by the Structures and Materials (S&M) Research Lab of Prince Sultan University. Furthermore, the authors acknowledge the support of Prince Sultan University for paying the article processing charges (APC) of this publication.

Institutional Review Board Statement: Not applicable.

Informed Consent Statement: Not applicable.

Data Availability Statement: Not applicable.

Acknowledgments: The authors acknowledge Seung-Bok Choi, Inha Distinguished Harlim Professor for the advice and suggestions on writing this review manuscript. The author Asrar Anjum and Nagma Parveen acknowledge the support of the TFW2020 scheme of Kulliyyah of Engineering, International Islamic University Malaysia.

Conflicts of Interest: The authors declare no conflict of interest.

References

1. PI Piezo Technology: "DuraAct Piezoelectric Transducers": PI Piezo Technology. 2017. Available online: https://www.piceramic.com/en/products/piezoceramic-actuators/patch-transducers/ (accessed on 31 March 2017).
2. Saraiva, F. Development of Press Forming Techniques for Thermoplastic Composites Investigation of a Multiple Step Forming Approach. Master's Thesis, TU Delft, Delft, The Netherlands, 2017.
3. Sun, D.; Wang, D. Distributed Piezoelectric Element Method for Vibration Control of Smart Plates. *AIAA J.* **1999**, *37*, 1459–1463. [CrossRef]
4. Wang, Q.; Quek, S.T. Repair of delaminated beams via piezoelectric patches. *Smart Mater. Struct.* **2004**, *13*, 1222–1229. [CrossRef]
5. Wang, Q.; Duan, W.H.; Quek, S.T. Repair of notched beam under dynamic load using piezoelectric patch. *Int. J. Mech. Sci.* **2004**, *46*, 1517–1533. [CrossRef]
6. Alaimo, A.; Milazzo, A.; Orlando, C. Boundary elements analysis of adhesively bonded piezoelectric active repair. *Eng. Fract. Mech.* **2009**, *76*, 500–511. [CrossRef]
7. Alaimo, A.; Milazzo, A.; Orlando, C. Numerical analysis of a piezoelectric structural health monitoring system for composite flange-skin delamination detection. *Compos. Struct.* **2013**, *100*, 343–355. [CrossRef]
8. Kapuria, S.; Yasin, M.Y.; Hagedorn, P. Active Vibration Control of Piezolaminated Composite Plates Considering Strong Electric Field Nonlinearity. *AIAA J.* **2015**, *53*, 603–616. [CrossRef]
9. Zhang, C.; Nanthakumar, S.S.; Lahmer, T.; Rabczuk, T. Multiple cracks identification for piezoelectric structures. *Int. J. Fract.* **2017**, *206*, 151–169. [CrossRef]
10. Krishna, P.; Mallik, S.; Rao, D.S. Vibration Control on Composite Beams With Multiple Piezoelectric Patches Using Finite Element Analysis. *Int. Res. J. Eng. Technol.* **2017**, *4*, 906–911.

11. Dineva, P.; Gross, D.; Müller, R.; Rangelov, T. *Dynamic Fracture of Piezoelectric Materials*; Springer International Publishing: Cham, Switzerland, 2014; Volume 212.

12. Holterman, J.; Groen, P. *An Introduction to Piezoelectric Materials and Applications*; Stichting Applied Piezo: Apeldoorn, The Netherlands, 2013; ISBN 978-9081936118.

13. Curie, J.; Curie, P. Développement, par pression, de l'électricité polaire dans les cristaux hémièdres à faces inclinées. *Comptes Rendus de l'Académie des Sciences* **1880**, *91*, 294–295.

14. Chee, C.Y.K.; Tong, L.; Steven, G.P. A review on the modelling of piezoelectric sensors and actuators incorporated in intelligent structures. *J. Intell. Mater. Syst. Struct.* **1998**, *9*, 3–19. [CrossRef]

15. De Jong, M.; Chen, W.; Geerlings, H.; Asta, M.; Persson, K.A. A database to enable discovery and design of piezoelectric materials. *Sci. Data* **2015**, *2*, 1–13.

16. Qin, Q.H. *Advanced Mechanics of Piezoelectricity*; Springer-Verlag: Berlin/Heidelberg, Germany, 2013; Volume 9783642297, ISBN 9783642297670.

17. Dahiya, A.; Thakur, O.P.; Juneja, J.K. Sensing and actuating applications of potassium sodium niobate: Use of potassium sodium niobate in sensor and actuator. *Proc. Int. Conf. Sens. Technol. ICST* **2013**, 383–386. [CrossRef]

18. Samal, M.K.; Seshu, P.; Dutta, B.K. Modeling and application of piezoelectric materials in smart structures. *Int. J. COMADEM* **2007**, *10*, 30.

19. Benjeddou, A. Shear-mode piezoceramic advanced materials and structures: A state of the art. *Mech. Adv. Mater. Struct.* **2007**, *14*, 263–275. [CrossRef]

20. Lang, S.B.; Muensit, S. Review of some lesser-known applications of piezoelectric and pyroelectric polymers. *Appl. Phys. A Mater. Sci. Process.* **2006**, *85*, 125–134. [CrossRef]

21. Hall, D.A. Nonlinearity in piezoelectric ceramics. *J. Mater. Sci.* **2001**, *36*, 4575–4601. [CrossRef]

22. Ramegowda, P.C.; Ishihara, D.; Takata, R.; Niho, T.; Horie, T. Hierarchically decomposed finite element method for a triply coupled piezoelectric, structure, and fluid fields of a thin piezoelectric bimorph in fluid. *Comput. Methods Appl. Mech. Eng.* **2020**, *365*, 113006. [CrossRef]

23. Lee, T.; Lakes, R. Damping properties of lead metaniobate. *IEEE Trans. Ultrason. Ferroelectr. Freq. Control* **2001**, *48*, 48–52.

24. Nogas-Ćwikiel, E. Fabrication of Mn Doped PZT for Ceramic-Polymer Composites. *Arch. Metall. Mater.* **2011**, *56*, 2–6. [CrossRef]

25. Arnau, A.; Soares, D. Fundamentals of piezoelectricity. In *Piezoelectric Transducers and Applications*; Springer-Verlag: Berlin/Heidelberg, Germany, 2008; pp. 1–38. ISBN 9783540775072.

26. Bafandeh, M.R.; Gharahkhani, R.; Lee, J.S. Dielectric and piezoelectric properties of sodium potassium niobate-based ceramics sintered in microwave furnace. *Mater. Chem. Phys.* **2015**, *156*, 254–260. [CrossRef]

27. Arian Nijmeijer, H.K. Synthesis and Properties of Lead Magnesium Niobate Zirconate. *J. Am. Ceram. Soc.* **1997**, *21*, 2717–2721.

28. Pmn, W. Lead Magnesium Niobate. 1958. Available online: http://research.physics.illinois.edu/Publications/theses/copies/fanning/4plmn.pdf (accessed on 5 April 2021).

29. Pasquali, M.; Gaudenzi, P. A nonlinear formulation of piezoelectric plates. *J. Intell. Mater. Syst. Struct.* **2012**, *23*, 1713–1723. [CrossRef]

30. Annamdas, V.G.M.; Soh, C.K. Application of electromechanical impedance technique for engineering structures: Review and future issues. *J. Intell. Mater. Syst. Struct.* **2010**, *21*, 41–59. [CrossRef]

31. Giurgiutiu, V. *Structural Health Monitoring: With Piezoelectric Wafer Active Sensors*; Elsevier: Amsterdam, The Netherlands, 2007.

32. Swigert, C.J.; Forward, R.L. Electronic damping of orthogonal bending modes in a cylindrical mast-theory. *J. Spacecr. Rockets* **1981**, *18*, 5–10. [CrossRef]

33. Iyengar, N.G.R.; Kamle, S. Development and application of flexible piezo patches for vibration control. In *Society of Expt. Mechanical Proceedin*; 2003; pp. 48–49.

34. Sharma, A.; Kumar, R.; Vaish, R.; Chauhan, V.S. Experimental and numerical investigation of active vibration control over wide range of operating temperature. *J. Intell. Mater. Syst. Struct.* **2016**, *27*, 1846–1860. [CrossRef]

35. Bao, B.; Guyomar, D.; Lallart, M. Vibration reduction for smart periodic structures via periodic piezoelectric arrays with nonlinear interleaved-switched electronic networks. *Mech. Syst. Signal Process.* **2016**, 1–29. [CrossRef]

36. Casagrande, D.; Gardonio, P.; Zilletti, M. Smart panel with time-varying shunted piezoelectric patch absorbers for broadband vibration control. *J. Sound Vib.* **2017**, *400*, 288–304. [CrossRef]

37. Bailey, T.; Hubbard, J.E. Distributed Piezoelectric-Polymer Active Vibration Control of a Cantilever Beam. *J. Guid. Control Dyn.* **1985**, *8*, 605–611. [CrossRef]

38. Yue, H.; Lu, Y.; Deng, Z.; Tzou, H. Experiments on vibration control of a piezoelectric laminated paraboloidal shell. *Mech. Syst. Signal Process.* **2016**, 1–17. [CrossRef]

39. Hagood, N.W.; Flotow, A. Von Damping of Structural Vibrations with Piezoelectric Materials and Passive Electrical Networks. *J. Sound Vib.* **1991**, *146*, 243–268. [CrossRef]

40. Clark, W.W. Vibration Control with State-Switched Piezoelectric Materials. *J. Intell. Mater. Syst. Struct.* **2000**, *11*, 263–271. [CrossRef]

41. Tsushima, N.; Su, W.; Introduction, I.; Member, S.; Member, S. Highly Flexible Piezoelectric Multifunctional Wings for. In Proceedings of the 58th AIAA/ASCE/AHS/ASC Structures, Structural Dynamics, and Materials Conference AIAA 2017-0624 Downloaded, Grapevine, TX, USA, 9–13 January 2017; pp. 1–15.

42. Na, W.S.; Baek, J. Impedance-Based Non-Destructive Testing Method Combined with Unmanned Aerial Vehicle for Structural Health Monitoring of Civil Infrastructures. *Appl. Sci.* **2017**, *7*, 15. [CrossRef]
43. Liu, T.; Li, C.; Wang, C.; Lai, J.W.; Cheong, K.H. A simple-fsdt-based isogeometric method for piezoelectric functionally graded plates. *Mathematics* **2020**, *8*, 1–24. [CrossRef]
44. Song, X.; Tan, S.; Wang, E.; Wu, S.; Wu, Z. Active shape control of an antenna reflector using piezoelectric actuators. *J. Intell. Mater. Syst. Struct.* **2019**, *30*, 2733–2747. [CrossRef]
45. Song, G.; Sethi, V.; Li, H.-N. Vibration control of civil structures using piezoceramic smart materials: A review. *Eng. Struct.* **2006**, *28*. [CrossRef]
46. Preumont, A.; Voltan, M.; Sangiovanni, A.; Bastaits, R.; Mokrani, B.; Alaluf, D. An investigation of the active damping of suspension bridges. *Adv. Inf. Knowl. Process.* **2015**, *3*, 1–36. [CrossRef]
47. Sethi, V.; Song, G. Optimal vibration control of a model frame structure using piezoceramic sensors and actuators. *JVC/Journal Vib. Control* **2005**, *11*. [CrossRef]
48. Sethi, V.; Song, G. Multimode vibration control of a smart model frame structure. *Smart Mater. Struct.* **2006**, *15*, 473–479. [CrossRef]
49. Sethi, V.; Song, G.; Franchek, M.A. Loop shaping control of a model-story building using smart materials. *J. Intell. Mater. Syst. Struct.* **2008**, *19*, 765–777. [CrossRef]
50. Aridogan, U.; Basdogan, I. A review of active vibration and noise suppression of plate-like structures with piezoelectric transducers. *J. Intell. Mater. Syst. Struct.* **2015**, *26*, 1455–1476. [CrossRef]
51. Search, H.; Journals, C.; Contact, A.; Iopscience, M.; Address, I.P.; Yang, A.M. Review on the use of piezoelectric materials for active vibration, noise, and flow control. *Smart Mater. Struct.* **2020**, *29*, 053001.
52. Gripp, J.A.B.; Rade, D.A. Vibration and noise control using shunted piezoelectric transducers: A review. *Mech. Syst. Signal Process.* **2018**, *112*, 359–383. [CrossRef]
53. Casadei, F.; Dozio, L.; Ruzzene, M.; Cunefare, K.A. Periodic shunted arrays for the control of noise radiation in an enclosure. *J. Sound Vib.* **2010**, *329*, 3632–3646. [CrossRef]
54. Ang, L.Y.L.; Koh, Y.K.; Lee, H.P. Acoustic metamaterials: A potential for cabin noise control in automobiles and armored vehicles. *Int. J. Appl. Mech.* **2016**, *8*, 1–35. [CrossRef]
55. Lai, S.C.; Mirshekarloo, M.S.; Yao, K. Effects of equivalent series resistance on the noise mitigation performance of piezoelectric shunt damping. *Smart Mater. Struct.* **2017**, *26*. [CrossRef]
56. Salvador, C.S.; Abas, M.C.A.; Teresa, J.A.; Castillo, M.; Dimaano, K.; Velasco, C.L.; Sangalang, J. Development of a traffic noise energy harvesting standalone system using piezoelectric transducers and super-capacitor. In Proceedings of the 2017 25th International Conference on Systems Engineering (ICSEng), Las Vegas, NV, USA, 22–24 August 2017; pp. 370–376.
57. Araújo, A.L.; Madeira, J.F.A. Multiobjective optimization solutions for noise reduction in composite sandwich panels using active control. *Compos. Struct.* **2020**, *247*, 112440. [CrossRef]
58. Al-Furjan, M.S.H.; Habibi, M.; Safarpour, H. Vibration Control of a Smart Shell Reinforced by Graphene Nanoplatelets. *Int. J. Appl. Mech.* **2020**, *12*. [CrossRef]
59. Li, S.; Liu, S.; Yang, L. Active Control of Vibration and Noise of Energy Equipment. *IOP Conf. Ser. Earth Environ. Sci.* **2020**, *446*. [CrossRef]
60. Kumamoto, M.; Kida, M.; Hirayama, R.; Kajikawa, Y.; Tani, T.; Kurumi, Y. Active noise control system for reducing MR noise. *IEICE Trans. Fundam. Electron. Commun. Comput. Sci.* **2011**, *E94*, 1479–1486. [CrossRef]
61. Timothy Zahl, "Car Body Cracked". 2015. Available online: https://www.carid.com/articles/what-type-of-body-kit-material-should-i-choose.html (accessed on 5 April 2021).
62. Fesharaki, J.J.; Madani, S.G.; Golabi, S. Effect of stiffness and thickness ratio of host plate and piezoelectric patches on reduction of the stress concentration factor. *Int. J. Adv. Struct. Eng.* **2016**, *8*, 229–242. [CrossRef]
63. Ihn, J.; Chang, F. Pitch-catch Active Sensing Methods in Structural Health Monitoring for Aircraft Structures. *Struct. Heal. Monit.* **2008**, *7*, 5–15. [CrossRef]
64. Rogers, C.A. Intelligent Material Systems—The Dawn of a New Materials Age. *J. Intell. Mater. Syst. Struct.* **1993**, *4*, 4–12. [CrossRef]
65. Kessler, S.S.; Johnson, C.E.; Dunn, C.T. Experimental Application of Optimized Lamb Wave Actuating/Sensing Patches for Health Monitoring of Composite Structures. In Proceedings of the 4th International Workshop, Stanford University, Stanford, CA, USA, 31 August–2 September 2003.
66. Kessler, S.S.; Spearing, S.M.; Soutis, C. Damage Detection in Composite Materials Using Lamb Wave Methods. *Smart Mater. […].* 2002. Available online: http://iopscience.iop.org/0964-1726/11/2/310 (accessed on 5 April 2021).
67. Bös, J.; Mayer, D. Comparison of various active vibration and noise reduction approaches applied to a planar test structure. In Proceedings of the 13th International Congress on Sound and Vibration, Wien, Österrcich, 2–6 July 2006.
68. Vergé, M.; Mechbal, N.; Coffignal, G. Active control of structures applied to an adaptable structure | Contrôle actif des structures appliqué à une structure souple. *J. Eur. Syst. Autom.* **2003**, *37*. [CrossRef]
69. Song, G.; Qiao, P.Z.; Binienda, W.K.; Zou, G.P. Active vibration damping of composite beam using smart sensors and actuators. *J. Aerosp. Eng.* **2002**, *15*, 97–103. [CrossRef]
70. Wu, N. Structural Repair using Smart Materials. *J. Aeronaut. Aerosp. Eng.* **2012**, *1*, 1–2. [CrossRef]

71. Rao, U.K.; Bangaru Babu, P.; Nagaraju, C. Active Repair of Engineering Structures Using Piezoelectric Patches. In Proceedings of the 17th ISME Conference ISME17, IIT Delhi, New Delhi, India, 3–4 October 2015; pp. 1–5.
72. Cheng, J.; Taheri, F. A novel smart adhesively bonded joint system. *Smart Mater. Struct.* **2005**, *14*, 971–981. [CrossRef]
73. Shaik Dawood, M.S.I.; Iannucci, L.; Greenhalgh, E.; Ariffin, A.K. Low Velocity Impact Induced Delamination Control Using MFC Actuator. *Appl. Mech. Mater.* **2012**, *165*, 346–351. [CrossRef]
74. Sohn, H.; Park, G.; Wait, J.R.; Limback, N.P.; Farrar, C.R. Wavelet-based active sensing for delamination detection in composite structures. *Smart Mater. Struct.* **2003**, *13*, 153–160. [CrossRef]
75. Liu, T.J.C. Fracture mechanics and crack contact analyses of the active repair of multi-layered piezoelectric patches bonded on cracked structures. *Theor. Appl. Fract. Mech.* **2007**, *47*, 120–132. [CrossRef]
76. Duan, W.H.; Quek, S.T.; Wang, Q. Finite element analysis of the piezoelectric-based repair of a delaminated beam. *Smart Mater. Struct.* **2008**, *17*, 015017. [CrossRef]
77. Alaimo, A.; Milazzo, A.; Orlando, C. Piezoelectric Patches for the Active Repair of Delaminated Structures. *J. Aerosp. Sci. Technol. Syst.* **2011**, *22*, 2137–2146.
78. Alaimo, A.; Milazzo, A.; Orlando, C.; Messineo, A. Numerical analysis of piezoelectric active repair in the presence of frictional contact conditions. *Sensors* **2013**, *13*, 4390–4403. [CrossRef]
79. Wu, N.; Wang, Q. Repair of a delaminated plate under static loading with piezoelectric patches. *Smart Mater. Struct.* **2010**, *19*, 105025. [CrossRef]
80. Muthu, N.; Maiti, S.K.; Falzon, B.G.; Yan, W. Crack propagation in non-homogenous materials: Evaluation of mixed-mode SIFs, T-stress and kinking angle using a variant of EFG Method. *Eng. Anal. Bound. Elem.* **2016**, *72*, 11–26. [CrossRef]
81. Barsoum, R.S. Triangular Quarter Point Elements as Elastic and Perfectly Plastic Crack Tip Elements. *Int. J. Numer. Meth. Engng.* **1977**, *11*, 85. [CrossRef]
82. Dally, J.W.; Sanford, R.J. Strain-gage methods for measuring the opening-mode stress-intensity factor, KI. *Exp. Mech.* **1987**, *27*, 381–388. [CrossRef]
83. Crawley, E.F.; De Luis, J. Use of piezoelectric actuators as elements of intelligent structures. *AIAA J.* **1987**, *25*, 1373–1385. [CrossRef]
84. Isaksson, P.; Hägglund, R. Crack-tip fields in gradient enhanced elasticity. *Eng. Fract. Mech.* **2013**, *97*, 186–192. [CrossRef]
85. Wang, Q.S. Active buckling control of beams using piezoelectric actuators and strain gauge sensors. *Smart Mater. Struct.* **2010**, *19*, 065022. [CrossRef]
86. Abuzaid, A.; Hrairi, M.; Dawood, M.S.I. Survey of Active Structural Control and Repair Using Piezoelectric Patches. *Actuators* **2015**, *4*, 77–98. [CrossRef]
87. Abuzaid, A.; Hrairi, M.; Dawood, M.S. Mode I Stress Intensity Factor for a Cracked Plate with an Integrated Piezoelectric Actuator. *Adv. Mater. Res.* **2015**, *1115*, 517–522. [CrossRef]
88. Abuzaid, A.; Hrairi, M.; Dawood, M. Evaluating the Reduction of Stress Intensity Factor in Center-Cracked Plates Using Piezoelectric Actuators. *Actuators* **2018**, *7*, 25. [CrossRef]
89. Aabid, A.; Hrairi, M.; Ali, J.S.M.; Abuzaid, A. Stress Concentration Analysis of a Composite Patch on a Hole in an Isotropic Plate. *Int. J. Mech. Prod. Eng. Res. Dev.* **2018**, *6*, 249–255.
90. Abuzaid, A.; Shaik Dawood, M.S.I.; Hrairi, M. The effect of piezoelectric actuation on stress distribution in aluminum plate with circular hole. *ARPN J. Eng. Appl. Sci.* **2015**, *10*, 9723–9729.
91. Abuzaid, A.; Hrairi, M.; Dawood, M.S. Modeling approach to evaluating reduction in stress intensity factor in center-cracked plate with piezoelectric actuator patches. *J. Intell. Mater. Syst. Struct.* **2017**, *28*, 1334–1345. [CrossRef]
92. Abuzaid, A.; Hrairi, M. Experimental and numerical analysis of piezoelectric active repair of edge-cracked plate. *J. Intell. Mater. Syst. Struct.* **2018**, *29*, 3656–3666. [CrossRef]
93. Abuzaid, A.; Dawood, M.S.; Hrairi, M. Effects of Adhesive Bond on Active Repair of Aluminium Plate Using Piezoelectric Patch. *Appl. Mech. Mater.* **2015**, *799–800*, 788–793. [CrossRef]
94. Aabid, A.; Hrairi, M.; Abuzaid, A.; Mohamed Ali, J.S. Estimation of stress intensity factor reduction for a center-cracked plate integrated with piezoelectric actuator and composite patch. *Thin-Walled Struct.* **2021**, *158*. [CrossRef]
95. Aabid, A.; Hrairi, M.; Dawood, M.S.I.S. Modeling Different Repair Configurations of an Aluminum Plate with a Hole. *Int. J. Recent Technol. Eng.* **2019**, *7*, 235–240.
96. Kang, K.; Chun, H.; Lee, J.A.; Byun, J.; Um, M.; Lee, S. Damage Detection of Composite Plates Using Finite Element Analysis Based on Structural Health Monitoring. *J. Mater. Sci. Eng.* **2011**, *1*, 14–21.
97. Martinez-Luengo, M.; Kolios, A.; Wang, L. Structural health monitoring of offshore wind turbines: A review through the Statistical Pattern Recognition Paradigm. *Renew. Sustain. Energy Rev.* **2016**, *64*, 91–105. [CrossRef]
98. Güemes, A.; Fernandez-Lopez, A.; Pozo, A.R.; Sierra-Pérez, J. Structural health monitoring for advanced composite structures: A review. *J. Compos. Sci.* **2020**, *4*, 15. [CrossRef]
99. Miorelli, R.; Kulakovskyi, A.; Chapuis, B.; D'Almeida, O.; Mesnil, O. Supervised learning strategy for classification and regression tasks applied to aeronautical structural health monitoring problems. *Ultrasonics* **2021**, *113*, 106372. [CrossRef] [PubMed]
100. Memmolo, V.; Ricci, F.; Boffa, N.D.; Maio, L.; Monaco, E. Structural Health Monitoring in Composites Based on Probabilistic Reconstruction Techniques. *Procedia Eng.* **2016**, *167*, 48–55. [CrossRef]
101. Anjum, A.; Syed, J.; Ali, M.; Zayan, J.M.; Aabid, A. Statistical Analysis of Adhesive Bond Parameters in a Single Lap Joint System. *J. Mod. Mech. Eng. Technol.* **2020**, *7*, 53–58.

102. Maio, L.; Memmolo, V.; Boccardi, S.; Meola, C.; Ricci, F.; Boffa, N.D.; Monaco, E. Ultrasonic and IR Thermographic Detection of a Defect in a Multilayered Composite Plate. *Procedia Eng.* **2016**, *167*, 71–79. [CrossRef]
103. Cot, L.D.; Wang, Y.; Bès, C.; Gogu, C. Scheduled and SHM structural airframe maintenance applications using a new probabilistic model. In Proceedings of the 17th European Workshop on Structural Health Monitoring (EWSHM 2014), Nantes, France, 8–11 July 2014; pp. 2306–2313.
104. Su, Z.; Ye, L.; Lu, Y. Guided Lamb waves for identification of damage in composite structures: A review. *J. Sound Vib.* **2006**, *295*, 753–780. [CrossRef]
105. Mitra, M.; Gopalakrishnan, S. Guided wave based structural health monitoring: A review. *Smart Mater. Struct.* **2016**, *25*. [CrossRef]
106. Memmolo, V.; Pasquino, N.; Ricci, F. Experimental characterization of a damage detection and localization system for composite structures. *Meas. J. Int. Meas. Confed.* **2018**, *129*, 381–388. [CrossRef]
107. Cawley, P. Structural health monitoring: Closing the gap between research and industrial deployment. *Struct. Heal. Monit.* **2018**, *17*, 1225–1244. [CrossRef]
108. Memmolo, V.; Monaco, E.; Boffa, N.D.; Maio, L.; Ricci, F. Guided wave propagation and scattering for structural health monitoring of stiffened composites. *Compos. Struct.* **2018**, *184*, 568–580. [CrossRef]
109. Salmanpour, M.S.; Sharif Khodaei, Z.; Aliabadi, M.H. Guided wave temperature correction methods in structural health monitoring. *J. Intell. Mater. Syst. Struct.* **2017**, *28*, 604–618. [CrossRef]
110. Miniaci, M.; Mazzotti, M.; Radzieński, M.; Kudela, P.; Kherraz, N.; Bosia, F.; Pugno, N.M.; Ostachowicz, W. Application of a laser-based time reversal algorithm for impact localization in a stiffened aluminum plate. *Front. Mater.* **2019**, *6*, 1–12. [CrossRef]
111. Cai, J.; Qiu, L.; Yuan, S.; Shi, S.; Liu, P.; Liang, D. Structural health monitoring for composite materials. In *Composites and Their Applications*; Hu, N., Ed.; IntechOpen: London, UK, 2012; pp. 37–58. ISBN 978-953-51-0706-4.
112. Hai, B.S.M.E.; Bause, M.; Kuberry, P. Finite element approximation of the extended fluid-structure interaction (eXFSI) problem. *Am. Soc. Mech. Eng. Fluids Eng. Div. FEDSM* **2016**, *1A-2016*. [CrossRef]
113. Kessler, S.; Mark, S.; Seth, S. Structural Health Monitoring in Composite Materials Using Lamb Wave Methods Structural Health Monitoring in Composite Materials Using Lamb Wave Methods. *ASC* **2001**, *43*, 1–4.
114. Duan, W.H.; Wang, Q.; Quek, S.T. Applications of piezoelectric materials in structural health monitoring and repair: Selected research examples. *Materials* **2010**, *3*, 5169–5194. [CrossRef] [PubMed]
115. Zhou, B.; Ma, X.; Wang, S.; Xue, S. Least-squares method for laminated beams with distributed braided piezoelectric composite actuators. *J. Intell. Mater. Syst. Struct.* **2020**, *31*, 2165–2176. [CrossRef]
116. Porchez, T.; Bencheikh, N.; Claeyssen, F. *Piezo-Composite Patches Applied to the Detection of Defects Using Lamb Wave Focusing*; Cedrat Technologies: Grenoble, France, 2013.
117. Shanker, R. An Integrated Approach for Structural Health Monitoring Rama Shanker. Ph.D. Thesis, Indian Institute of Technology Delhi, New Delhi, India, 2009.
118. Chen, C.-D.; Chiu, Y.-C.; Huang, Y.-H.; Wang, P.-H.; Chien, R.-D. Assessments of Structural Health Monitoring for Fatigue Cracks in Metallic Structures by Using Lamb Waves Driven by Piezoelectric Transducers. *J. Aerosp. Eng.* **2021**, *34*, 04020091. [CrossRef]
119. Venu, A.; Madhav, G.; John, P.; Lye, H.; Youxiang, C.; Jen, H.O.H.H.; Kun, Z.; Bin, S. Fatigue Monitoring of double surface defects using PZT based Electromechanical Impedance and Digital image correlation methods. *Adv. Mater. Res. Vols.* **2014**, *892*, 551–556.
120. Qing, X.; Li, W.; Wang, Y.; Sun, H. Piezoelectric transducer-based structural health monitoring for aircraft applications. *Sensors* **2019**, *19*, 1–27. [CrossRef] [PubMed]
121. Mewer, R.C. Analysis and Structural Health Monitoring of Composite Plates with Piezoelectric Sensors and Actuators. Master's Thesis, The University of Maine, Orono, ME, USA, 2003.
122. Bergmayr, T.; Kralovec, C.; Schagerl, M. Vibration-based thermal health monitoring for face layer debonding detection in aerospace sandwich structures. *Appl. Sci.* **2021**, *11*, 1–15.
123. Wang, Y.; Qiu, L.; Luo, Y.; Ding, R.; Jiang, F. A piezoelectric sensor network with shared signal transmission wires for structural health monitoring of aircraft smart skin. *Mech. Syst. Signal Process.* **2020**, *141*, 106730. [CrossRef]
124. Rocha, H.; Semprimoschnig, C.; Nunes, J.P. Sensors for process and structural health monitoring of aerospace composites: A review. *Eng. Struct.* **2021**, *237*, 112231. [CrossRef]
125. Ferri Aliabadi, M.H.; Khodaei, Z.S. Structural health monitoring for advanced composite structures. *Struct. Heal. Monit. Adv. Compos. Struct.* **2017**, 1–274. [CrossRef]
126. Giurgiutiu, V.; Zagrai, A.; Jing, B.J. Piezoelectric Wafer Embedded Active Sensors for Aging Aircraft Structural Health Monitoring. *Struct. Heal. Monit.* **2002**, *1*, 41–61. [CrossRef]
127. Aeronautics, S.M. Piezoelectric-Based In-Situ Damage Detection of Composite Materials for Structural Health Monitoring Systems. Ph.D Thesis, Massachusetts Institute of Technology, Cambridge, MA, USA, 2002.
128. Boukabache, H.; Escriba, C.; Fourniols, J.Y. Toward smart aerospace structures: Design of a piezoelectric sensor and its analog interface for flaw detection. *Sensors* **2014**, *14*, 20543–20561. [CrossRef]
129. Xu, Y.G.; Liu, G.R. A modified electro-mechanical impedance model of piezoelectric actuator-sensors for debonding detection of composite patches. *J. Intell. Mater. Syst. Struct.* **2002**, *13*, 389–396. [CrossRef]
130. Yang, Y.; Annamdas, V.G.M.; Wang, C.; Zhou, Y. Application of multiplexed FBG and PZT impedance sensors for health monitoring of rocks. *Sensors* **2008**, *8*, 271–289. [CrossRef]

131. Annamdas, V.G.M.; Yang, Y.; Soh, C.K. Influence of loading on the electromechanical admittance of piezoceramic transducers. *Smart Mater. Struct.* **2007**, *16*, 1888–1897. [CrossRef]

132. Yang, Y.; Lim, Y.Y.; Soh, C.K. Practical issues related to the application of the electromechanical impedance technique in the structural health monitoring of civil structures: I. Experiment. *Smart Mater. Struct.* **2008**, *17*. [CrossRef]

133. Yang, Y.; Hu, Y.; Lu, Y. Sensitivity of PZT impedance sensors for damage detection of concrete structures. *Sensors* **2008**, *8*, 327–346. [CrossRef]

134. Xu, B.; Jiang, F. Concrete-steel composite girder bolt loosening monitoring using electromechanical impedance measurements. In *Earth and Space 2012: Engineering, Science, Construction, and Operations in Challenging Environments*; American Society of Civil Engineers: Reston, VA, USA, 2012; pp. 629–634.

135. Hu, X.; Zhu, H.; Wang, D. A study of concrete slab damage detection based on the electromechanical impedance method. *Sensors* **2014**, *14*, 19897–19909. [CrossRef]

136. Gu, H.; Song, G.; Dhonde, H.; Mo, Y.L.; Yan, S. Concrete early-age strength monitoring using embedded piezoelectric transducers. *Smart Mater. Struct.* **2006**, *15*, 1837–1845. [CrossRef]

137. Ai, D.; Luo, H.; Wang, C.; Zhu, H. Monitoring of the load-induced RC beam structural tension / compression stress and damage using piezoelectric transducers. *Eng. Struct.* **2018**, *154*, 38–51. [CrossRef]

138. Song, G.; Gu, H.; Mo, Y.L.; Hsu, T.T.C.; Dhonde, H. Concrete structural health monitoring using embedded piezoceramic transducers. *Smart Mater. Struct.* **2015**, *16*, 959–968. [CrossRef]

139. Talakokula, V.; Bhalla, S.; Gupta, A. Corrosion assessment of reinforced concrete structures based on equivalent structural parameters using electro-mechanical impedance technique. *Intell. Mater. Syst. Struct.* **2014**, *25*, 484–500. [CrossRef]

140. Na, W.S.; Baek, J. A Review of the Piezoelectric Electromechanical Impedance Based Structural Health Monitoring Technique for Engineering Structures. *Sensors* **2018**, *18*, 18. [CrossRef]

141. Kim, H.; Liu, X.; Ahn, E.; Shin, M.; Shin, S.W.; Sim, S.H. Performance assessment method for crack repair in concrete using PZT-based electromechanical impedance technique. *NDT E Int.* **2019**, *104*, 90–97. [CrossRef]

142. Yan, S.; Ma, H.; Li, P.; Song, G.; Wu, J. Development and application of a structural health monitoring system based on wireless smart aggregates. *Sensors* **2017**, *17*, 1641. [CrossRef]

143. Sun, F.P.; Chaudhry, Z.; Liang, C.; Rogers, C.A. Truss Structure Integrity Identification Using PZT Sensor-Actuator. *J. Intell. Mater. Syst. Struct.* **1995**, *6*, 134–139. [CrossRef]

144. Park, G.; Kabeya, K.; Cudney, H.H.; Inman, D.J. Impednce-Based Structural Health Monitoring for Temperature Varying Applications. *Chem. Pharm. Bull.* **1992**, *40*, 1569–1572.

145. Grisso, B.L.; Inman, D.J. Temperature corrected sensor diagnostics for impedance-based SHM. *J. Sound Vib.* **2010**, *329*, 2323–2336. [CrossRef]

146. Wandowski, T.; Malinowski, P.H.; Ostachowicz, W.M. Temperature and damage influence on electromechanical impedance method used for carbon fibre-reinforced polymer panels. *J. Intell. Mater. Syst. Struct.* **2017**, *28*, 782–798. [CrossRef]

147. Na, W.S.; Lee, H. Experimental investigation for an isolation technique on conducting the electromechanical impedance method in high-temperature pipeline facilities. *J. Sound Vib.* **2016**, *383*, 210–220. [CrossRef]

148. Park, G.; Inman, D.J. Structural health monitoring using piezoelectric impedance measurements. *Philos. Trans. R. Soc. A Math. Phys. Eng. Sci.* **2007**, *365*, 373–392. [CrossRef]

149. Jiao, P.; Egbe, K.J.I.; Xie, Y.; Nazar, A.M.; Alavi, A.H. Piezoelectric sensing techniques in structural health monitoring: A state-of-the-art review. *Sensors* **2020**, *20*, 1–21. [CrossRef] [PubMed]

150. Spencer, B.F.; Ruiz-Sandoval, M.E.; Kurata, N. Smart sensing technology: Opportunities and challenges. *Struct. Control Heal. Monit.* **2004**, *11*, 349–368. [CrossRef]

151. Cusson, D.; Daigle, L. *Continuous Condition Assessment of Highway Bridges Using Field Monitoring NRCC-50863*; INFRA: Quebec City, QC, Canada, 2008; p. 14.

152. Albedah, A.; Khan, S.M.A.; Bouiadjra, B.B. Fatigue crack propagation in aluminum plates with composite patch including plasticity effect. *Proc. Inst. Mech. Eng. Part G* **2018**, *232*, 2122–2131. [CrossRef]

153. Oudad, W.; Belhadri, D.E.; Fekirini, H.; Khodja, M. Analysis of the Plastic Zone under Mixed Mode Fracture in Bonded Composite Repair of Aircraft Structures. *Aerosp. Sci. Technol.* **2017**. [CrossRef]

154. Hu, D.; Yang, Q.; Liu, H.; Mao, J.; Meng, F. Crack closure effect and crack growth behavior in GH2036 superalloy plates under combined high and low cycle fatigue. *Int. J. Fatigue* **2016**. [CrossRef]

155. Daryabor, P.; Safizadeh, M.S. Investigation of defect characteristics and heat transfer in step heating thermography of metal plates repaired with composite patches. *Infrared Phys. Technol.* **2016**, *76*, 608–620. [CrossRef]

MDPI

St. Alban-Anlage 66

4052 Basel

Switzerland

Tel. +41 61 683 77 34

Fax +41 61 302 89 18

www.mdpi.com

Actuators Editorial Office

E-mail: actuators@mdpi.com

www.mdpi.com/journal/actuators